An Introduction to Parallel Programming

An Introduction to Parallel Programming

Second Edition

Peter S. Pacheco
University of San Francisco

Matthew Malensek
University of San Francisco

MORGAN KAUFMANN PUBLISHERS

ELSEVIER AN IMPRINT OF ELSEVIER

Morgan Kaufmann is an imprint of Elsevier
50 Hampshire Street, 5th Floor, Cambridge, MA 02139, United States

Notices

Knowledge and best practice in this field are constantly changing. As new research and experience broaden our understanding, changes in research methods, professional practices, or medical treatment may become necessary.

Practitioners and researchers must always rely on their own experience and knowledge in evaluating and using any information, methods, compounds, or experiments described herein. In using such information or methods they should be mindful of their own safety and the safety of others, including parties for whom they have a professional responsibility.

To the fullest extent of the law, neither the Publisher nor the authors, contributors, or editors, assume any liability for any injury and/or damage to persons or property as a matter of products liability, negligence or otherwise, or from any use or operation of any methods, products, instructions, or ideas contained in the material herein.

Library of Congress Cataloging-in-Publication Data
A catalog record for this book is available from the Library of Congress

British Library Cataloguing-in-Publication Data
A catalogue record for this book is available from the British Library

ISBN: 978-0-12-804605-0

For information on all Morgan Kaufmann publications
visit our website at https://www.elsevier.com/books-and-journals

Publisher: Katey Birtcher
Acquisitions Editor: Stephen Merken
Content Development Manager: Meghan Andress
Publishing Services Manager: Shereen Jameel
Production Project Manager: Rukmani Krishnan
Designer: Victoria Pearson

Typeset by VTeX

Printed in United States of America

Last digit is the print number:
9 8 7 6 5 4 3 2 1

Working together
to grow libraries in
developing countries

www.elsevier.com • www.bookaid.org

To the memory of Robert S. Miller

Contents

Preface

Parallel hardware has been ubiquitous for some time now: it's difficult to find a laptop, desktop, or server that doesn't use a multicore processor. Cluster computing is nearly as common today as high-powered workstations were in the 1990s, and cloud computing is making distributed-memory systems as accessible as desktops. In spite of this, most computer science majors graduate with little or no experience in parallel programming. Many colleges and universities offer upper-division elective courses in parallel computing, but since most computer science majors have to take a large number of required courses, many graduate without ever writing a multithreaded or multiprocess program.

It seems clear that this state of affairs needs to change. Whereas many programs can obtain satisfactory performance on a single core, computer scientists should be made aware of the potentially vast performance improvements that can be obtained with parallelism, and they should be able to exploit this potential when the need arises.

Introduction to Parallel Programming was written to partially address this problem. It provides an introduction to writing parallel programs using MPI, Pthreads, OpenMP, and CUDA, four of the most widely used APIs for parallel programming. The intended audience is students and professionals who need to write parallel programs. The prerequisites are minimal: a college-level course in mathematics and the ability to write serial programs in C.

The prerequisites are minimal, because we believe that students should be able to start programming parallel systems *as early as possible*. At the University of San Francisco, computer science students can fulfill a requirement for the major by taking a course on which this text is based immediately after taking the "Introduction to Computer Science I" course that most majors take in the first semester of their freshman year. It has been our experience that there really is no reason for students to defer writing parallel programs until their junior or senior year. To the contrary, the course is popular, and students have found that using concurrency in other courses is much easier after having taken this course.

If second-semester freshmen can learn to write parallel programs by taking a class, then motivated computing professionals should be able to learn to write parallel programs through self-study. We hope this book will prove to be a useful resource for them.

The Second Edition

It has been nearly ten years since the first edition of *Introduction to Parallel Programming* was published. During that time much has changed in the world of parallel programming, but, perhaps surprisingly, much also remains the same. Our intent in writing this second edition has been to preserve the material from the first edition that continues to be generally useful, but also to add new material where we felt it was needed.

The most obvious addition is the inclusion of a new chapter on CUDA programming. When the first edition was published, CUDA was still very new. It was already clear that the use of GPUs in high-performance computing would become very widespread, but at that time we felt that GPGPU wasn't readily accessible to programmers with relatively little experience. In the last ten years, that has clearly changed. Of course, CUDA is not a standard, and features are added, modified, and deleted with great rapidity. As a consequence, authors who use CUDA must present a subject that changes much faster than a standard, such as MPI, Pthreads, or OpenMP. In spite of this, we hope that our presentation of CUDA will continue to be useful for some time.

Another big change is that Matthew Malensek has come onboard as a coauthor. Matthew is a relatively new colleague at the University of San Francisco, but he has extensive experience with both the teaching and application of parallel computing. His contributions have greatly improved the second edition.

About This Book

As we noted earlier, the main purpose of the book is to teach parallel programming in MPI, Pthreads, OpenMP, and CUDA to an audience with a limited background in computer science and no previous experience with parallelism. We also wanted to make the book as flexible as possible so that readers who have no interest in learning one or two of the APIs can still read the remaining material with little effort. Thus the chapters on the four APIs are largely independent of each other: they can be read in any order, and one or two of these chapters can be omitted. This independence has some cost: it was necessary to repeat some of the material in these chapters. Of course, repeated material can be simply scanned or skipped.

On the other hand, readers with no prior experience with parallel computing should read Chapter 1 first. This chapter attempts to provide a relatively nontechnical explanation of why parallel systems have come to dominate the computer landscape. It also provides a short introduction to parallel systems and parallel programming.

Chapter 2 provides technical background on computer hardware and software. Chapters 3 to 6 provide independent introductions to MPI, Pthreads, OpenMP, and CUDA, respectively. Chapter 7 illustrates the development of two different parallel programs using each of the four APIs. Finally, Chapter 8 provides a few pointers to additional information on parallel computing.

We use the C programming language for developing our programs, because all four API's have C-language interfaces, and, since C is such a small language, it is a relatively easy language to learn—especially for C++ and Java programmers, since they will already be familiar with C's control structures.

Classroom Use

This text grew out of a lower-division undergraduate course at the University of San Francisco. The course fulfills a requirement for the computer science major, and it also fulfills a prerequisite for the undergraduate operating systems, architecture, and networking courses. The course begins with a four-week introduction to C programming. Since most of the students have already written Java programs, the bulk of this introduction is devoted to the use pointers in C.[1] The remainder of the course provides introductions first to programming in MPI, then Pthreads and/or OpenMP, and it finishes with material covering CUDA.

We cover most of the material in Chapters 1, 3, 4, 5, and 6, and parts of the material in Chapters 2 and 7. The background in Chapter 2 is introduced as the need arises. For example, before discussing cache coherence issues in OpenMP (Chapter 5), we cover the material on caches in Chapter 2.

The coursework consists of weekly homework assignments, five programming assignments, a couple of midterms and a final exam. The homework assignments usually involve writing a very short program or making a small modification to an existing program. Their purpose is to insure that the students stay current with the coursework, and to give the students hands-on experience with ideas introduced in class. It seems likely that their existence has been one of the principle reasons for the course's success. Most of the exercises in the text are suitable for these brief assignments.

The programming assignments are larger than the programs written for homework, but we typically give the students a good deal of guidance: we'll frequently include pseudocode in the assignment and discuss some of the more difficult aspects in class. This extra guidance is often crucial: it's easy to give programming assignments that will take far too long for the students to complete.

The results of the midterms and finals and the enthusiastic reports of the professor who teaches operating systems suggest that the course is actually very successful in teaching students how to write parallel programs.

For more advanced courses in parallel computing, the text and its online supporting materials can serve as a supplement so that much of the material on the syntax and semantics of the four APIs can be assigned as outside reading.

The text can also be used as a supplement for project-based courses and courses outside of computer science that make use of parallel computation.

Support Materials

An online companion site for the book is located at www.elsevier.com/books-and-journals/book-companion/9780128046050.. This site will include errata and complete source for the longer programs we discuss in the text. Additional material for instructors, including downloadable figures and solutions to the exercises in the book, can be downloaded from https://educate.elsevier.com/9780128046050.

We would greatly appreciate readers' letting us know of any errors they find. Please send email to mmalensek@usfca.edu if you do find a mistake.

---

[1] Interestingly, a number of students have said that they found the use of C pointers more difficult than MPI programming.

Acknowledgments

In the course of working on this book we've received considerable help from many individuals. Among them we'd like to thank the reviewers of the second edition, Steven Frankel (Technion) and Il-Hyung Cho (Saginaw Valley State University), who read and commented on draft versions of the new CUDA chapter. We'd also like to thank the reviewers who read and commented on the initial proposal for the book: Fikret Ercal (Missouri University of Science and Technology), Dan Harvey (Southern Oregon University), Joel Hollingsworth (Elon University), Jens Mache (Lewis and Clark College), Don McLaughlin (West Virginia University), Manish Parashar (Rutgers University), Charlie Peck (Earlham College), Stephen C. Renk (North Central College), Rolfe Josef Sassenfeld (The University of Texas at El Paso), Joseph Sloan (Wofford College), Michela Taufer (University of Delaware), Pearl Wang (George Mason University), Bob Weems (University of Texas at Arlington), and Cheng-Zhong Xu (Wayne State University). We are also deeply grateful to the following individuals for their reviews of various chapters of the book: Duncan Buell (University of South Carolina), Matthias Gobbert (University of Maryland, Baltimore County), Krishna Kavi (University of North Texas), Hong Lin (University of Houston–Downtown), Kathy Liszka (University of Akron), Leigh Little (The State University of New York), Xinlian Liu (Hood College), Henry Tufo (University of Colorado at Boulder), Andrew Sloss (Consultant Engineer, ARM), and Gengbin Zheng (University of Illinois). Their comments and suggestions have made the book immeasurably better. Of course, we are solely responsible for remaining errors and omissions.

Slides and the solutions manual for the first edition were prepared by Kathy Liszka and Jinyoung Choi, respectively. Thanks to both of them.

The staff at Elsevier has been very helpful throughout this project. Nate McFadden helped with the development of the text. Todd Green and Steve Merken were the acquisitions editors. Meghan Andress was the content development manager. Rukmani Krishnan was the production editor. Victoria Pearson was the designer. They did a great job, and we are very grateful to all of them.

Our colleagues in the computer science and mathematics departments at USF have been extremely helpful during our work on the book. Peter would like to single out Prof. Gregory Benson for particular thanks: his understanding of parallel computing—especially Pthreads and semaphores—has been an invaluable resource. We're both very grateful to our system administrators, Alexey Fedosov and Elias Husary. They've patiently and efficiently dealt with all of the "emergencies" that cropped up while we were working on programs for the book. They've also done an amazing job of providing us with the hardware we used to do all program development and testing.

Peter would never have been able to finish the book without the encouragement and moral support of his friends Holly Cohn, John Dean, and Maria Grant. He will always be very grateful for their help and their friendship. He is especially grateful to Holly for allowing us to use her work, *seven notations*, for the cover.

Matthew would like to thank his colleagues in the USF Department of Computer Science, as well as Maya Malensek and Doyel Sadhu, for their love and support.

Most of all, he would like to thank Peter Pacheco for being a mentor and infallible source of advice and wisdom during the formative years of his career in academia.

Our biggest debt is to our students. As always, they showed us what was too easy and what was far too difficult. They taught us how to teach parallel computing. Our deepest thanks to all of them.

An Introduction to Parallel Programming

Why parallel computing

1

From 1986 to 2003, the performance of microprocessors increased, on average, more than 50% per year [28]. This unprecedented increase meant that users and software developers could often simply wait for the next generation of microprocessors to obtain increased performance from their applications. Since 2003, however, single-processor performance improvement has slowed to the point that in the period from 2015 to 2017, it increased at less than 4% per year [28]. This difference is dramatic: at 50% per year, performance will increase by almost a factor of 60 in 10 years, while at 4%, it will increase by about a factor of 1.5.

Furthermore, this difference in performance increase has been associated with a dramatic change in processor design. By 2005, most of the major manufacturers of microprocessors had decided that the road to rapidly increasing performance lay in the direction of parallelism. Rather than trying to continue to develop ever-faster monolithic processors, manufacturers started putting *multiple* complete processors on a single integrated circuit.

This change has a very important consequence for software developers: simply adding more processors will not magically improve the performance of the vast majority of **serial** programs, that is, programs that were written to run on a single processor. Such programs are unaware of the existence of multiple processors, and the performance of such a program on a system with multiple processors will be effectively the same as its performance on a single processor of the multiprocessor system.

All of this raises a number of questions:

- Why do we care? Aren't single-processor systems fast enough?
- Why can't microprocessor manufacturers continue to develop much faster single-processor systems? Why build **parallel systems**? Why build systems with multiple processors?
- Why can't we write programs that will automatically convert serial programs into **parallel programs**, that is, programs that take advantage of the presence of multiple processors?

Let's take a brief look at each of these questions. Keep in mind, though, that some of the answers aren't carved in stone. For example, the performance of many applications may already be more than adequate.

An Introduction to Parallel Programming. https://doi.org/10.1016/B978-0-12-804605-0.00008-7

1.1 Why we need ever-increasing performance

The vast increases in computational power that we've been enjoying for decades now have been at the heart of many of the most dramatic advances in fields as diverse as science, the Internet, and entertainment. For example, decoding the human genome, ever more accurate medical imaging, astonishingly fast and accurate Web searches, and ever more realistic and responsive computer games would all have been impossible without these increases. Indeed, more recent increases in computational power would have been difficult, if not impossible, without earlier increases. But we can never rest on our laurels. As our computational power increases, the number of problems that we can seriously consider solving also increases. Here are a few examples:

- *Climate modeling.* To better understand climate change, we need far more accurate computer models, models that include interactions between the atmosphere, the oceans, solid land, and the ice caps at the poles. We also need to be able to make detailed studies of how various interventions might affect the global climate.
- *Protein folding.* It's believed that misfolded proteins may be involved in diseases such as Huntington's, Parkinson's, and Alzheimer's, but our ability to study configurations of complex molecules such as proteins is severely limited by our current computational power.
- *Drug discovery.* There are many ways in which increased computational power can be used in research into new medical treatments. For example, there are many drugs that are effective in treating a relatively small fraction of those suffering from some disease. It's possible that we can devise alternative treatments by careful analysis of the genomes of the individuals for whom the known treatment is ineffective. This, however, will involve extensive computational analysis of genomes.
- *Energy research.* Increased computational power will make it possible to program much more detailed models of technologies, such as wind turbines, solar cells, and batteries. These programs may provide the information needed to construct far more efficient clean energy sources.
- *Data analysis.* We generate tremendous amounts of data. By some estimates, the quantity of data stored worldwide doubles every two years [31], but the vast majority of it is largely useless unless it's analyzed. As an example, knowing the sequence of nucleotides in human DNA is, by itself, of little use. Understanding how this sequence affects development and how it can cause disease requires extensive analysis. In addition to genomics, huge quantities of data are generated by particle colliders, such as the Large Hadron Collider at CERN, medical imaging, astronomical research, and Web search engines—to name a few.

These and a host of other problems won't be solved without tremendous increases in computational power.

1.2 Why we're building parallel systems

Much of the tremendous increase in single-processor performance was driven by the ever-increasing density of transistors—the electronic switches—on integrated circuits. As the size of transistors decreases, their speed can be increased, and the overall speed of the integrated circuit can be increased. However, as the speed of transistors increases, their power consumption also increases. Most of this power is dissipated as heat, and when an integrated circuit gets too hot, it becomes unreliable. In the first decade of the twenty-first century, air-cooled integrated circuits reached the limits of their ability to dissipate heat [28].

Therefore it is becoming impossible to continue to increase the speed of integrated circuits. Indeed, in the last few years, the increase in transistor density has slowed dramatically [36].

But given the potential of computing to improve our existence, there is a moral imperative to continue to increase computational power.

How then, can we continue to build ever more powerful computers? The answer is *parallelism*. Rather than building ever-faster, more complex, monolithic processors, the industry has decided to put multiple, relatively simple, complete processors on a single chip. Such integrated circuits are called **multicore** processors, and **core** has become synonymous with central processing unit, or CPU. In this setting a conventional processor with one CPU is often called a **single-core** system.

1.3 Why we need to write parallel programs

Most programs that have been written for conventional, single-core systems cannot exploit the presence of multiple cores. We can run multiple instances of a program on a multicore system, but this is often of little help. For example, being able to run multiple instances of our favorite game isn't really what we want—we want the program to run faster with more realistic graphics. To do this, we need to either rewrite our serial programs so that they're *parallel*, so that they can make use of multiple cores, or write translation programs, that is, programs that will automatically convert serial programs into parallel programs. The bad news is that researchers have had very limited success writing programs that convert serial programs in languages such as C, C++, and Java into parallel programs.

This isn't terribly surprising. While we can write programs that recognize common constructs in serial programs, and automatically translate these constructs into efficient parallel constructs, the sequence of parallel constructs may be terribly inefficient. For example, we can view the multiplication of two $n \times n$ matrices as a sequence of dot products, but parallelizing a matrix multiplication as a sequence of parallel dot products is likely to be fairly slow on many systems.

An efficient parallel implementation of a serial program may not be obtained by finding efficient parallelizations of each of its steps. Rather, the best parallelization may be obtained by devising an entirely new algorithm.

As an example, suppose that we need to compute n values and add them together. We know that this can be done with the following serial code:

```
sum = 0;
for (i = 0; i < n; i++) {
    x = Compute_next_value(. . .);
    sum += x;
}
```

Now suppose we also have p cores and $p \leq n$. Then each core can form a partial sum of approximately n/p values:

```
my_sum = 0;
my_first_i = . . . . ;
my_last_i = . . . . ;
for (my_i = my_first_i; my_i < my_last_i; my_i++) {
    my_x = Compute_next_value(. . .);
    my_sum += my_x;
}
```

Here the prefix `my_` indicates that each core is using its own, private variables, and each core can execute this block of code independently of the other cores.

After each core completes execution of this code, its variable `my_sum` will store the sum of the values computed by its calls to `Compute_next_value`. For example, if there are eight cores, $n = 24$, and the 24 calls to `Compute_next_value` return the values

$$1, 4, 3, \quad 9, 2, 8, \quad 5, 1, 1, \quad 6, 2, 7, \quad 2, 5, 0, \quad 4, 1, 8, \quad 6, 5, 1, \quad 2, 3, 9,$$

then the values stored in `my_sum` might be

Core	0	1	2	3	4	5	6	7
my_sum	8	19	7	15	7	13	12	14

Here we're assuming the cores are identified by nonnegative integers in the range $0, 1, \ldots, p - 1$, where p is the number of cores.

When the cores are done computing their values of `my_sum`, they can form a global sum by sending their results to a designated "master" core, which can add their results:

```
if (I'm the master core) {
    sum = my_sum;
    for each core other than myself {
        receive value from core;
        sum += value;
    }
} else {
    send my_sum to the master;
}
```

In our example, if the master core is core 0, it would add the values $8 + 19 + 7 + 15 + 7 + 13 + 12 + 14 = 95$.

But you can probably see a better way to do this—especially if the number of cores is large. Instead of making the master core do all the work of computing the final sum, we can pair the cores so that while core 0 adds in the result of core 1, core 2 can add in the result of core 3, core 4 can add in the result of core 5, and so on. Then we can repeat the process with only the even-ranked cores: 0 adds in the result of 2, 4 adds in the result of 6, and so on. Now cores divisible by 4 repeat the process, and so on. See Fig. 1.1. The circles contain the current value of each core's sum, and the lines with arrows indicate that one core is sending its sum to another core. The plus signs indicate that a core is receiving a sum from another core and adding the received sum into its own sum.

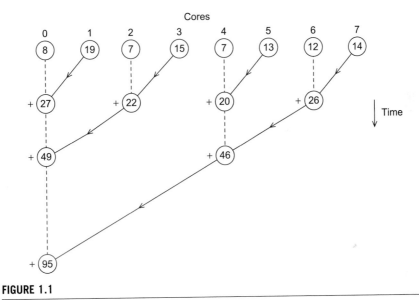

FIGURE 1.1

Multiple cores forming a global sum.

For both "global" sums, the master core (core 0) does more work than any other core, and the length of time it takes the program to complete the final sum should be the length of time it takes for the master to complete. However, with eight cores, the master will carry out seven receives and adds using the first method, while with the second method, it will only carry out three. So the second method results in an improvement of more than a factor of two. The difference becomes much more dramatic with large numbers of cores. With 1000 cores, the first method will require 999 receives and adds, while the second will only require 10—an improvement of almost a factor of 100!

The first global sum is a fairly obvious generalization of the serial global sum: divide the work of adding among the cores, and after each core has computed its part of the sum, the master core simply repeats the basic serial addition—if there are p cores, then it needs to add p values. The second global sum, on the other hand, bears little relation to the original serial addition.

The point here is that it's unlikely that a translation program would "discover" the second global sum. Rather, there would more likely be a predefined efficient global sum that the translation program would have access to. It could "recognize" the original serial loop and replace it with a precoded, efficient, parallel global sum.

We might expect that software could be written so that a large number of common serial constructs could be recognized and efficiently **parallelized**, that is, modified so that they can use multiple cores. However, as we apply this principle to ever more complex serial programs, it becomes more and more difficult to recognize the construct, and it becomes less and less likely that we'll have a precoded, efficient parallelization.

Thus we cannot simply continue to write serial programs; we must write parallel programs, programs that exploit the power of multiple processors.

1.4 How do we write parallel programs?

There are a number of possible answers to this question, but most of them depend on the basic idea of *partitioning* the work to be done among the cores. There are two widely used approaches: **task-parallelism** and **data-parallelism**. In task-parallelism, we partition the various tasks carried out in solving the problem among the cores. In data-parallelism, we partition the data used in solving the problem among the cores, and each core carries out more or less similar operations on its part of the data.

As an example, suppose that Prof P has to teach a section of "Survey of English Literature." Also suppose that Prof P has one hundred students in her section, so she's been assigned four teaching assistants (TAs): Mr. A, Ms. B, Mr. C, and Ms. D. At last the semester is over, and Prof P makes up a final exam that consists of five questions. To grade the exam, she and her TAs might consider the following two options: each of them can grade all one hundred responses to one of the questions; say, P grades question 1, A grades question 2, and so on. Alternatively, they can divide the one hundred exams into five piles of twenty exams each, and each of them can grade all the papers in one of the piles; P grades the papers in the first pile, A grades the papers in the second pile, and so on.

In both approaches the "cores" are the professor and her TAs. The first approach might be considered an example of task-parallelism. There are five tasks to be carried out: grading the first question, grading the second question, and so on. Presumably, the graders will be looking for different information in question 1, which is about Shakespeare, from the information in question 2, which is about Milton, and so on. So the professor and her TAs will be "executing different instructions."

On the other hand, the second approach might be considered an example of data-parallelism. The "data" are the students' papers, which are divided among the cores, and each core applies more or less the same grading instructions to each paper.

The first part of the global sum example in Section 1.3 would probably be considered an example of data-parallelism. The data are the values computed by Compute_next_value, and each core carries out roughly the same operations on its assigned elements: it computes the required values by calling Compute_next_value and adds them together. The second part of the first global sum example might be considered an example of task-parallelism. There are two tasks: receiving and adding the cores' partial sums, which is carried out by the master core; and giving the partial sum to the master core, which is carried out by the other cores.

When the cores can work independently, writing a parallel program is much the same as writing a serial program. Things get a great deal more complex when the cores need to coordinate their work. In the second global sum example, although the tree structure in the diagram is very easy to understand, writing the actual code is relatively complex. See Exercises 1.3 and 1.4. Unfortunately, it's much more common for the cores to need coordination.

In both global sum examples, the coordination involves **communication**: one or more cores send their current partial sums to another core. The global sum examples should also involve coordination through **load balancing**. In the first part of the global sum, it's clear that we want the amount of time taken by each core to be roughly the same as the time taken by the other cores. If the cores are identical, and each call to Compute_next_value requires the same amount of work, then we want each core to be assigned roughly the same number of values as the other cores. If, for example, one core has to compute most of the values, then the other cores will finish much sooner than the heavily loaded core, and their computational power will be wasted.

A third type of coordination is **synchronization**. As an example, suppose that instead of computing the values to be added, the values are read from stdin. Say, x is an array that is read in by the master core:

```
if (I'm the master core)
    for (my_i = 0; my_i < n; my_i++)
        scanf("%lf", &x[my_i]);
```

In most systems the cores are not automatically synchronized. Rather, each core works at its own pace. In this case, the problem is that we don't want the other cores to race ahead and start computing their partial sums before the master is done initializing x and making it available to the other cores. That is, the cores need to wait before starting execution of the code:

```
for (my_i = my_first_i; my_i < my_last_i; my_i++)
    my_sum += x[my_i];
```

We need to add in a point of synchronization between the initialization of x and the computation of the partial sums:

```
Synchronize_cores();
```

The idea here is that each core will wait in the function Synchronize_cores until all the cores have entered the function—in particular, until the master core has entered this function.

Currently, the most powerful parallel programs are written using *explicit* parallel constructs, that is, they are written using extensions to languages such as C, C++, and Java. These programs include explicit instructions for parallelism: core 0 executes task 0, core 1 executes task 1, ..., all cores synchronize, ..., and so on, so such programs are often extremely complex. Furthermore, the complexity of modern cores often makes it necessary to use considerable care in writing the code that will be executed by a single core.

There are other options for writing parallel programs—for example, higher level languages—but they tend to sacrifice performance to make program development somewhat easier.

1.5 What we'll be doing

We'll be focusing on learning to write programs that are explicitly parallel. Our purpose is to learn the basics of programming parallel computers using the C language and four different **APIs** or **application program interfaces**: the **Message-Passing Interface** or **MPI**, **POSIX threads** or **Pthreads**, **OpenMP**, and **CUDA**. MPI and Pthreads are libraries of type definitions, functions, and macros that can be used in C programs. OpenMP consists of a library and some modifications to the C compiler. CUDA consists of a library and modifications to the C++ compiler.

You may well wonder why we're learning about four different APIs instead of just one. The answer has to do with both the extensions and parallel systems. Currently, there are two main ways of classifying parallel systems: one is to consider the memory that the different cores have access to, and the other is to consider whether the cores can operate independently of each other.

In the memory classification, we'll be focusing on **shared-memory** systems and **distributed-memory** systems. In a shared-memory system, the cores can share access to the computer's memory; in principle, each core can read and write each memory location. In a shared-memory system, we can coordinate the cores by having them examine and update shared-memory locations. In a distributed-memory system, on the other hand, each core has its own, private memory, and the cores can communicate explicitly by doing something like sending messages across a network. Fig. 1.2 shows schematics of the two types of systems.

The second classification divides parallel systems according to the number of independent instruction streams and the number of independent data streams. In one type of system, the cores can be thought of as conventional processors, so they have their own control units, and they are capable of operating independently of each other. Each core can manage its own instruction stream and its own data stream, so this type of system is called a **Multiple-Instruction Multiple-Data** or **MIMD** system.

An alternative is to have a parallel system with cores that are not capable of managing their own instruction streams: they can be thought of as cores with no control unit. Rather, the cores share a single control unit. However, each core can access either its own private memory or memory that's shared among the cores. In this type of

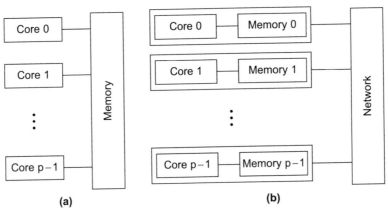

FIGURE 1.2

(a) A shared memory system and (b) a distributed memory system.

system, all the cores carry out the same instruction on their own data, so this type of system is called a **Single-Instruction Multiple-Data** or **SIMD** system.

In a MIMD system, it's perfectly feasible for one core to execute an addition while another core executes a multiply. In a SIMD system, two cores either execute the same instruction (on their own data) or, if they need to execute different instructions, one executes its instruction while the other is idle, and then the second executes its instruction while the first is idle. In a SIMD system, we couldn't have one core executing an addition while another core executes a multiplication. The system would have to do something like this:

Time	First core	Second core
1	Addition	Idle
2	Idle	Multiply

Since you're used to programming a processor with its own control unit, MIMD systems may seem more natural to you. However, as we'll see, there are many problems that are *very* easy to solve using a SIMD system. As a very simple example, suppose we have three arrays, each with n elements, and we want to add corresponding entries of the first two arrays to get the values in the third array. The serial pseudocode might look like this:

```
double x[n], y[n], z[n];
. . .
for (int i = 0; i < n; i++)
    z[i] = x[i] + y[i];
```

Now suppose we have n SIMD cores, and each core is assigned one element from each of the three arrays: core i is assigned elements x[i], y[i], and z[i]. Then our program can simply tell each core to add its x- and y-values to get the z value:

```
i = Compute_my_subscript ();
z[i] = x[i] + y[i];
```

This type of system is fundamental to modern Graphics Processing Units or GPUs, and since GPUs are extremely powerful parallel processors, it's important that we learn how to program them.

Our different APIs are used for programming different types of systems:

- MPI is an API for programming distributed memory MIMD systems.
- Pthreads is an API for programming shared memory MIMD systems.
- OpenMP is an API for programming both shared memory MIMD and shared memory SIMD systems, although we'll be focusing on programming MIMD systems.
- CUDA is an API for programming Nvidia GPUs, which have aspects of all four of our classifications: shared memory and distributed memory, SIMD, and MIMD. We will, however, be focusing on the shared memory SIMD and MIMD aspects of the API.

1.6 Concurrent, parallel, distributed

If you look at some other books on parallel computing or you search the Web for information on parallel computing, you're likely to also run across the terms **concurrent computing** and **distributed computing**. Although there isn't complete agreement on the distinction between the terms parallel, distributed, and concurrent, many authors make the following distinctions:

- In concurrent computing, a program is one in which multiple tasks can be *in progress* at any instant [5].
- In parallel computing, a program is one in which multiple tasks *cooperate closely* to solve a problem.
- In distributed computing, a program may need to cooperate with other programs to solve a problem.

So parallel and distributed programs are concurrent, but a program such as a multitasking operating system is also concurrent, even when it is run on a machine with only one core, since multiple tasks can be *in progress* at any instant. There isn't a clear-cut distinction between parallel and distributed programs, but a parallel program usually runs multiple tasks simultaneously on cores that are physically close to each other and that either share the same memory or are connected by a very high-speed network. On the other hand, distributed programs tend to be more "loosely coupled." The tasks may be executed by multiple computers that are separated by relatively large distances, and the tasks themselves are often executed by programs that were created independently. As examples, our two concurrent addition programs would be considered parallel by most authors, while a Web search program would be considered distributed.

But beware, there isn't general agreement on these terms. For example, many authors consider shared-memory programs to be "parallel" and distributed-memory programs to be "distributed." As our title suggests, we'll be interested in parallel programs—programs in which closely coupled tasks cooperate to solve a problem.

1.7 The rest of the book

How can we use this book to help us write parallel programs?

First, when you're interested in high performance, whether you're writing serial or parallel programs, you need to know a little bit about the systems you're working with—both hardware and software. So in Chapter 2, we'll give an overview of parallel hardware and software. In order to understand this discussion, it will be necessary to review some information on serial hardware and software. Much of the material in Chapter 2 won't be needed when we're getting started, so you might want to skim some of this material and refer back to it occasionally when you're reading later chapters.

The heart of the book is contained in Chapters 3–7. Chapters 3, 4, 5, and 6 provide a very elementary introduction to programming parallel systems using C and MPI, Pthreads, OpenMP, and CUDA, respectively. The only prerequisite for reading these chapters is a knowledge of C programming. We've tried to make these chapters independent of each other, and you should be able to read them in any order. However, to make them independent, we did find it necessary to repeat some material. So if you've read one of the three chapters, and you go on to read another, be prepared to skim over some of the material in the new chapter.

Chapter 7 puts together all we've learned in the preceding chapters. It develops two fairly large programs using each of the four APIs. However, it should be possible to read much of this even if you've only read one of Chapters 3, 4, 5, or 6. The last chapter, Chapter 8, provides a few suggestions for further study on parallel programming.

1.8 A word of warning

Before proceeding, a word of warning. It may be tempting to write parallel programs "by the seat of your pants," without taking the trouble to carefully design and incrementally develop your program. This will almost certainly be a mistake. Every parallel program contains at least one serial program. Since we almost always need to coordinate the actions of multiple cores, writing parallel programs is almost always more complex than writing a serial program that solves the same problem. In fact, it is often *far* more complex. All the rules about careful design and development are usually *far* more important for the writing of parallel programs than they are for serial programs.

1.9 Typographical conventions

We'll make use of the following typefaces in the text:

- Program text, displayed or within running text, will use the following typefaces:

```
/*   This  is  a  short  program  */
#include <stdio.h>

int main(int argc, char* argv[]) {
   printf("hello, world\n");

   return 0;
}
```

- Definitions are given in the body of the text, and the term being defined is printed in boldface type: A **parallel program** can make use of multiple cores.
- When we need to refer to the environment in which a program is being developed, we'll assume that we're using a UNIX shell, such as bash, and we'll use a $ to indicate the shell prompt:

```
$ gcc −g −Wall −o hello hello.c
```

- We'll specify the syntax of function calls with fixed argument lists by including a sample argument list. For example, the integer absolute value function, abs, in stdlib, might have its syntax specified with

```
int abs(int x);   /* Returns absolute value of int x */
```

For more complicated syntax, we'll enclose required content in angle brackets <> and optional content in square brackets []. For example, the C **if** statement might have its syntax specified as follows:

```
if ( <expression> )
   <statement1>
[else
   <statement2>]
```

This says that the **if** statement must include an expression enclosed in parentheses, and the right parenthesis must be followed by a statement. This statement can be followed by an optional **else** clause. If the **else** clause is present, it must include a second statement.

1.10 Summary

For many years we've reaped the benefits of having ever-faster processors. However, because of physical limitations, the rate of performance improvement in conventional processors has decreased dramatically. To increase the power of processors, chip-makers have turned to **multicore** integrated circuits, that is, integrated circuits with multiple conventional processors on a single chip.

Ordinary **serial** programs, which are programs written for a conventional single-core processor, usually cannot exploit the presence of multiple cores, and it's unlikely that translation programs will be able to shoulder all the work of converting serial programs into **parallel programs**—programs that can make use of multiple cores. As software developers, we need to learn to write parallel programs.

When we write parallel programs, we usually need to **coordinate** the work of the cores. This can involve **communication** among the cores, **load balancing**, and **synchronization** of the cores.

In this book we'll be learning to program parallel systems, so that we can maximize their performance. We'll be using the C language with four different **application program interfaces** or **APIs**: MPI, Pthreads, OpenMP, and CUDA. These APIs are used to program parallel systems that are classified according to how the cores access memory and whether the individual cores can operate independently of each other.

In the first classification, we distinguish between **shared-memory** and **distributed-memory** systems. In a shared-memory system, the cores share access to one large pool of memory, and they can coordinate their actions by accessing shared memory locations. In a distributed-memory system, each core has its own private memory, and the cores can coordinate their actions by sending messages across a network.

In the second classification, we distinguish between systems with cores that can operate independently of each other and systems in which the cores all execute the same instruction. In both types of system, the cores can operate on their own data stream. So the first type of system is called a **multiple-instruction multiple-data** or **MIMD** system, and the second type of system is called a **single-instruction multiple-data** or **SIMD** system.

MPI is used for programming distributed-memory MIMD systems. Pthreads is used for programming shared-memory MIMD systems. OpenMP can be used to program both shared-memory MIMD and shared-memory SIMD systems, although we'll be looking at using it to program MIMD systems. CUDA is used for programming Nvidia **graphics processing units** or **GPUs**. GPUs have aspects of all four types of system, but we'll be mainly interested in the shared-memory SIMD and shared-memory MIMD aspects.

Concurrent programs can have multiple tasks in progress at any instant. **Parallel** and **distributed** programs usually have tasks that execute simultaneously. There isn't a hard and fast distinction between parallel and distributed, although in parallel programs, the tasks are usually more tightly coupled.

Parallel programs are usually very complex. So it's even more important to use good program development techniques with parallel programs.

1.11 Exercises

1.1 Devise formulas for the functions that calculate `my_first_i` and `my_last_i` in the global sum example. Remember that each core should be assigned roughly the same number of elements of computations in the loop. *Hint*: First consider the case when n is evenly divisible by p.

1.2 We've implicitly assumed that each call to `Compute_next_value` requires roughly the same amount of work as the other calls. How would you change your answer to the preceding question if call $i = k$ requires $k + 1$ times as much work as the call with $i = 0$? How would you change your answer if the first call ($i = 0$) requires 2 milliseconds, the second call ($i = 1$) requires 4, the third ($i = 2$) requires 6, and so on?

1.3 Try to write pseudocode for the tree-structured global sum illustrated in Fig. 1.1. Assume the number of cores is a power of two (1, 2, 4, 8, ...). *Hints*: Use a variable `divisor` to determine whether a core should send its sum or receive and add. The `divisor` should start with the value 2 and be doubled after each iteration. Also use a variable `core_difference` to determine which core should be partnered with the current core. It should start with the value 1 and also be doubled after each iteration. For example, in the first iteration 0 % `divisor` = 0 and 1 % `divisor` = 1, so 0 receives and adds, while 1 sends. Also in the first iteration 0 + `core_difference` = 1 and 1 − `core_difference` = 0, so 0 and 1 are paired in the first iteration.

1.4 As an alternative to the approach outlined in the preceding problem, we can use C's bitwise operators to implement the tree-structured global sum. To see how this works, it helps to write down the binary (base 2) representation of each of the core ranks and note the pairings during each stage:

| | Stages | | |
Cores	1	2	3
$0_{10} = 000_2$	$1_{10} = 001_2$	$2_{10} = 010_2$	$4_{10} = 100_2$
$1_{10} = 001_2$	$0_{10} = 000_2$	×	×
$2_{10} = 010_2$	$3_{10} = 011_2$	$0_{10} = 000_2$	×
$3_{10} = 011_2$	$2_{10} = 010_2$	×	×
$4_{10} = 100_2$	$5_{10} = 101_2$	$6_{10} = 110_2$	$0_{10} = 000_2$
$5_{10} = 101_2$	$4_{10} = 100_2$	×	×
$6_{10} = 110_2$	$7_{10} = 111_2$	$4_{10} = 100_2$	×
$7_{10} = 111_2$	$6_{10} = 110_2$	×	×

From the table, we see that during the first stage each core is paired with the core whose rank differs in the rightmost or first bit. During the second stage, cores that continue are paired with the core whose rank differs in the second bit; and during the third stage, cores are paired with the core whose rank differs in the third bit. Thus if we have a binary value `bitmask` that is 001_2 for the first stage, 010_2 for the second, and 100_2 for the third, we can get the rank of the core we're paired with by "inverting" the bit in our rank that is nonzero in `bitmask`. This can be done using the bitwise exclusive or ∧ operator. Implement this algorithm in pseudocode using the bitwise exclusive or and the left-shift operator.

1.5 What happens if your pseudocode in Exercise 1.3 or Exercise 1.4 is run when the number of cores is *not* a power of two (e.g., 3, 5, 6, 7)? Can you modify the pseudocode so that it will work correctly regardless of the number of cores?

1.6 Derive formulas for the number of receives and additions that core 0 carries out using

   a. the original pseudocode for a global sum, and

   b. the tree-structured global sum.

   Make a table showing the numbers of receives and additions carried out by core 0 when the two sums are used with 2, 4, 8, ..., 1024 cores.

1.7 The first part of the global sum example—when each core adds its assigned computed values—is usually considered to be an example of data-parallelism, while the second part of the first global sum—when the cores send their partial sums to the master core, which adds them—could be considered to be an example of task-parallelism. What about the second part of the second global sum—when the cores use a tree structure to add their partial sums? Is this an example of data- or task-parallelism? Why?

1.8 Suppose the faculty members are throwing a party for the students in the department.

   a. Identify tasks that can be assigned to the faculty members that will allow them to use task-parallelism when they prepare for the party. Work out a schedule that shows when the various tasks can be performed.

   b. We might hope that one of the tasks in the preceding part is cleaning the house where the party will be held. How can we use data-parallelism to partition the work of cleaning the house among the faculty?

   c. Use a combination of task- and data-parallelism to prepare for the party. (If there's too much work for the faculty, you can use TAs to pick up the slack.)

1.9 Write an essay describing a research problem in your major that would benefit from the use of parallel computing. Provide a rough outline of how parallelism would be used. Would you use task- or data-parallelism?

Parallel hardware and parallel software

2

It's perfectly feasible for specialists in disciplines other than computer science and computer engineering to write parallel programs. However, to write *efficient* parallel programs, we often need some knowledge of the underlying hardware and system software. It's also very useful to have some knowledge of different types of parallel software, so in this chapter we'll take a brief look at a few topics in hardware and software. We'll also take a brief look at evaluating program performance and a method for developing parallel programs. We'll close with a discussion of what kind of environment we might expect to be working in, and a few rules and assumptions we'll make in the rest of the book.

This is a long, broad chapter, so it may be a good idea to skim through some of the sections on a first reading so that you have a good idea of what's in the chapter. Then, when a concept or term in a later chapter isn't quite clear, it may be helpful to refer back to this chapter. In particular, you may want to skim over most of the material in "Modifications to the von Neumann Model," except "The Basics of Caching." Also, in the "Parallel Hardware" section, you can safely skim the material on "Interconnection Networks." You can also skim the material on "SIMD Systems" unless you're planning to read the chapter on CUDA programming.

2.1 Some background

Parallel hardware and software have grown out of conventional **serial** hardware and software: hardware and software that runs (more or less) a single job at a time. So to better understand the current state of parallel systems, let's begin with a brief look at a few aspects of serial systems.

2.1.1 The von Neumann architecture

The "classical" **von Neumann architecture** consists of **main memory**, a **central-processing unit** (CPU) or **processor** or **core**, and an **interconnection** between the memory and the CPU. Main memory consists of a collection of locations, each of which is capable of storing both instructions and data. Every location has an *address* and the location's *contents*. The address is used to access the location, and the contents of the location is the instruction or data stored in the location.

An Introduction to Parallel Programming. https://doi.org/10.1016/B978-0-12-804605-0.00009-9

The central processing unit is logically divided into a **control unit** and a **datapath**. The control unit is responsible for deciding which instructions in a program should be executed, and the datapath is responsible for executing the actual instructions. Data in the CPU and information about the state of an executing program are stored in special, very fast storage, called **registers**. The control unit has a special register called the **program counter**. It stores the address of the next instruction to be executed.

Instructions and data are transferred between the CPU and memory via the interconnect. This has traditionally been a **bus**, which consists of a collection of parallel wires and some hardware controlling access to the wires. More recent systems use more complex interconnects. (See Section 2.3.4.) A von Neumann machine executes a single instruction at a time, and each instruction operates on only a few pieces of data. See Fig. 2.1.

When data or instructions are transferred from memory to the CPU, we sometimes say the data or instructions are **fetched** or **read** from memory. When data are transferred from the CPU to memory, we sometimes say the data are **written to mem-**

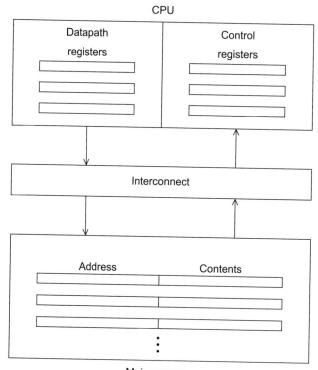

FIGURE 2.1

The von Neumann architecture.

ory or **stored**. The separation of memory and CPU is often called the **von Neumann bottleneck**, since the interconnect determines the rate at which instructions and data can be accessed. The potentially vast quantity of data and instructions needed to run a program is effectively isolated from the CPU. In 2021, CPUs are capable of executing instructions more than one hundred times faster than they can fetch items from main memory.

To better understand this problem, imagine that a large company has a single factory (the CPU) in one town and a single warehouse (main memory) in another. Furthermore, imagine that there is a single two-lane road joining the warehouse and the factory. All the raw materials used in manufacturing the products are stored in the warehouse. Also, all the finished products are stored in the warehouse before being shipped to customers. If the rate at which products can be manufactured is much larger than the rate at which raw materials and finished products can be transported, then it's likely that there will be a huge traffic jam on the road, and the employees and machinery in the factory will either be idle for extended periods or will have to reduce the rate at which they produce finished products.

To address the von Neumann bottleneck and, more generally, improve computer performance, computer engineers and computer scientists have experimented with many modifications to the basic von Neumann architecture. Before discussing some of these modifications, let's first take a moment to discuss some aspects of the software that is used in both von Neumann systems and more modern systems.

2.1.2 Processes, multitasking, and threads

Recall that the **operating system**, or OS, is a major piece of software, whose purpose is to manage hardware and software resources on a computer. It determines which programs can run and when they can run. It also controls the allocation of memory to running programs and access to peripheral devices, such as hard disks and network interface cards.

When a user runs a program, the operating system creates a **process**—an instance of a computer program that is being executed. A process consists of several entities:

- The executable machine language program.
- A block of memory, which will include the executable code, a **call stack** that keeps track of active functions, a **heap** that can be used for memory explicitly allocated by the user program, and some other memory locations.
- Descriptors of resources that the operating system has allocated to the process, for example, file descriptors.
- Security information—for example, information specifying which hardware and software resources the process can access.
- Information about the state of the process, such as whether the process is ready to run or is waiting on some resource, the content of the registers, and information about the process's memory.

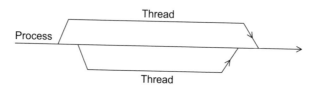

FIGURE 2.2

A process with three threads.

Most modern operating systems are **multitasking**. This means that the operating system provides support for the apparent simultaneous execution of multiple programs. This is possible even on a system with a single core, since each process runs for a small interval of time (typically a few milliseconds), often called a **time slice**. After one running program has executed for a time slice, the operating system can run a different program. A multitasking OS may change the running process many times a minute, even though changing the running process can take a long time.

In a multitasking OS, if a process needs to wait for a resource—for example, it needs to read data from external storage—it will **block**. This means that it will stop executing and the operating system can run another process. However, many programs can continue to do useful work even though the part of the program that is currently executing must wait on a resource. For example, an airline reservation system that is blocked waiting for a seat map for one user could provide a list of available flights to another user. **Threading** provides a mechanism for programmers to divide their programs into more or less independent tasks, with the property that when one thread is blocked, another thread can be run. Furthermore, in most systems it's possible to switch between threads much faster than it's possible to switch between processes. This is because threads are "lighter weight" than processes. Threads are contained within processes, so they can use the same executable, and they usually share the same memory and the same I/O devices. In fact, two threads belonging to one process can share most of the process's resources. The two most important exceptions are that they'll need a record of their own program counters and they'll need their own call stacks, so that they can execute independently of each other.

If a process is the "master" thread of execution and threads are started and stopped by the process, then we can envision the process and its subsidiary threads as lines: when a thread is started, it **forks** off the process; when a thread terminates, it **joins** the process. See Fig. 2.2.

2.2 Modifications to the von Neumann model

As we noted earlier, since the first electronic digital computers were developed back in the 1940s, computer scientists and computer engineers have made many improvements to the basic von Neumann architecture. Many are targeted at reducing the problem of the von Neumann bottleneck, but many are also targeted at simply making CPUs faster. In this section, we'll look at three of these improvements: caching, virtual memory, and low-level parallelism.

2.2.1 The basics of caching

Caching is one of the most widely used methods of addressing the von Neumann bottleneck. To understand the ideas behind caching, recall our example. A company has a factory (CPU) in one town and a warehouse (main memory) in another, and there is a single, two-lane road joining the factory and the warehouse. There are a number of possible solutions to the problem of transporting raw materials and finished products between the warehouse and the factory. One is to widen the road. Another is to move the factory and/or the warehouse, or to build a unified factory and warehouse. Caching exploits both of these ideas. Rather than transporting a single instruction or data item, we can use an effectively wider interconnection that can transport more data or more instructions in a single memory access. Also, rather than storing all data and instructions exclusively in main memory, we can store blocks of data and instructions in special memory that is effectively closer to the registers in the CPU.

In general, a **cache** is a collection of memory locations that can be accessed in less time than some other memory locations. In our setting, when we talk about caches we'll usually mean a **CPU cache**, which is a collection of memory locations that the CPU can access more quickly than it can access main memory. A CPU cache can either be located on the same chip as the CPU or it can be located on a separate chip that can be accessed much faster than an ordinary memory chip. Memory that can be accessed faster than main memory is more expensive. So caches are much smaller than main memory.

Once we have a cache, an obvious problem is deciding which data and instructions should be stored in the cache. The universally used principle is based on the idea that programs tend to use data and instructions that are physically close to recently used data and instructions. After executing an instruction, programs typically execute the next instruction; branching tends to be relatively rare. Similarly, after a program has accessed data in one memory location, it often accesses data in a memory location that is physically nearby. An extreme example of this is in the use of arrays. Consider the loop

```
float  z[1000];
. . .
sum = 0.0;
for (i = 0; i < 1000; i++)
    sum += z[i];
```

Arrays are allocated as blocks of contiguous memory locations. So, for example, the location storing $z[1]$ immediately follows the location $z[0]$. Thus as long as $i < 999$, the read of $z[i]$ is immediately followed by a read of $z[i+1]$.

The principle that an access of one location is followed by an access of a nearby location is often called **locality**. After accessing one memory location (instruction or data), a program will typically access a nearby location (**spatial** locality) in the near future (**temporal** locality).

To exploit the principle of locality, the system uses an effectively *wider* interconnect to access data and instructions. That is, a memory access will effectively operate on blocks of data and instructions instead of individual instructions and individual data items. These blocks are called **cache blocks** or **cache lines**. A typical cache line stores 8 to 16 times as much information as a single memory location. In our example, if a cache line stores 16 floats, then when we first go to add sum += z[0], the system might read the first 16 elements of z, z[0], z[1], ..., z[15] from memory into cache. So the next 15 additions will use elements of z that are already in the cache.

Conceptually, it's often convenient to think of a CPU cache as a single monolithic structure. In practice, though, the cache is usually divided into **levels**: the first level (L1) is the smallest and the fastest, and other levels (L2, L3, ...) are larger and slower. In 2021, most systems have at least two levels, and having three levels is quite common. Caches often store copies of information in slower memory, and, if we think of a lower-numbered (faster, smaller) cache as a cache that's closer to the registers, this often applies. So, for example, a variable stored in a level 1 cache will often also be stored in level 2. However, some multilevel caches don't duplicate information that's available in another level. For these caches, a variable in a level 1 cache might not be stored in any other level of the cache, but it *would* be stored in main memory.

When the CPU needs to access an instruction or data, it works its way down the cache hierarchy: First it checks the level 1 cache, then the level 2, and so on. Finally, if the information needed isn't in any of the caches, it accesses main memory. When a cache is checked for information and the information is available, it's called a **cache hit** or just a **hit**. If the information isn't available, it's called a **cache miss** or a **miss**. Hit or miss is often modified by the level. For example, when the CPU attempts to access a variable, it might have an L1 miss and an L2 hit.

Note that the memory access terms **read** and **write** are also used for caches. For example, we might read an instruction from an L2 cache, and we might write data to an L1 cache.

When the CPU attempts to read data or instructions and there's a cache read-miss, it will read from memory the cache line that contains the needed information and store it in the cache. This may stall the processor while it waits for the slower memory: the processor may stop executing statements from the current program until the required data or instructions have been fetched from memory. So in our example, when we read z[0], the processor may stall while the cache line containing z[0] is transferred from memory into the cache.

When the CPU writes data to a cache, the value in the cache and the value in main memory are different or **inconsistent**. There are two basic approaches to dealing with the inconsistency. In **write-through** caches, the line is written to main memory when it is written to the cache. In **write-back** caches, the data isn't written immediately. Rather, the updated data in the cache is marked *dirty*, and when the cache line is replaced by a new cache line from memory, the dirty line is written to memory.

Table 2.1 Assignments of a 16-line main memory to a 4-line cache.

Memory Index	Cache Location		
	Fully Assoc	**Direct Mapped**	**2-way**
0	0, 1, 2, or 3	0	0 or 1
1	0, 1, 2, or 3	1	2 or 3
2	0, 1, 2, or 3	2	0 or 1
3	0, 1, 2, or 3	3	2 or 3
4	0, 1, 2, or 3	0	0 or 1
5	0, 1, 2, or 3	1	2 or 3
6	0, 1, 2, or 3	2	0 or 1
7	0, 1, 2, or 3	3	2 or 3
8	0, 1, 2, or 3	0	0 or 1
9	0, 1, 2, or 3	1	2 or 3
10	0, 1, 2, or 3	2	0 or 1
11	0, 1, 2, or 3	3	2 or 3
12	0, 1, 2, or 3	0	0 or 1
13	0, 1, 2, or 3	1	2 or 3
14	0, 1, 2, or 3	2	0 or 1
15	0, 1, 2, or 3	3	2 or 3

2.2.2 Cache mappings

Another issue in cache design is deciding where lines should be stored. That is, if we fetch a cache line from main memory, where in the cache should it be placed? The answer to this question varies from system to system. At one extreme is a **fully associative** cache, in which a new line can be placed at any location in the cache. At the other extreme is a **direct mapped** cache, in which each cache line has a unique location in the cache to which it will be assigned. Intermediate schemes are called n-**way set associative**. In these schemes, each cache line can be placed in one of n different locations in the cache. For example, in a two-way set associative cache, each line can be mapped to one of two locations.

As an example, suppose our main memory consists of 16 lines with indexes 0–15, and our cache consists of 4 lines with indexes 0–3. In a fully associative cache, line 0 can be assigned to cache location 0, 1, 2, or 3. In a direct mapped cache, we might assign lines by looking at their remainder after division by 4. So lines 0, 4, 8, and 12 would be mapped to cache index 0, lines 1, 5, 9, and 13 would be mapped to cache index 1, and so on. In a two-way set associative cache, we might group the cache into two sets: indexes 0 and 1 form one set—set 0—and indexes 2 and 3 form another—set 1. So we could use the remainder of the main memory index modulo 2, and cache line 0 would be mapped to either cache index 0 or cache index 1. See Table 2.1.

When more than one line in memory can be mapped to several different locations in a cache (fully associative and n-way set associative), we also need to be able to decide which line should be replaced or **evicted**. In our preceding example, if, for example, line 0 is in location 0 and line 2 is in location 1, where would we store line 4? The idea behind most commonly used approaches is called **least recently used**. As the name suggests, the cache has a record of the relative order in which the blocks have been used, and if line 0 were used more recently than line 2, then line 2 would be evicted and replaced by line 4.

2.2.3 Caches and programs: an example

It's important to remember that the workings of the CPU cache are controlled by the system hardware, and we, the programmers, don't directly determine which data and which instructions are in the cache. However, knowing the principle of spatial and temporal locality allows us to have some indirect control over caching. As an example, C stores two-dimensional arrays in "row-major" order. That is, although we think of a two-dimensional array as a rectangular block, memory is effectively a huge one-dimensional array. So in row-major storage, we store row 0 first, then row 1, and so on. In the following two-code segments, we would expect the first pair of nested loops to have much better performance than the second, since it's accessing the data in the two-dimensional array in contiguous blocks:

```
double  A[MAX][MAX],  x[MAX],  y[MAX];
.  .  .
/*  Initialize  A  and  x,  assign  y  =  0  */
.  .  .
/*  First  pair  of  loops  */
for  (i  =  0;  i  <  MAX;  i++)
    for  (j  =  0;  j  <  MAX;  j++)
        y[i]  +=  A[i][j]*x[j];
.  .  .
/*  Assign  y  =  0  */
.  .  .
/*  Second  pair  of  loops  */
for  (j  =  0;  j  <  MAX;  j++)
    for  (i  =  0;  i  <  MAX;  i++)
        y[i]  +=  A[i][j]*x[j];
```

To better understand this, suppose MAX is four, a cache line stores four doubles, and the elements of A are stored in memory as follows:

Cache Line	Elements of A			
0	A[0][0]	A[0][1]	A[0][2]	A[0][3]
1	A[1][0]	A[1][1]	A[1][2]	A[1][3]
2	A[2][0]	A[2][1]	A[2][2]	A[2][3]
3	A[3][0]	A[3][1]	A[3][2]	A[3][3]

So, for example, A[0][1] is stored immediately after A[0][0], and A[1][0] is stored immediately after A[0][3].

Let's suppose that none of the elements of A are in the cache when each pair of loops starts executing. Let's also suppose that the cache is direct mapped, and it can only store eight elements of A, or two cache lines. (We won't worry about x and y.)

Both pairs of loops attempt to first access A[0][0]. Since it's not in the cache, this will result in a cache miss, and the system will read the line consisting of the first row of A, A[0][0], A[0][1], A[0][2], A[0][3] into the cache. The first pair of loops then accesses A[0][1], A[0][2], A[0][3], all of which are in the cache, and the next miss in the first pair of loops will occur when the code accesses A[1][0]. Continuing in this fashion, we see that the first pair of loops will result in a total of four misses when it accesses elements of A, one for each row. Note that since our hypothetical cache can only store two lines or eight elements of A, when we read the first element of row two and the first element of row three, one of the lines that's already in the cache will have to be evicted from the cache, but once a line is evicted, the first pair of loops won't need to access the elements of that line again.

After reading the first row into the cache, the second pair of loops needs to then access A[1][0], A[2][0], A[3][0], none of which are in the cache. So the next three accesses of A will also result in misses. Furthermore, because the cache is small, the reads of A[2][0] and A[3][0] will require that lines already in the cache be evicted. Since A[2][0] is stored in cache line 2, reading its line will evict line 0, and reading A[3][0] will evict line 1. After finishing the first pass through the outer loop, we'll next need to access A[0][1], which was evicted with the rest of the first row. So we see that *every* time we read an element of A, we'll have a miss, and the second pair of loops results in 16 misses.

Thus we'd expect the first pair of nested loops to be much faster than the second. In fact, if we run the code on one of our systems with MAX = 1000, the first pair of nested loops is approximately three times faster than the second pair.

2.2.4 Virtual memory

Caches make it possible for the CPU to quickly access instructions and data that are in main memory. However, if we run a very large program or a program that accesses very large data sets, all of the instructions and data may not fit into main memory. This is especially true with multitasking operating systems; to switch between programs and create the illusion that multiple programs are running simultaneously, the instructions and data that will be used during the next time slice should be in main memory. Thus in a multitasking system, even if the main memory is very large, many running programs must share the available main memory. Furthermore, this sharing must be done in such a way that each program's data and instructions are protected from corruption by other programs.

Virtual memory was developed so that main memory can function as a cache for secondary storage. It exploits the principle of spatial and temporal locality by keeping in main memory only the active parts of the many running programs; those parts that are idle can be kept in a block of secondary storage, called **swap space**. Like CPU caches, virtual memory operates on blocks of data and instructions. These blocks are commonly called **pages**, and since secondary storage access can be hundreds of thousands of times slower than main memory access, pages are relatively large—most systems have a fixed page size that currently ranges from 4 to 16 kilobytes.

Table 2.2 Virtual address divided into virtual page number and byte offset.

Virtual Address								
Virtual Page Number					Byte Offset			
31	30	...	13	12	11	10	...	1	0
1	0	...	1	1	0	0	...	1	1

We may run into trouble if we try to assign physical memory addresses to pages when we compile a program. If we do this, then each page of the program can only be assigned to one block of memory, and with a multitasking operating system, we're likely to have many programs wanting to use the same block of memory. To avoid this problem, when a program is compiled, its pages are assigned *virtual* page numbers. Then, when the program is run, a table is created that maps the virtual page numbers to physical addresses. When the program is run and it refers to a virtual address, this **page table** is used to translate the virtual address into a physical address. If the creation of the page table is managed by the operating system, it can ensure that the memory used by one program doesn't overlap the memory used by another.

A drawback to the use of a page table is that it can double the time needed to access a location in main memory. Suppose, for example, that we want to execute an instruction in main memory. Then our executing program will have the *virtual* address of this instruction, but before we can find the instruction in memory, we'll need to translate the virtual address into a physical address. To do this, we'll need to find the page in memory that contains the instruction. Now the virtual page number is stored as a part of the virtual address. As an example, suppose our addresses are 32 bits and our pages are 4 kilobytes = 4096 bytes. Then each byte in the page can be identified with 12 bits, since $2^{12} = 4096$. Thus we can use the low-order 12 bits of the virtual address to locate a byte within a page, and the remaining bits of the virtual address can be used to locate an individual page. See Table 2.2. Observe that the virtual page number can be computed from the virtual address without going to memory. However, once we've found the virtual page number, we'll need to access the page table to translate it into a physical page address. If the required part of the page table isn't in cache, we'll need to load it from memory. After it's loaded, we can translate our virtual address to a physical address, and get the required instruction.

This is clearly a problem. Although multiple programs can use main memory at more or less the same time, using a page table has the potential to significantly increase each program's overall run-time. To address this issue, processors have a special address translation cache, called a **translation-lookaside buffer**, or TLB. It caches a small number of entries (typically 16–512) from the page table in very fast memory. Using the principle of spatial and temporal locality, we would expect that most of our memory references will be to pages whose physical address is stored in the TLB, and the number of memory references that require accesses to the page table in main memory will be substantially reduced.

The terminology for the TLB is the same as the terminology for caches. When we look for an address and the virtual page number is in the TLB, it's called a TLB *hit*. If it's not in the TLB, it's called a *miss*. The terminology for the page table, however, has an important difference from the terminology for caches. If we attempt to access a page that's not in memory—that is, the page table doesn't have a valid physical address for the page and the page is only stored on disk—then the attempted access is called a **page fault**.

The relative slowness of disk accesses has a couple of additional consequences for virtual memory. First, with CPU caches, we could handle write-misses with either a write-through or write-back scheme. With virtual memory, however, disk accesses are so expensive that they should be avoided whenever possible, so virtual memory always uses a write-back scheme. This can be handled by keeping a bit on each page in memory that indicates whether the page has been updated. If it has been updated, when it is evicted from main memory, it will be written to disk. Second, since disk accesses are so slow, management of the page table and the handling of disk accesses can be done by the operating system. Thus, even though we as programmers don't directly control virtual memory, unlike CPU caches, which are handled by system hardware, virtual memory is usually controlled by a combination of system hardware and operating system software.

2.2.5 Instruction-level parallelism

Instruction-level parallelism, or ILP, attempts to improve processor performance by having multiple processor components or **functional units** simultaneously executing instructions. There are two main approaches to ILP: **pipelining**, in which functional units are arranged in stages; and **multiple issue**, in which multiple instructions can be simultaneously initiated. Both approaches are used in virtually all modern CPUs.

Pipelining

The principle of pipelining is similar to a factory assembly line: while one team is bolting a car's engine to the chassis, another team can connect the transmission to the engine and the driveshaft of a car that's already been processed by the first team, and a third team can bolt the body to the chassis in a car that's been processed by the first two teams. As an example involving computation, suppose we want to add the floating point numbers 9.87×10^4 and 6.54×10^3. Then we can use the following steps:

Time	Operation	Operand 1	Operand 2	Result
1	Fetch operands	9.87×10^4	6.54×10^3	
2	Compare exponents	9.87×10^4	6.54×10^3	
3	Shift one operand	9.87×10^4	0.654×10^4	
4	Add	9.87×10^4	0.654×10^4	10.524×10^4
5	Normalize result	9.87×10^4	0.654×10^4	1.0524×10^5
6	Round result	9.87×10^4	0.654×10^4	1.05×10^5
7	Store result	9.87×10^4	0.654×10^4	1.05×10^5

Table 2.3 Pipelined addition. Numbers in the table are subscripts of operands/results.

Time	Fetch	Compare	Shift	Add	Normalize	Round	Store
0	0						
1	1	0					
2	2	1	0				
3	3	2	1	0			
4	4	3	2	1	0		
5	5	4	3	2	1	0	
6	6	5	4	3	2	1	0
⋮	⋮	⋮	⋮	⋮	⋮	⋮	⋮
999	999	998	997	996	995	994	993
1000		999	998	997	996	995	994
1001			999	998	997	996	995
1003				999	998	997	
1004					999	998	
1005						999	

Here we're using base 10 and a three-digit mantissa or significand with one digit to the left of the decimal point. Thus, in the example, normalizing shifts the decimal point one unit to the left, and rounding rounds to three digits.

Now if each of the operations takes one nanosecond (10^{-9} seconds), the addition operation will take seven nanoseconds. So if we execute the code

```
float x[1000], y[1000], z[1000];
. . .
for (i = 0; i < 1000; i++)
    z[i] = x[i] + y[i];
```

the **for** loop will take something like 7000 nanoseconds.

As an alternative, suppose we divide our floating point adder into seven separate pieces of hardware or functional units. The first unit will fetch two operands, the second will compare exponents, and so on. Also, suppose that the output of one functional unit is the input to the next. So, for example, the output of the functional unit that adds the two values is the input to the unit that normalizes the result. Then a single floating point addition will still take 7 nanoseconds. However, when we execute the **for** loop, we can fetch $x[1]$ and $y[1]$ while we're comparing the exponents of $x[0]$ and $y[0]$. More generally, it's possible for us to simultaneously execute seven different stages in seven different additions. See Table 2.3. From the table, we see that after time 5, the pipelined loop produces a result every nanosecond, instead of every seven nanoseconds, so the total time to execute the **for** loop has been reduced from 7000 nanoseconds to 1006 nanoseconds—an improvement of almost a factor of seven.

In general, a pipeline with k stages won't get a k-fold improvement in performance. For example, if the times required by the various functional units are different, then the stages will effectively run at the speed of the slowest functional unit. Furthermore, delays—such as waiting for an operand to become available—can cause the pipeline to stall. See Exercise 2.1 for more details on the performance of pipelines.

Multiple issue

Pipelines improve performance by taking individual pieces of hardware or functional units and connecting them in sequence. Multiple issue processors replicate functional units and try to simultaneously execute different instructions in a program. For example, if we have two complete floating point adders, we can approximately halve the time it takes to execute the loop

```
for (i = 0; i < 1000; i++)
    z[i] = x[i] + y[i];
```

While the first adder is computing $z[0]$, the second can compute $z[1]$; while the first is computing $z[2]$, the second can compute $z[3]$; and so on.

If the functional units are scheduled at compile time, the multiple issue system is said to use **static** multiple issue. If they're scheduled at run-time, the system is said to use **dynamic** multiple issue. A processor that supports dynamic multiple issue is sometimes said to be **superscalar**.

Of course, to make use of multiple issue, the system must find instructions that can be executed simultaneously. One of the most important techniques is **speculation**. In speculation, the compiler or the processor makes a guess about an instruction, and then executes the instruction on the basis of the guess. As a simple example, in the following code, the system might predict that the outcome of $z = x + y$ will give z a positive value, and, as a consequence, it will assign $w = x$:

```
z = x + y;
if (z > 0)
    w = x;
else
    w = y;
```

As another example, in the code

```
z = x + y;
w = *a_p;   /* a_p is a pointer */
```

the system might predict that a_p does not refer to z, and hence it can simultaneously execute the two assignments.

As both examples make clear, speculative execution must allow for the possibility that the predicted behavior is incorrect. In the first example, we will need to go back and execute the assignment $w = y$ if the assignment $z = x + y$ results in a value that's not positive. In the second example, if a_p does point to z, we'll need to re-execute the assignment $w = *a\_p$.

If the compiler does the speculation, it will usually insert code that tests whether the speculation was correct, and, if not, takes corrective action. If the hardware does the speculation, the processor usually stores the result(s) of the speculative execution in a buffer. When it's known that the speculation was correct, the contents of the buffer are transferred to registers or memory. If the speculation was incorrect, the contents of the buffer are discarded, and the instruction is re-executed.

While dynamic multiple issue systems can execute instructions out of order, in current generation systems the instructions are still loaded in order, and the results of the instructions are also committed in order. That is, the results of instructions are written to registers and memory in the program-specified order.

Optimizing compilers, on the other hand, can reorder instructions. This, as we'll see later, can have important consequences for shared-memory programming.

2.2.6 Hardware multithreading

ILP can be very difficult to exploit: a program with a long sequence of dependent statements offers few opportunities. For example, in a direct calculation of the Fibonacci numbers

```
f[0] = f[1] = 1;
for (i = 2; i <= n; i++)
    f[i] = f[i-1] + f[i-2];
```

there's essentially no opportunity for simultaneous execution of instructions.

Thread-level parallelism, or TLP, attempts to provide parallelism through the simultaneous execution of different threads, so it provides a **coarser-grained** parallelism than ILP, that is, the program units that are being simultaneously executed—threads—are larger or coarser than the **finer-grained** units—individual instructions.

Hardware multithreading provides a means for systems to continue doing useful work when the task being currently executed has stalled—for example, if the current task has to wait for data to be loaded from memory. Instead of looking for parallelism in the currently executing thread, it may make sense to simply run another thread. Of course, for this to be useful, the system must support very rapid switching between threads. For example, in some older systems, threads were simply implemented as processes, and in the time it took to switch between processes, thousands of instructions could be executed.

In **fine-grained** multithreading, the processor switches between threads after each instruction, skipping threads that are stalled. While this approach has the potential to avoid wasted machine time due to stalls, it has the drawback that a thread that's ready to execute a long sequence of instructions may have to wait to execute every instruction. **Coarse-grained** multithreading attempts to avoid this problem by only switching threads that are stalled waiting for a time-consuming operation to complete (e.g., a load from main memory). This has the virtue that switching threads doesn't need to be nearly instantaneous. However, the processor can be idled on shorter stalls, and thread switching will also cause delays.

Simultaneous multithreading, or SMT, is a variation on fine-grained multithreading. It attempts to exploit superscalar processors by allowing multiple threads to make use of the multiple functional units. If we designate "preferred" threads—threads that have many instructions ready to execute—we can somewhat reduce the problem of thread slowdown.

2.3 Parallel hardware

Multiple issue and pipelining can clearly be considered to be parallel hardware, since different functional units can be executed simultaneously. However, since this form of parallelism isn't usually visible to the programmer, we're treating both of them as extensions to the basic von Neumann model, and for our purposes, parallel hardware will be limited to hardware that's visible to the programmer. In other words, if she can readily modify her source code to exploit it, or if she must modify her source code to exploit it, then we'll consider the hardware to be parallel.

2.3.1 Classifications of parallel computers

We'll make use of two, independent classifications of parallel computers. The first, **Flynn's taxonomy** [20], classifies a parallel computer according to the number of instruction streams and the number of data streams (or datapaths) it can simultaneously manage. A classical von Neumann system is therefore a **single instruction stream, single data stream**, or SISD system, since it executes a single instruction at a time and, in most cases, computes a single data value at a time. The parallel systems in the classification can always manage multiple data streams, and we differentiate between systems that support only a single instruction stream (SIMD) and systems that support multiple instruction streams (MIMD).

We already encountered the alternative classification in Chapter 1: it has to do with how the cores access memory. In **shared memory** systems, the cores can share access to memory locations, and the cores coordinate their work by modifying shared memory locations. In **distributed memory** systems, each core has its own, private memory, and the cores coordinate their work by communicating across a network.

2.3.2 SIMD systems

Single instruction, multiple data, or SIMD, systems are parallel systems. As the name suggests, SIMD systems operate on multiple data streams by applying the same instruction to multiple data items, so an abstract SIMD system can be thought of as having a single control unit and multiple datapaths. An instruction is broadcast from the control unit to the datapaths, and each datapath either applies the instruction to the current data item, or it is idle. As an example, suppose we want to carry out a "vector addition." That is, suppose we have two arrays x and y, each with n elements, and we want to add the elements of y to the elements of x:

```
for (i = 0; i < n; i++)
    x[i] += y[i];
```

Suppose further that our SIMD system has n datapaths. Then we could load $x[i]$ and $y[i]$ into the ith datapath, have the ith datapath add $y[i]$ to $x[i]$, and store the result in $x[i]$. If the system has m datapaths and $m < n$, we can simply execute the additions in blocks of m elements at a time. For example, if $m = 4$ and $n = 15$, we can first add elements 0 to 3, then elements 4 to 7, then elements 8 to 11, and finally elements 12 to 14. Note that in the last group of elements in our example—elements 12 to 14—we're only operating on three elements of x and y, so one of the four datapaths will be idle.

The requirement that all the datapaths execute the same instruction or are idle can seriously degrade the overall performance of a SIMD system. For example, suppose we only want to carry out the addition if $y[i]$ is positive:

```
for (i = 0; i < n; i++)
    if (y[i] > 0.0) x[i] += y[i];
```

In this setting, we must load each element of y into a datapath and determine whether it's positive. If $y[i]$ is positive, we can proceed to carry out the addition. Otherwise, the datapath storing $y[i]$ will be idle while the other datapaths carry out the addition.

Note also that in a "classical" SIMD system, the datapaths must operate synchronously, that is, each datapath must wait for the next instruction to be broadcast before proceeding. Furthermore, the datapaths have no instruction storage, so a datapath can't delay execution of an instruction by storing it for later execution.

Finally, as our first example shows, SIMD systems are ideal for parallelizing simple loops that operate on large arrays of data. Parallelism that's obtained by dividing data among the processors and having the processors all apply (more or less) the same instructions to their subsets of the data is called **data-parallelism**. SIMD parallelism can be very efficient on large data parallel problems, but SIMD systems often don't do very well on other types of parallel problems.

SIMD systems have had a somewhat checkered history. In the early 1990s, a maker of SIMD systems (Thinking Machines) was one of the largest manufacturers of parallel supercomputers. However, by the late 1990s, the only widely produced SIMD systems were **vector processors**. More recently, graphics processing units, or GPUs, and desktop CPUs are making use of aspects of SIMD computing.

Vector processors

Although what constitutes a vector processor has changed over the years, their key characteristic is that they can operate on arrays or *vectors* of data, while conventional CPUs operate on individual data elements or *scalars*. Typical recent systems have the following characteristics:

- *Vector registers.* These are registers capable of storing a vector of operands and operating simultaneously on their contents. The vector length is fixed by the system, and can range from 4 to 256 64-bit elements.

- *Vectorized and pipelined functional units.* Note that the same operation is applied to each element in the vector, or, in the case of operations like addition, the same operation is applied to each pair of corresponding elements in the two vectors. Thus vector operations are SIMD.
- *Vector instructions.* These are instructions that operate on vectors rather than scalars. If the vector length is `vector_length`, these instructions have the great virtue that a simple loop such as

```
for (i = 0; i < n; i++)
    x[i] += y[i];
```

requires only a single load, add, and store for each block of `vector_length` elements, while a conventional system requires a load, add, and store for each element.
- *Interleaved memory.* The memory system consists of multiple "banks" that can be accessed more or less independently. After accessing one bank, there will be a delay before it can be reaccessed, but a different bank can be accessed much sooner. So if the elements of a vector are distributed across multiple banks, there can be little to no delay in loading/storing successive elements.
- *Strided memory access and hardware scatter/gather.* In *strided memory access*, the program accesses elements of a vector located at fixed intervals. For example, accessing the first element, the fifth element, the ninth element, and so on would be strided access with a stride of four. Scatter/gather (in this context) is writing (scatter) or reading (gather) elements of a vector located at irregular intervals— for example, accessing the first element, the second element, the fourth element, the eighth element, and so on. Typical vector systems provide special hardware to accelerate strided access and scatter/gather.

Vector processors have the virtue that for many applications, they are very fast and very easy to use. Vectorizing compilers are quite good at identifying code that can be vectorized. Furthermore, they identify loops that cannot be vectorized, and they often provide information about why a loop couldn't be vectorized. The user can thereby make informed decisions about whether it's possible to rewrite the loop so that it will vectorize. Vector systems have very high memory bandwidth, and every data item that's loaded is actually used, unlike cache-based systems that may not make use of every item in a cache line. On the other hand, they don't handle irregular data structures as well as other parallel architectures, and there seems to be a very finite limit to their **scalability**, that is, their ability to handle ever larger problems. It's difficult to see how systems could be created that would operate on ever longer vectors. Current generation systems usually scale by increasing the number of vector processors, not the vector length. Current commodity systems provide limited support for operations on very short vectors, while processors that operate on long vectors are custom manufactured, and, consequently, very expensive.

Graphics processing units

Real-time graphics application programming interfaces, or APIs, use points, lines, and triangles to internally represent the surface of an object. They use a **graphics processing pipeline** to convert the internal representation into an array of pixels that can be sent to a computer screen. Several of the stages of this pipeline are programmable. The behavior of the programmable stages is specified by functions called **shader functions**. The shader functions are typically quite short—often just a few lines of C code. They're also implicitly parallel, since they can be applied to multiple elements (e.g., vertices) in the graphics stream. Since the application of a shader function to nearby elements often results in the same flow of control, GPUs can optimize performance by using SIMD parallelism, and in the current generation all GPUs use SIMD parallelism. This is obtained by including a large number of datapaths (e.g., 128) on each GPU processing core.

Processing a single image can require very large amounts of data—hundreds of megabytes of data for a single image is not unusual. GPUs therefore need to maintain very high rates of data movement, and to avoid stalls on memory accesses, they rely heavily on hardware multithreading; some systems are capable of storing the state of more than a hundred suspended threads for each executing thread. The actual number of threads depends on the amount of resources (e.g., registers) needed by the shader function. A drawback here is that many threads processing a lot of data are needed to keep the datapaths busy, and GPUs may have relatively poor performance on small problems.

It should be stressed that GPUs are not pure SIMD systems. Although the datapaths on a given core can use SIMD parallelism, current generation GPUs can run more than one instruction stream on a single core. Furthermore, typical GPUs can have dozens of cores, and these cores are also capable of executing independent instruction streams. So GPUs are neither purely SIMD nor purely MIMD.

Also note that GPUs can use shared or distributed memory. The largest systems often use both: several cores may have access to a common block of memory, but other SIMD cores may have access to a different block of shared memory, and two cores with access to different blocks of shared memory may communicate over a network. However, in the remainder of this text, we'll limit our discussion to GPUs that use shared memory.

GPUs are becoming increasingly popular for general, high-performance computing, and several languages have been developed that allow users to exploit their power. We'll go into more detail on the architecture of Nvidia processors and how to program them in Chapter 6. Also see [33].

2.3.3 MIMD systems

Multiple instruction, multiple data, or MIMD, systems support multiple simultaneous instruction streams operating on multiple data streams. Thus MIMD systems typically consist of a collection of fully independent processing units or cores, each of which has its own control unit and its own datapath. Furthermore, unlike SIMD

systems, MIMD systems are usually **asynchronous**, that is, the processors can operate at their own pace. In many MIMD systems, there is no global clock, and there may
be no relation between the system times on two different processors. In fact, unless
the programmer imposes some synchronization, even if the processors are executing
exactly the same sequence of instructions, at any given instant they may be executing
different statements.

As we noted in Chapter 1, there are two principal types of MIMD systems: shared-
memory systems and distributed-memory systems. In a **shared-memory system** a
collection of autonomous processors is connected to a memory system via an interconnection network, and each processor can access each memory location. In a
shared-memory system, the processors usually communicate implicitly by accessing
shared data structures. In a **distributed-memory system**, each processor is paired
with its own *private* memory, and the processor-memory pairs communicate over
an interconnection network. So in distributed-memory systems, the processors usually communicate explicitly by sending messages or by using special functions that
provide access to the memory of another processor. See Figs. 2.3 and 2.4.

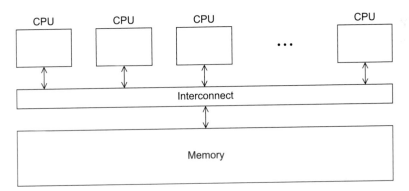

FIGURE 2.3

A shared-memory system.

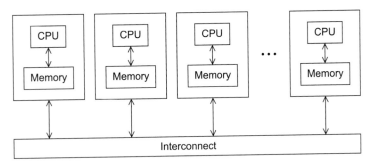

FIGURE 2.4

A distributed-memory system.

Shared-memory systems

The most widely available shared-memory systems use one or more **multicore** processors. As we discussed in Chapter 1, a multicore processor has multiple CPUs or cores on a single chip. Typically, the cores have private level 1 caches, while other caches may or may not be shared between the cores.

In shared-memory systems with multiple multicore processors, the interconnect can either connect all the processors directly to main memory, or each processor can have a direct connection to a block of main memory, and the processors can access each other's blocks of main memory through special hardware built into the processors. (See Figs. 2.5 and 2.6.) In the first type of system, the time to access all the memory locations will be the same for all the cores, while in the second type, a memory location to which a core is directly connected, can be accessed more quickly than a memory location that must be accessed through another chip. Thus the first type of system is called a **uniform memory access**, or UMA, system, while the second type is called a **nonuniform memory access**, or NUMA, system. UMA systems are

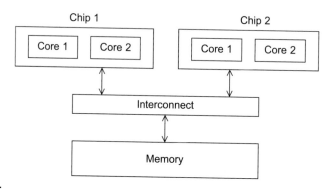

FIGURE 2.5

A UMA multicore system.

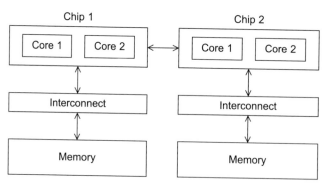

FIGURE 2.6

A NUMA multicore system.

usually easier to program, since the programmer doesn't need to worry about different access times for different memory locations. This advantage can be offset by the faster access to the directly connected memory in NUMA systems. Furthermore, NUMA systems have the potential to use larger amounts of memory than UMA systems.

Distributed-memory systems

The most widely available distributed-memory systems are called **clusters**. They are composed of a collection of commodity systems—for example, PCs—connected by a commodity interconnection network—for example, Ethernet. In fact, the **nodes** of these systems, the individual computational units joined together by the communication network, are usually shared-memory systems with one or more multicore processors. To distinguish such systems from pure distributed-memory systems, they are sometimes called **hybrid systems**. Nowadays, it's usually understood that a cluster has shared-memory nodes.

The **grid** provides the infrastructure necessary to turn large networks of geographically distributed computers into a unified distributed-memory system. In general, such a system is *heterogeneous*, that is, the individual nodes are built from different types of hardware.

2.3.4 Interconnection networks

The interconnect plays a decisive role in the performance of both distributed- and shared-memory systems: even if the processors and memory have virtually unlimited performance, a slow interconnect will seriously degrade the overall performance of all but the simplest parallel program. See, for example, Exercise 2.11.

Although some of the interconnects have a great deal in common, there are enough differences to make it worthwhile to treat interconnects for shared-memory and distributed-memory separately.

Shared-memory interconnects

In the past, it was common for shared memory systems to use a **bus** to connect processors and memory. Originally, a **bus** was a collection of parallel communication wires together with some hardware that controls access to the bus. The key characteristic of a bus is that the communication wires are shared by the devices that are connected to it. Buses have the virtue of low cost and flexibility; multiple devices can be connected to a bus with little additional cost. However, since the communication wires are shared, as the number of devices connected to the bus increases, the likelihood that there will be contention for use of the bus increases, and the expected performance of the bus decreases. Therefore if we connect a large number of processors to a bus, we would expect that the processors would frequently have to wait for access to main memory. So, as the size of shared-memory systems has increased, buses are being replaced by *switched* interconnects.

As the name suggests, **switched** interconnects use switches to control the routing of data among the connected devices. A **crossbar** is a relatively simple and powerful switched interconnect. The diagram in Fig. 2.7(a) shows a simple crossbar. The lines are bidirectional communication links, the squares are cores or memory modules, and the circles are switches.

The individual switches can assume one of the two configurations shown in Fig. 2.7(b). With these switches and at least as many memory modules as processors, there will only be a conflict between two cores attempting to access memory if the two cores attempt to simultaneously access the same memory module. For example, Fig. 2.7(c) shows the configuration of the switches if P_1 writes to M_4, P_2 reads from M_3, P_3 reads from M_1, and P_4 writes to M_2.

Crossbars allow simultaneous communication among different devices, so they are much faster than buses. However, the cost of the switches and links is relatively high. A small bus-based system will be much less expensive than a crossbar-based system of the same size.

Distributed-memory interconnects

Distributed-memory interconnects are often divided into two groups: direct interconnects and indirect interconnects. In a **direct interconnect** each switch is directly connected to a processor-memory pair, and the switches are connected to each other. Fig. 2.8 shows a **ring** and a two-dimensional **toroidal mesh**. As before, the circles are switches, the squares are processors, and the lines are bidirectional links.

One of the simplest measures of the power of a direct interconnect is the number of links. When counting links in a direct interconnect, it's customary to count only switch-to-switch links. This is because the speed of the processor-to-switch links may be very different from the speed of the switch-to-switch links. Furthermore, to get the total number of links, we can usually just add the number of processors to the number of switch-to-switch links. So in the diagram for a ring (Fig. 2.8a), we would ordinarily count 3 links instead of 6, and in the diagram for the toroidal mesh (Fig. 2.8b), we would count 18 links instead of 27.

A ring is superior to a simple bus, since it allows multiple simultaneous communications. However, it's easy to devise communication schemes, in which some of the processors must wait for other processors to complete their communications. The toroidal mesh will be more expensive than the ring, because the switches are more complex—they must support five links instead of three—and if there are p processors, the number of links is $2p$ in a toroidal mesh, while it's only p in a ring. However, it's not difficult to convince yourself that the number of possible simultaneous communications patterns is greater with a mesh than with a ring.

One measure of "number of simultaneous communications" or "connectivity" is **bisection width**. To understand this measure, imagine that the parallel system is divided into two halves, and each half contains half of the processors or nodes. How many simultaneous communications can take place "across the divide" between the halves? In Fig. 2.9(a) we've divided a ring with eight nodes into two groups of four nodes, and we can see that only two communications can take place between the

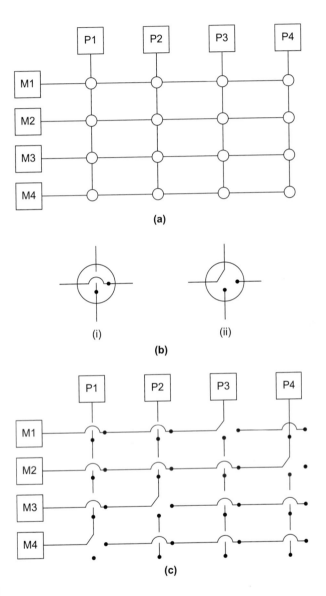

FIGURE 2.7

(a) A crossbar switch connecting four processors (P_i) and four memory modules (M_j); (b) configuration of internal switches in a crossbar; (c) simultaneous memory accesses by the processors.

halves. (To make the diagrams easier to read, we've grouped each node with its switch in this and subsequent diagrams of direct interconnects.) However, in Fig. 2.9(b) we've divided the nodes into two parts so that four simultaneous communications

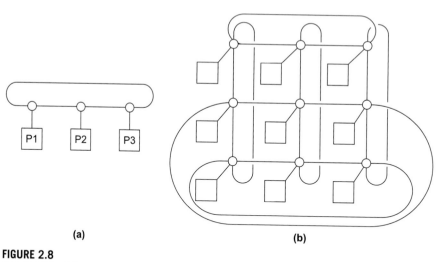

(a)   (b)

FIGURE 2.8

(a) A ring and (b) a toroidal mesh.

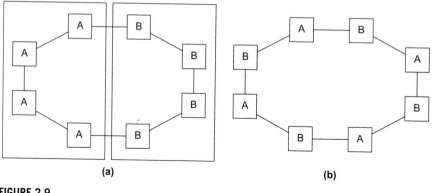

(a)   (b)

FIGURE 2.9

Two bisections of a ring: (a) only two communications can take place between the halves and (b) four simultaneous connections can take place.

can take place, so what's the bisection width? The bisection width is supposed to give a "worst-case" estimate, so the bisection width is two—not four.

An alternative way of computing the bisection width is to remove the minimum number of links needed to split the set of nodes into two equal halves. The number of links removed is the bisection width. If we have a square two-dimensional toroidal mesh with $p = q^2$ nodes (where q is even), then we can split the nodes into two halves by removing the "middle" horizontal links and the "wraparound" horizontal links. (See Fig. 2.10.) This suggests that the bisection width is at most $2q = 2\sqrt{p}$. In

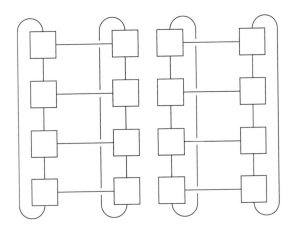

FIGURE 2.10

A bisection of a toroidal mesh.

fact, this is the smallest possible number of links, and the bisection width of a square two-dimensional toroidal mesh is $2\sqrt{p}$.

The **bandwidth** of a link is the rate at which it can transmit data. It's usually given in megabits or megabytes per second. **Bisection bandwidth** is often used as a measure of network quality. It's similar to bisection width. However, instead of counting the number of links joining the halves, it sums the bandwidth of the links. For example, if the links in a ring have a bandwidth of one billion bits per second, then the bisection bandwidth of the ring will be two billion bits per second or 2000 megabits per second.

The ideal direct interconnect is a **fully connected network**, in which each switch is directly connected to every other switch. See Fig. 2.11. Its bisection width is $p^2/4$. However, it's impractical to construct such an interconnect for systems with more than a few nodes, since it requires a total of $p^2/2 - p/2$ links, and each switch must be capable of connecting to p links. It is therefore more a "theoretical best possible" interconnect than a practical one, and it is used as a basis for evaluating other interconnects.

The **hypercube** is a highly connected direct interconnect that has been used in actual systems. Hypercubes are built inductively: A one-dimensional hypercube is a fully connected system with two processors. A two-dimensional hypercube is built from two one-dimensional hypercubes by joining "corresponding" switches. Similarly, a three-dimensional hypercube is built from two two-dimensional hypercubes. (See Fig. 2.12.) Thus a hypercube of dimension d has $p = 2^d$ nodes, and a switch in a d-dimensional hypercube is directly connected to a processor and d switches. The bisection width of a hypercube is $p/2$, so it has more connectivity than a ring or toroidal mesh, but the switches must be more powerful, since they must support $1 + d = 1 + \log_2(p)$ wires, while the mesh switches only require five wires. So a hypercube with p nodes is more expensive to construct than a toroidal mesh.

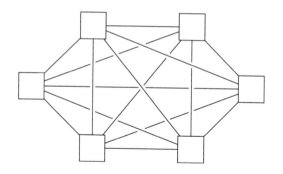

FIGURE 2.11

A fully connected network.

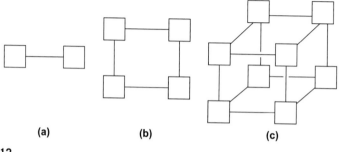

FIGURE 2.12

(a) One-, (b) two-, and (c) three-dimensional hypercubes.

Indirect interconnects provide an alternative to direct interconnects. In an indirect interconnect, the switches may not be directly connected to a processor. They're often shown with unidirectional links and a collection of processors, each of which has an outgoing and an incoming link, and a switching network. (See Fig. 2.13.)

The **crossbar** and the **omega network** are relatively simple examples of indirect networks. We saw a shared-memory crossbar with bidirectional links earlier (Fig. 2.7). The diagram of a distributed-memory crossbar in Fig. 2.14 has unidirectional links. Notice that as long as two processors don't attempt to communicate with the same processor, all the processors can simultaneously communicate with another processor.

An omega network is shown in Fig. 2.15. The switches are two-by-two crossbars (see Fig. 2.16). Observe that unlike the crossbar, there are communications that cannot occur simultaneously. For example, in Fig. 2.15, if processor 0 sends a message to processor 6, then processor 1 cannot simultaneously send a message to processor 7. On the other hand, the omega network is less expensive than the crossbar. The omega network uses $\frac{1}{2}p \log_2(p)$ of the 2×2 crossbar switches, so it uses a total of $2p \log_2(p)$ switches, while the crossbar uses p^2.

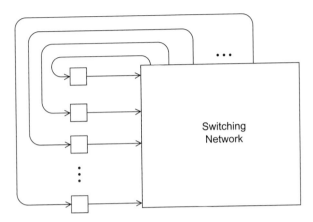

FIGURE 2.13

A generic indirect network.

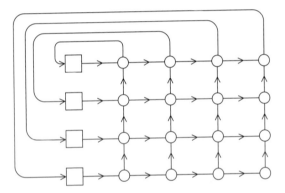

FIGURE 2.14

A crossbar interconnect for distributed-memory.

It's a little bit more complicated to define bisection width for indirect networks. However, the principle is the same: we want to divide the nodes into two groups of equal size and determine how much communication can take place between the two halves, or alternatively, the minimum number of links that need to be removed so that the two groups can't communicate. The bisection width of a $p \times p$ crossbar is p, and the bisection width of an omega network is $p/2$.

Latency and bandwidth

Any time data is transmitted, we're interested in how long it will take for the data to reach its destination. This is true whether we're talking about transmitting data between main memory and cache, cache and register, hard disk and memory, or between two nodes in a distributed-memory or hybrid system. There are two figures

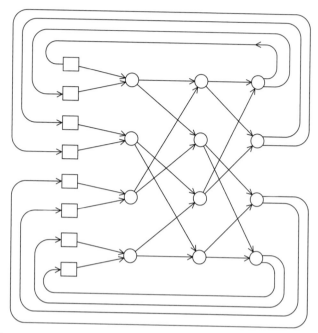

FIGURE 2.15

An omega network.

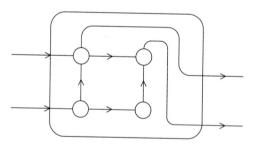

FIGURE 2.16

A switch in an omega network.

that are often used to describe the performance of an interconnect (regardless of what it's connecting): the **latency** and the **bandwidth**. The latency is the time that elapses between the source's beginning to transmit the data and the destination's starting to receive the first byte. The bandwidth is the rate at which the destination receives data after it has started to receive the first byte. So if the latency of an interconnect is l seconds and the bandwidth is b bytes per second, then the time it takes to transmit a message of n bytes is

$$\text{message transmission time} = l + n/b.$$

Beware, however, that these terms are often used in different ways. For example, latency is sometimes used to describe total message transmission time. It's also often used to describe the time required for any fixed overhead involved in transmitting data. For example, if we're sending a message between two nodes in a distributed-memory system, a message is not just raw data. It might include the data to be transmitted, a destination address, some information specifying the size of the message, some information for error correction, and so on. So in this setting, latency might be the time it takes to assemble the message on the sending side—the time needed to combine the various parts—and the time to disassemble the message on the receiving side (the time needed to extract the raw data from the message and store it in its destination).

2.3.5 Cache coherence

Recall that CPU caches are managed by system hardware: programmers don't have direct control over them. This has several important consequences for shared-memory systems. To understand these issues, suppose we have a shared-memory system with two cores, each of which has its own private data cache. (See Fig. 2.17.) As long as the two cores only read shared data, there is no problem. For example, suppose that

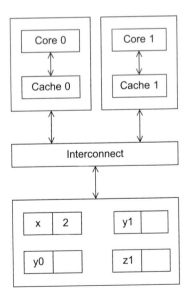

FIGURE 2.17

A shared-memory system with two cores and two caches.

x is a shared variable that has been initialized to 2, y0 is private and owned by core 0, and y1 and z1 are private and owned by core 1. Now suppose the following statements are executed at the indicated times:

Time	Core 0	Core 1
0	y0 = x;	y1 = 3*x;
1	x = 7;	Statement(s) not involving x
2	Statement(s) not involving x	z1 = 4*x;

Then the memory location for y0 will eventually get the value 2, and the memory location for y1 will eventually get the value 6. However, it's not so clear what value z1 will get. It might at first appear that since core 0 updates x to 7 before the assignment to z1, z1 will get the value $4 \times 7 = 28$. However, at time 0, x is in the cache of core 1. So unless for some reason x is evicted from core 0's cache and then reloaded into core 1's cache, it actually appears that the original value x = 2 may be used, and z1 will get the value $4 \times 2 = 8$.

Note that this unpredictable behavior will occur regardless of whether the system is using a write-through or a write-back policy. If it's using a write-through policy, the main memory will be updated by the assignment x = 7. However, this will have no effect on the value in the cache of core 1. If the system is using a write-back policy, the new value of x in the cache of core 0 probably won't even be available to core 1 when it updates z1.

Clearly, this is a problem. The programmer doesn't have direct control over when the caches are updated, so her program cannot execute these apparently innocuous statements and know what will be stored in z1. There are several problems here, but the one we want to look at right now is that the caches we described for single processor systems provide no mechanism for ensuring that when the caches of multiple processors store the same variable, an update by one processor to the cached variable is "seen" by the other processors. That is, that the cached value stored by the other processors is also updated. This is called the **cache coherence** problem.

Snooping cache coherence

There are two main approaches to ensuring cache coherence: **snooping cache coherence** and **directory-based cache coherence**. The idea behind snooping comes from bus-based systems: When the cores share a bus, any signal transmitted on the bus can be "seen" by all the cores connected to the bus. Thus when core 0 updates the copy of x stored in its cache, if it also broadcasts this information across the bus, and if core 1 is "snooping" the bus, it will see that x has been updated, and it can mark its copy of x as invalid. This is more or less how snooping cache coherence works. The principal difference between our description and the actual snooping protocol is that the broadcast only informs the other cores that the *cache line* containing x has been updated, not that x has been updated.

A couple of points should be made regarding snooping. First, it's not essential that the interconnect be a bus, only that it support broadcasts from each processor to all the other processors. Second, snooping works with both write-through and write-back caches. In principle, if the interconnect is shared—as with a bus—with write-through caches, there's no need for additional traffic on the interconnect, since each core can simply "watch" for writes. With write-back caches, on the other hand, an extra communication *is* necessary, since updates to the cache don't get immediately sent to memory.

Directory-based cache coherence

Unfortunately, in large networks, broadcasts are expensive, and snooping cache coherence requires a broadcast every time a variable is updated. So snooping cache coherence isn't scalable, because for larger systems it will cause performance to degrade. For example, suppose we have a system with the basic distributed-memory architecture (Fig. 2.4). However, the system provides a single address space for all the memories. So, for example, core 0 can access the variable x stored in core 1's memory, by simply executing a statement such as y = x. (Of course, accessing the memory attached to another core will be slower than accessing "local" memory, but that's another story.) Such a system can, in principle, scale to very large numbers of cores. However, snooping cache coherence is clearly a problem, since a broadcast across the interconnect will be very slow relative to the speed of accessing local memory.

Directory-based cache coherence protocols attempt to solve this problem through the use of a data structure called a **directory**. The directory stores the status of each cache line. Typically, this data structure is distributed; in our example, each core/memory pair might be responsible for storing the part of the structure that specifies the status of the cache lines in its local memory. Thus when a line is read into, say, core 0's cache, the directory entry corresponding to that line would be updated, indicating that core 0 has a copy of the line. When a variable is updated, the directory is consulted, and the cache controllers of the cores that have that variable's cache line in their caches will invalidate those lines.

Clearly, there will be substantial additional storage required for the directory, but when a cache variable is updated, only the cores storing that variable need to be contacted.

False sharing

It's important to remember that CPU caches are implemented in hardware, so they operate on cache lines, not individual variables. This can have disastrous consequences for performance. As an example, suppose we want to repeatedly call a function f(i,j) and add the computed values into a vector:

```
int  i, j, m, n;
double  y[m];

/* Assign y = 0 */
. . .
```

```
for (i = 0; i < m; i++)
    for (j = 0; j < n; j++)
        y[i] += f(i,j);
```

We can parallelize this by dividing the iterations in the outer loop among the cores. If we have core_count cores, we might assign the first m/core_count iterations to the first core, the next m/core_count iterations to the second core, and so on.

```
/* Private variables */
int i, j, iter_count;

/* Shared variables initialized by one core */
int m, n, core_count
double y[m];

iter_count = m/core_count

/* Core 0 does this */
for (i = 0; i < iter_count; i++)
    for (j = 0; j < n; j++)
        y[i] += f(i,j);

/* Core 1 does this */
for (i = iter_count; i < 2*iter_count; i++)
    for (j = 0; j < n; j++)
        y[i] += f(i,j);

. . .
```

Now suppose our shared-memory system has two cores, m = 8, doubles are eight bytes, cache lines are 64 bytes, and y[0] is stored at the beginning of a cache line. A cache line can store eight doubles, and y takes one full cache line. What happens when core 0 and core 1 simultaneously execute their codes? Since all of y is stored in a single cache line, each time one of the cores executes the statement y[i] += f(i,j), the line will be invalidated, and the next time the other core tries to execute this statement, it will have to fetch the updated line from memory! So if n is large, we would expect that a large percentage of the assignments y[i] += f(i,j) will access main memory—in spite of the fact that core 0 and core 1 never access each other's elements of y. This is called **false sharing**, because the system is behaving *as if* the elements of y were being shared by the cores.

Note that false sharing does not cause incorrect results. However, it can ruin the performance of a program by causing many more accesses to main memory than necessary. We can reduce its effect by using temporary storage that is local to the thread or process, and then copying the temporary storage to the shared storage. We'll return to the subject of false sharing in Chapters 4 and 5.

2.3.6 Shared-memory vs. distributed-memory

Newcomers to parallel computing sometimes wonder why all MIMD systems aren't shared-memory, since most programmers find the concept of implicitly coordinating the work of the processors through shared data structures more appealing than explicitly sending messages. There are several issues, some of which we'll discuss when we talk about software for distributed- and shared-memory. However, the principal hardware issue is the cost of scaling the interconnect. As we add processors to a bus, the chance that there will be conflicts over access to the bus increase dramatically, so buses are suitable for systems with only a few processors. Large crossbars are very expensive, so it's also unusual to find systems with large crossbar interconnects. On the other hand, distributed-memory interconnects, such as the hypercube and the toroidal mesh, are relatively inexpensive, and distributed-memory systems with thousands of processors that use these and other interconnects have been built. Thus distributed-memory systems are often better suited for problems requiring vast amounts of data or computation.

2.4 Parallel software

Parallel hardware has arrived. Virtually all desktop and server systems use multicore processors. Nowadays even mobile phones and tablets make use of multicore processors. The first edition (2011) asserted, "The same cannot be said for parallel software." Currently (2021) the situation for parallel software is in flux. Most system software makes some use of parallelism, and many widely used application programs (e.g., Excel, Photoshop, Chrome) can also use multiple cores. However, there are still many programs that can only make use of a single core, and there are many programmers with no experience writing parallel programs. This is a problem, because we can no longer rely on hardware and compilers to provide a steady increase in application performance. If we're to continue to have routine increases in application performance and application power, software developers must learn to write applications that exploit shared- and distributed-memory architectures and MIMD and SIMD systems. In this section, we'll take a look at some of the issues involved in writing software for parallel systems.

First, some terminology. Typically when we run our shared-memory programs, we'll start a single process and fork multiple threads. So when we discuss shared-memory programs, we'll talk about *threads* carrying out tasks. On the other hand, when we run distributed-memory programs, we'll start multiple processes, and we'll talk about *processes* carrying out tasks. When the discussion applies equally well to shared-memory and distributed-memory systems, we'll talk about *processes/threads* carrying out tasks.

2.4.1 Caveats

Before proceeding, we need to stress some of the limitations of this section. First, we stress that our coverage in this chapter is only meant to give some idea of the issues: there is no attempt to be comprehensive.

Second, we'll mainly focus on what's often called **single program, multiple data**, or SPMD, programs. Instead of running a different program on each core, SPMD programs consist of a single executable that can behave as if it were multiple different programs through the use of conditional branches. For example,

```
if (I'm thread/process 0)
    do this;
else
    do that;
```

Observe that SPMD programs can readily implement data-parallelism. For example,

```
if (I'm thread/process 0)
    operate on the first half of the array;
else /* I'm thread/process 1 */
    operate on the second half of the array;
```

Recall that a program is **task parallel** if it obtains its parallelism by dividing tasks among the threads or processes. The first example makes it clear that SPMD programs can also implement **task-parallelism**.

2.4.2 Coordinating the processes/threads

In a very few cases, obtaining excellent parallel performance is trivial. For example, suppose we have two arrays and we want to add them:

```
double x[n], y[n];
 . . .
for (int i = 0; i < n; i++)
    x[i] += y[i];
```

To parallelize this, we only need to assign elements of the arrays to the processes/threads. For example, if we have p processes/threads, we might make process/thread 0 responsible for elements $0, \ldots, n/p - 1$, process/thread 1 would be responsible for elements $n/p, \ldots, 2n/p - 1$, and so on.

So for this example, the programmer only needs to

1. Divide the work among the processes/threads in such a way that
   a. each process/thread gets roughly the same amount of work, and
   b. the amount of communication required is minimized.

   Recall that the process of dividing the work among the processes/threads so that (a) is satisfied is called **load balancing**. The two qualifications on dividing the work are obvious, but nonetheless important. In many cases it won't be necessary to give much thought to them; they typically become concerns in situations in which the amount of work isn't known in advance by the programmer, but rather the work is generated as the program runs.

Although we might wish for a term that's a little easier to pronounce, recall that the process of converting a serial program or algorithm into a parallel program is often called **parallelization**. Programs that can be parallelized by simply dividing the work among the processes/threads are sometimes said to be **embarrassingly parallel**. This is a bit unfortunate, since it suggests that programmers should be embarrassed to have written an embarrassingly parallel program, when, to the contrary, successfully devising a parallel solution to *any* problem is a cause for great rejoicing.

Alas, the vast majority of problems are much more determined to resist our efforts to find a parallel solution. As we saw in Chapter 1, for these problems, we need to coordinate the work of the processes/threads. In these programs, we also usually need to

2. Arrange for the processes/threads to synchronize.
3. Arrange for communication among the processes/threads.

These last two problems are often interrelated. For example, in distributed-memory programs, we often implicitly synchronize the processes by communicating among them, and in shared-memory programs, we often communicate among the threads by synchronizing them. We'll say more about both issues below.

2.4.3 Shared-memory

As we noted earlier, in shared-memory programs, variables can be **shared** or **private**. Shared variables can be read or written by any thread, and private variables can ordinarily only be accessed by one thread. Communication among the threads is usually done through shared variables, so communication is implicit rather than explicit.

Dynamic and static threads

In many environments, shared-memory programs use **dynamic threads**. In this paradigm, there is often a master thread and at any given instant a (possibly empty) collection of worker threads. The master thread typically waits for work requests—for example, over a network—and when a new request arrives, it forks a worker thread, the thread carries out the request, and when the thread completes the work, it terminates and joins the master thread. This paradigm makes efficient use of system resources, since the resources required by a thread are only being used while the thread is actually running.

An alternative to the dynamic paradigm is the **static thread** paradigm. In this paradigm, all of the threads are forked after any needed setup by the master thread and the threads run until all the work is completed. After the threads join the master thread, the master thread may do some cleanup (e.g., free memory), and then it also terminates. In terms of resource usage, this may be less efficient: if a thread is idle, its resources (e.g., stack, program counter, and so on) can't be freed. However, forking and joining threads can be fairly time-consuming operations. So if the necessary

resources are available, the static thread paradigm has the potential for better performance than the dynamic paradigm. It also has the virtue that it's closer to the most
widely used paradigm for distributed-memory programming, so part of the mindset
that is used for one type of system is preserved for the other. Hence, we'll often use
the static thread paradigm.

Nondeterminism

In any MIMD system in which the processors execute asynchronously it is likely that
there will be **nondeterminism**. A computation is nondeterministic if a given input
can result in different outputs. If multiple threads are executing independently, the
relative rate at which they'll complete statements varies from run to run, and hence
the results of the program may be different from run to run. As a very simple example,
suppose we have two threads, one with ID or rank 0 and the other with ID or rank 1.
Suppose also that each is storing a private variable my_x, thread 0's value for my_x is 7,
and thread 1's is 19. Further, suppose both threads execute the following code:

```
.  .  .
printf("Thread %d > my_x = %d\n", my_rank, my_x);
.  .  .
```

Then the output could be

```
Thread 0 > my_x = 7
Thread 1 > my_x = 19
```

but it could also be

```
Thread 1 > my_x = 19
Thread 0 > my_x = 7
```

In fact, things could be even worse: the output of one thread could be broken up by
the output of the other thread. However, the point here is that because the threads are
executing independently and interacting with the operating system, the time it takes
for one thread to complete a block of statements varies from execution to execution,
so the order in which these statements complete can't be predicted.

In many cases nondeterminism isn't a problem. In our example, since we've labeled the output with the thread's rank, the order in which the output appears probably
doesn't matter. However, there are also many cases in which nondeterminism—
especially in shared-memory programs—can be disastrous, because it can easily
result in program errors. Here's a simple example with two threads.

Suppose each thread computes an **int**, which it stores in a private variable my_val.
Suppose also that we want to add the values stored in my_val into a shared-memory
location x that has been initialized to 0. Both threads therefore want to execute code
that looks something like this:

```
my_val = Compute_val(my_rank);
x += my_val;
```

Now recall that an addition typically requires loading the two values to be added into registers, adding the values, and finally storing the result. To keep things relatively simple, we'll assume that values are loaded from main memory directly into registers and stored in main memory directly from registers. Here is one possible sequence of events:

Time	Core 0	Core 1
0	Finish assignment to my_val	In call to Compute_val
1	Load x = 0 into register	Finish assignment to my_val
2	Load my_val = 7 into register	Load x = 0 into register
3	Add my_val = 7 to x	Load my_val = 19 into register
4	Store x = 7	Add my_val to x
5	Start other work	Store x = 19

Clearly this is not what we want, and it's easy to imagine other sequences of events that result in an incorrect value for x. The nondeterminism here is a result of the fact that two threads are attempting to more or less simultaneously update the memory location x. When threads or processes attempt to simultaneously access a shared resource, and the accesses can result in an error, we often say the program has a **race condition**, because the threads or processes are in a "race" to carry out an operation. That is, the outcome of the computation depends on which thread wins the race. In our example, the threads are in a race to execute x += my_val. In this case, unless one thread completes x += my_val before the other thread starts, the result will be incorrect. So we need for each thread's operation x += my_val to be **atomic**. An operation that writes to a memory location is atomic if, after a thread has completed the operation, it appears that no *other* thread has modified the memory location. There are usually several ways to ensure that an operation is atomic. One possibility is to ensure that only one thread executes the update x += my_val at a time. A block of code that can only be executed by one thread at a time is called a **critical section**, and it's usually our job as programmers to ensure **mutually exclusive** access to a critical section. In other words, we need to ensure that if one thread is executing the code in the critical section, then the other threads are excluded.

The most commonly used mechanism for ensuring mutual exclusion is a **mutual exclusion lock** or **mutex**, or simply **lock**. A mutex is a special type of object that has support in the underlying hardware. The basic idea is that each critical section is *protected* by a lock. Before a thread can execute the code in the critical section, it must "obtain" the mutex by calling a mutex function, and when it's done executing the code in the critical section, it should "relinquish" the mutex by calling an unlock function. While one thread "owns" the lock—that is, has returned from a call to the lock function but hasn't yet called the unlock function—any other thread attempting to execute the code in the critical section will wait in its call to the lock function.

Thus to ensure that our code functions correctly, we might modify it so that it looks something like this:

```
my_val = Compute_val(my_rank);
Lock(&add_my_val_lock);
x += my_val;
Unlock(&add_my_val_lock);
```

This ensures that only one thread at a time can execute the statement x += my_val. Note that the code does *not* impose any predetermined order on the threads. Either thread 0 or thread 1 can execute x += my_val first.

Also note that the use of a mutex enforces **serialization** of the critical section. Since only one thread at a time can execute the code in the critical section, this code is effectively serial. Thus we want our code to have as few critical sections as possible, and we want our critical sections to be as short as possible.

There are alternatives to mutexes. In **busy-waiting**, a thread enters a loop, whose sole purpose is to test a condition. In our example, suppose there is a shared variable ok_for_1 that has been initialized to false. Then something like the following code can ensure that thread 1 won't update x until after thread 0 has updated it:

```
my_val = Compute_val(my_rank);
if (my_rank == 1)
    while (!ok_for_1);     /* Busy-wait loop */
x += my_val;              /* Critical section */
if (my_rank == 0)
    ok_for_1 = true;      /* Let thread 1 update x */
```

So until thread 0 executes ok_for_1 = true, thread 1 will be stuck in the loop **while** (!ok_for_1). This loop is called a "busy-wait," because the thread can be very busy waiting for the condition. This has the virtue that it's simple to understand and implement. However, it can be very wasteful of system resources, because even when a thread is doing no useful work, the core running the thread will be repeatedly checking to see if the critical section can be entered. **Semaphores** are similar to mutexes, although the details of their behavior are slightly different, and there are some types of thread synchronization that are easier to implement with semaphores than mutexes. A **monitor** provides mutual exclusion at a somewhat higher level: it is an object whose methods can only be executed by one thread at a time. We'll discuss busy-waiting and semaphores in Chapter 4.

Thread safety

In many, if not most, cases, parallel programs can call functions developed for use in serial programs, and there won't be any problems. However, there are some notable exceptions. The most important exception for C programmers occurs in functions that make use of *static* local variables. Recall that ordinary C local variables—variables declared inside a function—are allocated from the system stack. Since each thread has its own stack, ordinary C local variables are private. However, recall that a static variable that's declared in a function persists from one call to the next. Thus static variables are effectively shared among any threads that call the function, and this can have unexpected and unwanted consequences.

For example, the C string library function strtok splits an input string into substrings. When it's first called, it's passed a string, and on subsequent calls it returns successive substrings. This can be arranged through the use of a static **char∗** variable that refers to the string that was passed on the first call. Now suppose two threads are splitting strings into substrings. Clearly, if, for example, thread 0 makes its first call to strtok, and then thread 1 makes its first call to strtok before thread 0 has completed splitting its string, then thread 0's string will be lost or overwritten, and, on subsequent calls, it may get substrings of thread 1's strings.

A function such as strtok is not **thread safe**. This means that if it is used in a multithreaded program, there may be errors or unexpected results. When a block of code isn't thread safe, it's usually because different threads are accessing shared data. Thus, as we've seen, even though many serial functions can be used safely in multithreaded programs—that is, they're *thread safe*—programmers need to be wary of functions that were written exclusively for use in serial programs. We'll take a closer look at thread safety in Chapters 4 and 5.

2.4.4 Distributed-memory

In distributed-memory programs, the cores can directly access only their own, private memories. There are several APIs that are used. However, by far the most widely used is message-passing. So we'll devote most of our attention in this section to message-passing. Then we'll take a brief look at a couple of other, less widely used, APIs.

Perhaps the first thing to note regarding distributed-memory APIs is that they can be used with shared-memory hardware. It's perfectly feasible for programmers to logically partition shared-memory into private address spaces for the various threads, and a library or compiler can implement the communication that's needed.

As we noted earlier, distributed-memory programs are usually executed by starting multiple processes rather than multiple threads. This is because typical "threads of execution" in a distributed-memory program may run on independent CPUs with independent operating systems, and there may be no software infrastructure for starting a single "distributed" process and having that process fork one or more threads on each node of the system.

Message-passing

A message-passing API provides (at a minimum) a send and a receive function. Processes typically identify each other by ranks in the range $0, 1, \ldots, p - 1$, where p is the number of processes. So, for example, process 1 might send a message to process 0 with the following pseudocode:

```
char message[100];
. . .
my_rank = Get_rank();
if (my_rank == 1) {
```

```
        sprintf(message, "Greetings from process 1");
        Send(message, MSG_CHAR, 100, 0);
   } else if (my_rank == 0) {
        Receive(message, MSG_CHAR, 100, 1);
        printf("Process 0 > Received: %s\n", message);
   }
```

Here the `Get_rank` function returns the calling process's rank. Then the processes branch depending on their ranks. Process 1 creates a message with `sprintf` from the standard C library and then sends it to process 0 with the call to `Send`. The arguments to the call are, in order, the message, the type of the elements in the message (`MSG_CHAR`), the number of elements in the message (100), and the rank of the destination process (0). On the other hand, process 0 calls `Receive` with the following arguments: the variable into which the message will be received (`message`), the type of the message elements, the number of elements available for storing the message, and the rank of the process sending the message. After completing the call to `Receive`, process 0 prints the message.

Several points are worth noting here. First, note that the program segment is SPMD. The two processes are using the same executable, but carrying out different actions. In this case, what they do depends on their ranks. Second, note that the variable `message` refers to different blocks of memory on the different processes. Programmers often stress this by using variable names, such as `my_message` or `local_message`. Finally, note that we're assuming that process 0 can write to `stdout`. This is usually the case: most implementations of message-passing APIs allow all processes access to `stdout` and `stderr`—even if the API doesn't explicitly provide for this. We'll talk a little more about I/O later on.

There are several possibilities for the exact behavior of the `Send` and `Receive` functions, and most message-passing APIs provide several different send and/or receive functions. The simplest behavior is for the call to `Send` to **block** until the call to `Receive` starts receiving the data. This means that the process calling `Send` won't return from the call until the matching call to `Receive` has started. Alternatively, the `Send` function may copy the contents of the message into storage that it owns, and then it will return as soon as the data are copied. The most common behavior for the `Receive` function is for the receiving process to block until the message is received. There are other possibilities for both `Send` and `Receive`, and we'll discuss some of them in Chapter 3.

Typical message-passing APIs also provide a wide variety of additional functions. For example, there may be functions for various "collective" communications, such as a **broadcast**, in which a single process transmits the same data to all the processes, or a **reduction**, in which results computed by the individual processes are combined into a single result—for example, values computed by the processes are added. There may also be special functions for managing processes and communicating complicated data structures. The most widely used API for message-passing is the **Message-Passing Interface** or MPI. We'll take a closer look at it in Chapter 3.

Message-passing is a very powerful and versatile API for developing parallel programs. Virtually all of the programs that are run on the most powerful computers in the world use message-passing. However, it is also very low level. That is, there is a huge amount of detail that the programmer needs to manage. For example, to parallelize a serial program, it is usually necessary to rewrite the vast majority of the program. The data structures in the program may have to either be replicated by each process or be explicitly distributed among the processes. Furthermore, the rewriting usually can't be done incrementally. For example, if a data structure is used in many parts of the program, distributing it for the parallel parts and collecting it for the serial (unparallelized) parts will probably be prohibitively expensive. Therefore message-passing is sometimes called "the assembly language of parallel programming," and there have been many attempts to develop other distributed-memory APIs.

One-sided communication

In message-passing, one process must call a send function, and the send must be matched by another process's call to a receive function. Any communication requires the explicit participation of two processes. In **one-sided communication**, or **remote memory access**, a single process calls a function, which updates either local memory with a value from another process or remote memory with a value from the calling process. This can simplify communication, since it only requires the active participation of a single process. Furthermore, it can significantly reduce the cost of communication by eliminating the overhead associated with synchronizing two processes. It can also reduce overhead by eliminating the overhead of one of the function calls (send or receive).

It should be noted that some of these advantages may be hard to realize in practice. For example, if process 0 is copying a value into the memory of process 1, 0 must have some way of knowing when it's safe to copy, since it will overwrite some memory location. Process 1 must also have some way of knowing when the memory location has been updated. The first problem can be solved by synchronizing the two processes before the copy, and the second problem can be solved by another synchronization, or by having a "flag" variable that process 0 sets after it has completed the copy. In the latter case, process 1 may need to **poll** the flag variable to determine that the new value is available. That is, it must repeatedly check the flag variable until it gets the value indicating 0 has completed its copy. Clearly, these problems can considerably increase the overhead associated with transmitting a value. A further difficulty is that since there is no explicit interaction between the two processes, remote memory operations can introduce errors that are very hard to track down.

Partitioned global address space languages

Since many programmers find shared-memory programming more appealing than message-passing or one-sided communication, a number of groups are developing parallel programming languages that allow the user to use some shared-memory techniques for programming distributed-memory hardware. This isn't quite as simple as it sounds. If we simply wrote a compiler that treated the collective memories in a distributed-memory system as a single large memory, our programs would have poor, or, at best, unpredictable performance, since each time a running process accessed memory, it might access local memory—that is, memory belonging to the core on which it was executing—or remote memory, memory belonging to another core. Accessing remote memory can take hundreds or even thousands of times longer than accessing local memory. As an example, consider the following pseudocode for a shared-memory vector addition:

```
shared int n = . . . ;
shared double x[n], y[n];
private int i, my_first_element, my_last_element;
my_first_element = . . . ;
my_last_element = . . . ;

/* Initialize x and y */
. . .

for (i = my_first_element; i <= my_last_element; i++)
    x[i] += y[i];
```

We first declare two shared arrays. Then, on the basis of the process's rank, we determine which elements of the array "belong" to which process. After initializing the arrays, each process adds its assigned elements. If the assigned elements of x and y have been allocated so that the elements assigned to each process are in the memory attached to the core the process is running on, then this code should be very fast. However, if, for example, all of x is assigned to core 0 and all of y is assigned to core 1, then the performance is likely to be terrible, since each time the assignment x[i] += y[i] is executed, the process will need to refer to remote memory.

Partitioned global address space, or PGAS, languages provide some of the mechanisms of shared-memory programs. However, they provide the programmer with tools to avoid the problem we just discussed. Private variables are allocated in the local memory of the core on which the process is executing, and the distribution of the data in shared data structures is controlled by the programmer. So, for example, she knows which elements of a shared array are in which process's local memory.

There are several projects currently working on the development of PGAS languages. See, for example, [8,54].

2.4.5 GPU programming

GPUs are usually not "standalone" processors. They don't ordinarily run an operating system and system services, such as direct access to secondary storage. So programming a GPU also involves writing code for the CPU "host" system, which runs on an ordinary CPU. The memory for the CPU host and the GPU memory are usually separate. So the code that runs on the host typically allocates and initializes storage on both the CPU and the GPU. It will start the program on the GPU, and it is responsible for the output of the results of the GPU program. Thus GPU programming is really *heterogeneous* programming, since it involves programming two different types of processors.

The GPU itself will have one or more processors. Each of these processors is capable of running hundreds or thousands of threads. In the systems we'll be using, the processors share a large block of memory, but each individual processor has a small block of much faster memory that can only be accessed by threads running on that processor. These blocks of faster memory can be thought of as a programmer-managed cache.

The threads running on a processor are typically divided into groups: the threads within a group use the SIMD model, and two threads in different groups can run independently. The threads in a SIMD group may not run in lockstep. That is, they may not all execute the same instruction at the same time. However, no thread in the group will execute the next instruction until all the threads in the group have completed executing the current instruction. If the threads in a group are executing a branch, it may be necessary to idle some of the threads. For example, suppose there are 32 threads in a SIMD group, and each thread has a private variable rank_in_gp that ranges from 0 to 31. Suppose also that the threads are executing the following code:

```
// Thread private variables
int rank_in_gp, my_x;
. . .
if (rank_in_gp < 16)
    my_x += 1;
else
    my_x -= 1;
```

Then the threads with rank < 16 will execute the first assignment, while the threads with rank ≥ 16 are idle. After the threads with rank < 16 are done, the roles will be reversed: the threads with rank < 16 will be idle, while the threads with rank ≥ 16 will execute the second assignment. Of course, idling half the threads for two instructions isn't a very efficient use of the available resources. So it's up to the programmer to minimize branching, where the threads within a SIMD group take different branches.

Another issue in GPU programming that's different from CPU programming is how the threads are scheduled to execute. GPUs use a hardware scheduler (unlike CPUs, which use software to schedule threads and processes), and this hardware scheduler uses very little overhead. However, the scheduler will choose to execute an instruction when all the threads in the SIMD group are ready. In the preceding example, before executing the test, we would want the variable rank_in_gp stored in a register by each thread. So, to maximize use of the hardware, we usually create a large number of SIMD groups. When this is the case, groups that aren't ready to execute (e.g., they're waiting for data from memory, or waiting for the completion of a previous instruction) can be idled, and the scheduler can choose a SIMD group that is ready.

2.4.6 Programming hybrid systems

Before moving on, we should note that it is possible to program systems such as clusters of multicore processors using a combination of a shared-memory API on the nodes and a distributed-memory API for internode communication. However, this is usually only done for programs that require the highest possible levels of performance, since the complexity of this "hybrid" API makes program development much more difficult. See, for example, [45]. Rather, such systems are often programmed using a single, distributed-memory API for both inter- and intra-node communication.

2.5 Input and output

2.5.1 MIMD systems

We've generally avoided the issue of input and output. There are a couple of reasons. First and foremost, parallel I/O, in which multiple cores access multiple disks or other devices, is a subject to which one could easily devote a book. See, for example, [38]. Second, the vast majority of the programs we'll develop do very little in the way of I/O. The amount of data they read and write is quite small and easily managed by the standard C I/O functions `printf`, `fprintf`, `scanf`, and `fscanf`. However, even the limited use we make of these functions can potentially cause some problems. Since these functions are part of standard C, which is a serial language, the standard says nothing about what happens when they're called by different processes. On the other hand, threads that are forked by a single process *do* share `stdin`, `stdout`, and `stderr`. However, as we've seen, when multiple threads attempt to access one of these, the outcome is nondeterministic, and it's impossible to predict what will happen.

When we call `printf` from multiple processes/threads, we, as developers, usually want the output to appear on the console of a single system, the system on which we started the program. In fact, this is what the vast majority of systems do. However, with processes, there is no guarantee, and we need to be aware that it is possible for a system to do something else: for example, only one process has access to `stdout` or `stderr`, or even *no* processes have access to `stdout` or `stderr`.

What *should* happen with calls to `scanf` when we're running multiple processes/threads is a little less obvious. Should the input be divided among the processes/threads? Or should only a single process/thread be allowed to call `scanf`? The vast majority of systems allow at least one process to call `scanf`—usually process 0—while most allow multiple threads to call `scanf`. Once again, there are some systems that don't allow any processes to call `scanf`.

When multiple processes/threads *can* access `stdout`, `stderr`, or `stdin`, as you might guess, the distribution of the input and the sequence of the output are usually nondeterministic. For output, the data will probably appear in a different order each time the program is run, or, even worse, the output of one process/thread may be broken up by the output of another process/thread. For input, the data read by each process/thread may be different on each run, even if the same input is used.

To partially address these issues, we'll be making these assumptions and following these rules when our parallel programs need to do I/O:

- In distributed-memory programs, only process 0 will access stdin. In shared-memory programs, only the master thread or thread 0 will access stdin.
- In both distributed-memory and shared-memory programs, all the processes/threads can access stdout and stderr.
- However, because of the nondeterministic order of output to stdout, in most cases only a single process/thread will be used for all output to stdout. The principal exception will be output for debugging a program. In this situation, we'll often have multiple processes/threads writing to stdout.
- Only a single process/thread will attempt to access any single file other than stdin, stdout, or stderr. So, for example, each process/thread can open its own, private file for reading or writing, but no two processes/threads will open the same file.
- Debug output should always include the rank or ID of the process/thread that's generating the output.

2.5.2 GPUs

In most cases, the host code in our GPU programs will carry out all I/O. Since we'll only be running one process/thread on the host, the standard C I/O functions should behave as they do in ordinary serial C programs.

The exception to the rule that we use the host for I/O is that when we are debugging our GPU code, we'll want to be able to write to stdout and/or stderr. In the systems we use, each thread can write to stdout, and, as with MIMD programs, the order of the output is nondeterministic. Also in the systems we use, *no* GPU thread has access to stderr, stdin, or secondary storage.

2.6 Performance

Of course our main purpose in writing parallel programs is usually increased performance. So what can we expect? And how can we evaluate our programs? In this section, we'll start by looking at the performance of homogeneous MIMD systems. So we'll assume that all of the cores have the same architecture. Since this is not the case for GPUs, we'll talk about the performance of GPUs in a separate subsection.

2.6.1 Speedup and efficiency in MIMD systems

Usually the best our parallel program can do is to divide the work equally among the cores while at the same time introducing no additional work for the cores. If we succeed in doing this, and we run our program with p cores, one thread or process on each core, then our parallel program will run p times faster than the serial program runs on a single core of the same design. If we call the serial run-time T_{serial} and our parallel run-time $T_{parallel}$, then it's usually the case that the best possible run-time

Table 2.4 Speedups and efficiencies of a parallel program.

p	1	2	4	8	16
S	1.0	1.9	3.6	6.5	10.8
$E = S/p$	1.0	0.95	0.90	0.81	0.68

of our parallel program is $T_{\text{parallel}} = T_{\text{serial}}/p$. When this happens, we say that our parallel program has **linear speedup**.

In practice, we usually don't get perfect linear speedup, because the use of multiple processes/threads almost invariably introduces some overhead. For example, shared-memory programs will almost always have critical sections, which will require that we use some mutual exclusion mechanism, such as a mutex. The calls to the mutex functions are the overhead that's not present in the serial program, and the use of the mutex forces the parallel program to serialize execution of the critical section. Distributed-memory programs will almost always need to transmit data across the network, which is usually much slower than local memory access. Serial programs, on the other hand, won't have these overheads. Thus it will be unusual for us to find that our parallel programs get linear speedup. Furthermore, it's likely that the overheads will increase as we increase the number of processes or threads. For example, more threads will probably mean more threads need to access a critical section, and more processes will probably mean more data needs to be transmitted across the network.

So if we define the **speedup** of a parallel program to be

$$S = \frac{T_{\text{serial}}}{T_{\text{parallel}}},$$

then linear speedup has $S = p$. Furthermore, since as p increases we expect the parallel overhead to increase, we also expect S to become a smaller and smaller fraction of the ideal, linear speedup p. Another way of saying this is that S/p will probably get smaller and smaller as p increases. Table 2.4 shows an example of the changes in S and S/p as p increases.[1] This value, S/p, is sometimes called the **efficiency** of the parallel program. If we substitute the formula for S, we see that the efficiency is

$$E = \frac{S}{p} = \frac{\left(\frac{T_{\text{serial}}}{T_{\text{parallel}}}\right)}{p}$$

$$= \frac{T_{\text{serial}}}{p \cdot T_{\text{parallel}}}.$$

If the serial run-time has been taken on the same type of core that the parallel system is using, we can think of efficiency as the average utilization of the parallel

---

[1] These data are taken from Chapter 3. See Tables 3.6 and 3.7.

Table 2.5 Speedups and efficiencies of parallel program on different problem sizes.

	p	1	2	4	8	16
Half	S	1.0	1.9	3.1	4.8	6.2
	E	1.0	0.95	0.78	0.60	0.39
Original	S	1.0	1.9	3.6	6.5	10.8
	E	1.0	0.95	0.90	0.81	0.68
Double	S	1.0	1.9	3.9	7.5	14.2
	E	1.0	0.95	0.98	0.94	0.89

cores on solving the problem. That is, the efficiency can be thought of as the fraction of the parallel run-time that's spent, on average, by each core working on solving the original problem. The remainder of the parallel run-time is the parallel overhead. This can be seen by simply multiplying the efficiency and the parallel run-time:

$$E \cdot T_{\text{parallel}} = \frac{T_{\text{serial}}}{p \cdot T_{\text{parallel}}} \cdot T_{\text{parallel}} = \frac{T_{\text{serial}}}{p}.$$

For example, suppose we have $T_{\text{serial}} = 24$ ms, $p = 8$, and $T_{\text{parallel}} = 4$ ms. Then

$$E = \frac{24}{8 \cdot 4} = \frac{3}{4},$$

and, on average, each process/thread spends $3/4 \cdot 4 = 3$ ms on solving the original problem, and $4 - 3 = 1$ ms in parallel overhead.

Many parallel programs are developed by explicitly dividing the work of the serial program among the processes/threads and adding in the necessary "parallel overhead," such as mutual exclusion or communication. Therefore if T_{overhead} denotes this parallel overhead, it's often the case that

$$T_{\text{parallel}} = T_{\text{serial}}/p + T_{\text{overhead}}.$$

When this formula applies, it's clear that the parallel efficiency is just the fraction of the parallel run-time spent by the parallel program in the original problem, because the formula divides the parallel run-time into the time on the original problem, T_{serial}/p, and the time spent in parallel overhead, T_{overhead}.

We've already seen that T_{parallel}, S, and E depend on p, the number of processes or threads. We also need to keep in mind that T_{parallel}, S, E, and T_{serial} all depend on the problem size. For example, if we halve and double the problem size of the program, whose speedups are shown in Table 2.4, we get the speedups and efficiencies shown in Table 2.5. The speedups are plotted in Fig. 2.18, and the efficiencies are plotted in Fig. 2.19.

We see that in this example, when we increase the problem size, the speedups and the efficiencies increase, while they decrease when we decrease the problem size.

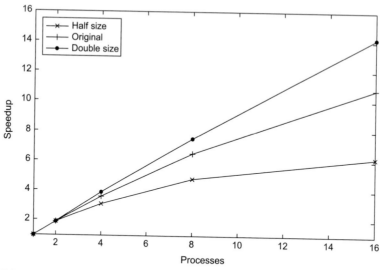

FIGURE 2.18

Speedups of parallel program on different problem sizes.

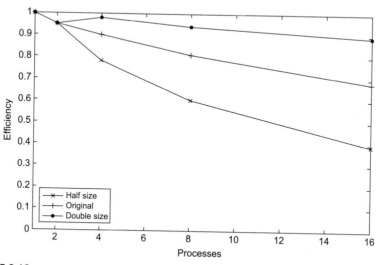

FIGURE 2.19

Efficiencies of parallel program on different problem sizes.

This behavior is quite common, because in many parallel programs, as the problem size is increased but the number of processes/threads is fixed, the parallel overhead grows much more slowly than the time spent in solving the original problem. That is, if we think of T_{serial} and T_{overhead} as functions of the problem size, T_{serial} grows much faster as the problem size is increased. Exercise 2.15 goes into more detail.

A final issue to consider is what values of T_{serial} should be used when reporting speedups and efficiencies. Some authors say that T_{serial} should be the run-time of the fastest program on the fastest processor available. However, since it's often the case that we think of efficiency as the utilization of the cores on the parallel system, in practice, most authors use a serial program, on which the parallel program was based and run it on a single processor of the parallel system. So if we were studying the performance of a parallel shell sort program, authors in the first group might use a serial radix sort or quicksort on a single core of the fastest system available, while authors in the second group would use a serial shell sort on a single processor of the parallel system. We'll generally use the second approach.

2.6.2 Amdahl's law

Back in the 1960s, Gene Amdahl made an observation [3] that's become known as **Amdahl's law**. It says, roughly, that unless virtually all of a serial program is parallelized, the possible speedup is going to be very limited—regardless of the number of cores available. Suppose, for example, that we're able to parallelize 90% of a serial program. Furthermore, suppose that the parallelization is "perfect," that is, regardless of the number of cores p we use, the speedup of this part of the program will be p. If the serial run-time is $T_{\text{serial}} = 20$ seconds, then the run-time of the parallelized part will be $0.9 \times T_{\text{serial}}/p = 18/p$ and the run-time of the "unparallelized" part will be $0.1 \times T_{\text{serial}} = 2$. The overall parallel run-time will be

$$T_{\text{parallel}} = 0.9 \times T_{\text{serial}}/p + 0.1 \times T_{\text{serial}} = 18/p + 2,$$

and the speedup will be

$$S = \frac{T_{\text{serial}}}{0.9 \times T_{\text{serial}}/p + 0.1 \times T_{\text{serial}}} = \frac{20}{18/p + 2}.$$

Now as p gets larger and larger, $0.9 \times T_{\text{serial}}/p = 18/p$ gets closer and closer to 0, so the total parallel run-time can't be smaller than $0.1 \times T_{\text{serial}} = 2$. That is, the denominator in S can't be smaller than $0.1 \times T_{\text{serial}} = 2$. The fraction S must therefore satisfy the inequality

$$S \leq \frac{T_{\text{serial}}}{0.1 \times T_{\text{serial}}} = \frac{20}{2} = 10.$$

That is, $S \leq 10$. This is saying that even though we've done a perfect job in parallelizing 90% of the program, and even if we have, say, 1000 cores, we'll never get a speedup better than 10.

More generally, if a fraction r of our serial program remains unparallelized, then Amdahl's law says we can't get a speedup better than $1/r$. In our example, $r = 1 - 0.9 = 1/10$, so we couldn't get a speedup better than 10. Therefore if a fraction r of our serial program is "inherently serial," that is, cannot possibly be parallelized, then we can't possibly get a speedup better than $1/r$. Thus even if r is quite small—say, 1/100—and we have a system with thousands of cores, we can't possibly get a speedup better than 100.

This is pretty daunting. Should we give up and go home? Well, no. There are several reasons not to be too worried by Amdahl's law. First, it doesn't take into consideration the problem size. For many problems, as we increase the problem size, the "inherently serial" fraction of the program decreases in size; a more mathematical version of this statement is known as **Gustafson's law** [25]. Second, there are thousands of programs used by scientists and engineers that routinely obtain huge speedups on large distributed-memory systems. Finally, is a small speedup so awful? In many cases, obtaining a speedup of 5 or 10 is more than adequate, especially if the effort involved in developing the parallel program wasn't very large.

2.6.3 Scalability in MIMD systems

The word "scalable" has a wide variety of informal uses. Indeed, we've used it several times already. Roughly speaking, a program is scalable if, by increasing the power of the system it's run on (e.g., increasing the number of cores), we can obtain speedups over the program when it's run on a less powerful system (e.g., a system with fewer cores). However, in discussions of MIMD parallel program performance, scalability has a somewhat more formal definition. Suppose we run a parallel program with a fixed number of processes/threads and a fixed input size, and we obtain an efficiency E. Suppose we now increase the number of processes/threads that are used by the program. If we can find a corresponding rate of increase in the problem size so that the program always has efficiency E, then the program is **scalable**.

As an example, suppose that $T_{serial} = n$, where the units of T_{serial} are in microseconds, and n is also the problem size. Also suppose that $T_{parallel} = n/p + 1$. Then

$$E = \frac{n}{p(n/p + 1)} = \frac{n}{n + p}.$$

To see if the program is scalable, we increase the number of processes/threads by a factor of k, and we want to find the factor x that we need to increase the problem size by, so that E is unchanged. The number of processes/threads will be kp; the problem size will be xn, and we want to solve the following equation for x:

$$E = \frac{n}{n + p} = \frac{xn}{xn + kp}.$$

Well, if $x = k$, there will be a common factor of k in the denominator $xn + kp = kn + kp = k(n + p)$, and we can reduce the fraction to get

$$\frac{xn}{xn + kp} = \frac{kn}{k(n + p)} = \frac{n}{n + p}.$$

In other words, if we increase the problem size at the same rate that we increase the number of processes/threads, then the efficiency will be unchanged, and our program is scalable.

There are a couple of cases that have special names. If, when we increase the number of processes/threads, we can keep the efficiency fixed *without* increasing the problem size, the program is said to be *strongly scalable*. If we can keep the efficiency fixed by increasing the problem size at the same rate as we increase the number of processes/threads, then the program is said to be *weakly scalable*. The program in our example would be weakly scalable.

2.6.4 Taking timings of MIMD programs

You may have been wondering how we find T_{serial} and $T_{parallel}$. There are a *lot* of different approaches, and with parallel programs the details may depend on the API. However, there are a few general observations we can make that may make things a little easier.

The first thing to note is that there are at least two different reasons for taking timings. During program development, we may take timings to determine if the program is behaving as we intend. For example, in a distributed-memory program, we might be interested in finding out how much time the processes are spending waiting for messages, because if this value is large, there is almost certainly something wrong either with our design or our implementation. On the other hand, once we've completed development of the program, we're often interested in determining how good its performance is. Perhaps surprisingly, the way we take these two timings is usually different. For the first timing, we usually need very detailed information: How much time did the program spend in this part of the program? How much time did it spend in that part? For the second, we usually report a single value. Right now we'll talk about the second type of timing. See Exercise 2.21 for a brief discussion of one issue in taking the first type of timing.

Second, we're usually *not* interested in the time that elapses between the program's start and the program's finish. We're usually interested only in some part of the program. For example, if we write a program that implements bubble sort, we're probably only interested in the time it takes to sort the keys, not the time it takes to read them in and print them out. So we probably can't use something like the Unix shell command `time`, which reports the time taken to run a program from start to finish.

Third, we're usually *not* interested in "CPU time." This is the time reported by the standard C function `clock`. It's the total time the program spends in code executed as part of the program. It would include the time for code we've written; it would include the time we spend in library functions, such as `pow` or `sin`; and it would include the time the operating system spends in functions we call, such as `printf` and `scanf`. It would not include time the program was idle, and this could be a problem. For example, in a distributed-memory program, a process that calls a receive function may have to wait for the sending process to execute the matching send, and the operating system might put the receiving process to sleep while it waits. This idle time wouldn't be counted as CPU time, since no function that's been called by the process is active. However, it should count in our evaluation of the overall run-time, since it may be a real cost in our program. If each time the program is run, the process has to wait, ignoring the time it spends waiting would give a misleading picture of the actual run-time of the program.

Thus when you see an article reporting the run-time of a parallel program, the reported time is usually "wall clock" time. That is, the authors of the article report the time that has elapsed between the start and finish of execution of the code that the user is interested in. If the user could see the execution of the program, she would hit the start button on her stopwatch when it begins execution and hit the stop button when it stops execution. Of course, she can't see her code executing, but she can modify the source code so that it looks something like this:

```
double start, finish;
.  .  .
start = Get_current_time();
/* Code that we want to time */
.  .  .
finish = Get_current_time();
printf("The elapsed time = %e seconds\n", finish-start);
```

The function Get_current_time() is a hypothetical function that's supposed to return the number of seconds that have elapsed since some fixed time in the past. It's just a placeholder. The actual function that is used will depend on the API. For example, MPI has a function MPI_Wtime that could be used here, and the OpenMP API for shared-memory programming has a function omp_get_wtime. Both functions return wall clock time instead of CPU time.

There may be an issue with the **resolution** of the timer function. The resolution is the unit of measurement on the timer. It's the duration of the shortest event that can have a nonzero time. Some timer functions have resolutions in milliseconds (10^{-3} seconds), and when instructions can take times that are less than a nanosecond (10^{-9} seconds), a program may have to execute millions of instructions before the timer reports a nonzero time. Many APIs provide a function that reports the resolution of the timer. Other APIs specify that a timer must have a given resolution. In either case we, as the programmers, need to check these values.

When we're timing parallel programs, we need to be a little more careful about how the timings are taken. In our example, the code that we want to time is probably being executed by multiple processes or threads, and our original timing will result in the output of p elapsed times:

```
private double start, finish;
.  .  .
start = Get_current_time();
/* Code that we want to time */
.  .  .
finish = Get_current_time();
printf("The elapsed time = %e seconds\n", finish-start);
```

However, what we're usually interested in is a single time: the time that has elapsed from when the first process/thread began execution of the code to the time the last process/thread finished execution of the code. We often can't obtain this exactly, since there may not be any correspondence between the clock on one node and the clock on another node. We usually settle for a compromise that looks something like this:

```
shared double global_elapsed;
private double my_start, my_finish, my_elapsed;
.  .  .
/* Synchronize all processes/threads */
Barrier();
my_start = Get_current_time();

/* Code that we want to time */
.  .  .

my_finish = Get_current_time();
my_elapsed = my_finish - my_start;

/* Find the max across all processes/threads */
global_elapsed = Global_max(my_elapsed);
if (my_rank == 0)
    printf("The elapsed time = %e seconds\n",
            global_elapsed);
```

Here, we first execute a **barrier** function that approximately synchronizes all of the processes/threads. We would like for all the processes/threads to return from the call simultaneously, but such a function usually can only guarantee that all the processes/threads have started the call when the first process/thread returns. We then execute the code as before, and each process/thread finds the time it took. Then all the processes/threads call a global maximum function, which returns the largest of the elapsed times, and process/thread 0 prints it out.

We also need to be aware of the *variability* in timings. When we run a program several times, it's extremely likely that the elapsed time will be different for each run. This will be true, even if each time we run the program we use the same input and the same systems. It might seem that the best way to deal with this would be to report either a mean or a median run-time. However, it's unlikely that some outside event could actually make our program run faster than its best possible run-time. So instead of reporting the mean or median time, we usually report the *minimum* time.

Running more than one thread per core can cause dramatic increases in the variability of timings. More importantly, if we run more than one thread per core, the system will have to take extra time to schedule and deschedule cores, and this will add to the overall run-time. Therefore we rarely run more than one thread per core.

Finally, as a practical matter, since our programs won't be designed for high-performance I/O, we'll usually not include I/O in our reported run-times.

2.6.5 GPU performance

In our discussion of MIMD performance, there is an assumption that we can evaluate a parallel program by comparing its performance to the performance of a serial program. We can, of course, compare the performance of a GPU program to the performance of a serial program, and it's quite common to see reported *speedups* of GPU programs over serial programs or parallel MIMD programs.

However, as we noted in our discussion of efficiency of MIMD programs, we often assume that the serial program is run on the same type of core that's used by the parallel computer. Since GPUs use cores that are inherently parallel, this type of comparison usually doesn't make sense for GPUs. So efficiency is ordinarily not used in discussions of the performance of GPUs.

Similarly, since the cores on the GPU are fundamentally different from conventional CPUs, it doesn't make sense to talk about linear speedup of a GPU program relative to a serial CPU program.

Note that since efficiency of a GPU program relative to a CPU program doesn't make sense, the formal definition of the scalability of a MIMD program can't be applied to a GPU program. However, the informal usage of scalability is routinely applied to GPUs: a GPU program is scalable if we can increase the size of the GPU and obtain speedups over the performance of the program on a smaller GPU.

If we run the inherently serial part of a GPU program on a conventional, serial processor, then Amdahl's law can be applied to GPU programs, and the resulting upper bound on the possible speedup will be the same as the upper bound on the possible speedup for a MIMD program. That is, if a fraction r of the original serial program isn't parallelized, and this fraction is run on a conventional serial processor, then the best possible speedup of the program running on the GPU and the serial processor will be less than $1/r$.

It should be noted that the same caveats that apply to Amdahl's law on MIMD systems also apply to Amdahl's law on GPUs: It's likely that the "inherently serial" fraction will depend on the problem size, and if it gets smaller as the problem size increases, the bound on the best possible speedup will increase. Also, *many* GPU programs obtain *huge* speedups, and, finally, a relatively small speedup may be perfectly adequate.

The same basic ideas about timing that we discussed for MIMD programs also apply to GPU programs. However, since a GPU program is ordinarily started and finished on a conventional CPU, as long as we're interested in the performance of the entirety of the program running on the GPU, we can usually just use the timer on the CPU, starting it before the GPU part(s) of the program are started, and stopping it after the GPU part(s) are done. There are more complicated scenarios—e.g., running a program on multiple CPU-GPU pairs—that require more care, but we won't be dealing with these types of programs. If we only want to time a subset of the code running on the GPU, we'll need to use a timer defined by the API for the GPU.

2.7 Parallel program design

So we've got a serial program. How do we parallelize it? We know that in general we need to divide the work among the processes/threads so that each process/thread gets roughly the same amount of work and any parallel overhead is minimized. In most cases, we also need to arrange for the processes/threads to synchronize and communicate. Unfortunately, there isn't some mechanical process we can follow; if there were, we could write a program that would convert any serial program into a parallel program, but, as we noted in Chapter 1, in spite of a tremendous amount of work and some progress, this seems to be a problem that has no universal solution.

However, Ian Foster provides an outline of steps in his online book *Designing and Building Parallel Programs* [21]:

1. *Partitioning.* Divide the computation to be performed and the data operated on by the computation into small tasks. The focus here should be on identifying tasks that can be executed in parallel.
2. *Communication.* Determine what communication needs to be carried out among the tasks identified in the previous step.
3. *Agglomeration or aggregation.* Combine tasks and communications identified in the first step into larger tasks. For example, if task A must be executed before task B can be executed, it may make sense to aggregate them into a single composite task.
4. *Mapping.* Assign the composite tasks identified in the previous step to processes/threads. This should be done so that communication is minimized, and each process/thread gets roughly the same amount of work.

This is sometimes called **Foster's methodology**.

2.7.1 An example

Let's look at a small example. Suppose we have a program that generates large quantities of floating point data that it stores in an array. To get some feel for the distribution of the data, we can make a histogram of the data. Recall that to make a histogram, we simply divide the range of the data up into equal-sized subintervals, or *bins*, determine the number of measurements in each bin, and plot a bar graph showing the relative sizes of the bins. As a very small example, suppose our data are

$$1.3, 2.9, 0.4, 0.3, 1.3, 4.4, 1.7, 0.4, 3.2, 0.3, 4.9, 2.4, 3.1, 4.4, 3.9, 0.4, 4.2, 4.5, 4.9, 0.9.$$

Then the data lie in the range 0–5, and if we choose to have five bins, the histogram might look something like Fig. 2.20.

A serial program

It's usually a good idea to start the design of a parallel program by first thinking about the design of a serial program. Although it may turn out that the algorithm used in the serial design is unsuitable for parallelization, just thinking through the problems

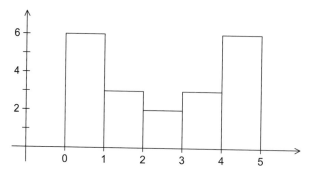

FIGURE 2.20

A histogram.

involved in a serial solution will give us a much better understanding of what needs to be done to solve the problem. Also, it's often the case that the design of the parallel program is based directly on the design of the serial program.

It's pretty straightforward to write a serial program that generates a histogram. We need to decide what the bins are, determine the number of measurements in each bin, and print the bars of the histogram. Since we're not focusing on I/O, we'll limit ourselves to just the first two steps, so the input will be

1. the number of measurements, `data_count`;
2. an array of `data_count` floats, `data`;
3. the minimum value for the bin containing the smallest values, `min_meas`;
4. the maximum value for the bin containing the largest values, `max_meas`;
5. the number of bins, `bin_count`.

The output will be an array containing the number of elements of data that lie in each bin. To make things precise, we'll make use of the following data structures:

```
- bin_maxes.   An array of bin_count floats
- bin_counts.  An array of bin_count ints
```

The array `bin_maxes` will store the upper bound for each bin, and `bin_counts` will store the number of data elements in each bin. To be explicit, we can define

```
bin_width = (max_meas - min_meas)/bin_count.
```

Then `bin_maxes` will be initialized by

```
for (b = 0; b < bin_count; b++)
    bin_maxes[b] = min_meas + bin_width*(b+1);
```

We'll adopt the convention that bin b will be all the measurements in the range

```
bin_maxes[b-1] <= measurement < bin_maxes[b]
```

Of course, this doesn't make sense if $b = 0$, and in this case we'll use the rule that bin 0 will be the measurements in the range

2.9 Assumptions

As we noted earlier, we'll be focusing on homogeneous MIMD systems—systems in which all of the nodes have the same architecture—and heterogeneous GPU systems. All of our programs will be SPMD. Thus we'll write a single program that can use branching to have multiple different behaviors.

For the MIMD systems, we'll assume the cores are identical but that they operate asynchronously. We'll usually assume that we run at most one process or thread of our program on a single core, and we'll often use static processes or threads. In other words, we'll often start all of our processes or threads at more or less the same time, and when they're done executing, we'll terminate them at more or less the same time.

For the GPUs, we'll write a single program, but it will contain some code that runs on the *host* or CPU and some code that runs on the *device* or GPU. The CPU code will start the code that runs on the GPU, and we'll let the system map the GPU threads to the GPU cores.

Some application programming interfaces or APIs for parallel MIMD systems define new programming languages. However, most extend existing languages, either through a library of functions—for example, functions to pass messages—or extensions to the compiler for the serial language. This latter approach will be our focus for the MIMD programs. We'll be using parallel extensions to the C language.

Since the GPU programs contain code that runs on two different types of processors, we will need a new language for these programs, CUDA. However, it is very similar to C++, and we'll usually only use the subset of C++ that is common to the C language. There are a few instances, however, where we'll need to use features of C++ that are different from C, but we'll be careful to note these.

When we want to be explicit about compiling and running programs, we'll use the command line of a Unix shell, the `gcc` compiler, some extension of it (e.g., `mpicc`), or the CUDA compiler (`nvcc`), and we'll start programs from the command line. For example, if we wanted to show compilation and execution of the "hello, world" program from Kernighan and Ritchie [32], we might show something like this:

```
$ gcc -g -Wall -o hello hello.c
$ ./hello
hello, world
```

The $-sign is the prompt from the shell. We will usually use the following options for the `gcc`-based compilers:

- `-g`. Create information that allows us to use a debugger
- `-Wall`. Issue lots of warnings
- `-o <outfile>`. Put the executable in the file named `outfile`
- When we're timing programs, we often tell the compiler to optimize the code by using the `-O2` option.

In most systems, user directories or folders are not, by default, in the user's execution path, so we'll usually start jobs by giving the path to the executable by adding `./` to its name.

For the CUDA compiler `nvcc`, we'll normally use just the `-o <outfile>` command line option.

2.10 Summary

There's a *lot* of material in this chapter, and a complete summary would run on for many pages, so we'll be very terse.

2.10.1 Serial systems

We started with a discussion of conventional serial hardware and software. The standard model of computer hardware has been the **von Neumann architecture**, which consists of a **central processing unit** that carries out the computations, and **main memory** that stores data and instructions. Each location in memory consists of an **address** and the **contents** of the location. The separation of CPU and memory is often called the **von Neumann bottleneck**, since it limits the rate at which instructions can be executed. The CPU is further divided into the **control unit**, which determines which instruction to execute, and the **datapath**, which carries out the instructions. Both the control unit and the datapath have very fast memory locations, called **registers**. An instruction is loaded into a register for execution, and data is loaded into registers to be used in the execution of instructions. Computed results are also stored in registers.

Perhaps the most important software on a computer is the **operating system** (OS). It manages the computer's resources. Most modern operating systems are **multitasking**. Even if the hardware doesn't have multiple processors or cores, by rapidly switching among executing programs, the OS creates the illusion that multiple jobs are running simultaneously. A running program is called a **process**. Since a running process is more or less autonomous, it has a lot of associated data and information. A **thread** is started by a process. A thread doesn't need as much associated data and information, and threads can be stopped and started much faster than processes.

In computer science, **caches** are memory locations that can be accessed faster than other memory locations. CPU caches are memory locations that are intermediate between the CPU registers and main memory. Their purpose is to reduce the delays associated with accessing main memory. Data are stored in cache using the principle of **temporal and spatial locality**. That is, items that are close to recently accessed items (spatial locality) are more likely to be accessed in the near future (temporal locality). Thus instead of transferring individual data items and individual instructions between main memory and caches, the system transfers **blocks** or **lines** of contiguous data items and contiguous instructions. When an instruction or data item is accessed and it's already in cache, it's called a cache **hit**. If the item isn't in cache, it's called a cache **miss**. Caches are directly managed by the computer hardware, so programmers can only indirectly control caching.

Main memory can also function as a cache for secondary storage. This is managed by the hardware and the operating system through **virtual memory**. Rather than storing all of a program's instructions and data in main memory, only the active parts are stored in main memory, and the remaining parts are stored in secondary storage called **swap space**. Like CPU caches, virtual memory operates on blocks of contiguous data and instructions, which in this setting are called **pages**. Note that instead of addressing the memory used by a program with physical addresses, virtual memory uses **virtual addresses**, which are independent of actual physical addresses. The correspondence between physical addresses and virtual addresses is stored in main memory in a **page table**. The combination of virtual addresses and a page table provides the system with the flexibility to store a program's data and instructions anywhere in memory. Thus it won't matter if two different programs use the same virtual addresses. With the page table stored in main memory, it could happen that every time a program needed to access a main memory location, it would need two memory accesses: one to get the appropriate page table entry so that it could find the location in main memory, and one to actually access the desired memory. To avoid this problem, CPUs have a special page table cache, called the **translation lookaside buffer**, which stores the most recently used page table entries.

Instruction-level parallelism (ILP) allows a single processor to execute multiple instructions simultaneously. There are two main types of ILP: **pipelining** and **multiple issue**. With pipelining, some of the functional units of a processor are ordered in sequence, with the output of one being the input to the next. Thus while one piece of data is being processed by, say, the second functional unit, another piece of data can be processed by the first. With multiple issue, the functional units are replicated, and the processor attempts to simultaneously execute different instructions in a single program.

Rather than attempting to simultaneously execute individual instructions, **hardware multithreading** attempts to simultaneously execute different threads. There are several different approaches to implementing hardware multithreading. However, all of them attempt to keep the processor as busy as possible by rapidly switching between threads. This is especially important when a thread **stalls** and has to wait (e.g., for a memory access to complete) before it can execute an instruction. In **simultaneous multithreading**, the different threads can simultaneously use the multiple functional units in a multiple issue processor. Since threads consist of multiple instructions, we sometimes say that **thread-level parallelism**, or TLP, is **coarser-grained** than ILP.

2.10.2 Parallel hardware

ILP and TLP provide parallelism at a very low level; they're typically controlled by the processor and the compiler, and their use isn't directly controlled by the programmer. For our purposes, hardware is parallel if the parallelism is visible to the programmer, and she can modify her source code to exploit it.

We use two main classifications of parallel hardware: Flynn's Taxonomy and shared vs. distributed memory. **Flynn's taxonomy** distinguishes between the number of instruction streams and the number of data streams a system can simultaneously manage.

Shared-memory and distributed-memory distinguish between how each core accesses main memory. In a shared-memory system each core has access to the same "shared" block of main memory. In a distributed-memory system each core has a private block of memory, to which only it has direct access. If a core needs to access a memory location in another core's block, it will communicate over the network.

A von Neumann system has a single instruction stream and a single data stream so it is classified as a **single instruction, single data**, or **SISD**, system. Since there is only a single core in a von Neumann system, it is neither shared- nor distributed-memory.

A **single instruction, multiple data**, or **SIMD**, system executes a single instruction at a time, but the instruction can operate on multiple data items. These systems often execute their instructions in lockstep: the first instruction is applied to all of the data elements simultaneously, then the second is applied, and so on. This type of parallel system is usually employed in **data parallel** programs, programs in which the data are divided among the processors, and each data item is subjected to more or less the same sequence of instructions. **Vector processors** and **graphics processing units** are often classified as SIMD systems, although current generation GPUs also have characteristics of multiple instruction, multiple data stream systems.

Branching in SIMD systems is handled by idling those processors that might operate on a data item to which the instruction doesn't apply. This behavior often makes SIMD systems poorly suited for **task parallelism**, in which each processor executes a different task, or even data-parallelism, with many conditional branches.

SIMD systems can use shared- and/or distributed-memory. However, when we look at writing programs for GPUs, we'll only consider shared-memory systems.

As the name suggests, **multiple instruction, multiple data**, or **MIMD**, systems execute multiple independent instruction streams, each of which can have its own data stream. In practice, MIMD systems are collections of autonomous processors that can execute at their own pace. The principal distinction between different MIMD systems is whether they are **shared-memory** or **distributed-memory** systems. Most of the larger MIMD systems are **hybrid** systems, in which a number of relatively small shared-memory systems are connected by an interconnection network. In such systems, the individual shared-memory systems are sometimes called **nodes**. Some MIMD systems are **heterogeneous** systems, in which the processors have different capabilities. For example, a system with a conventional CPU and a GPU is a heterogeneous system. A system in which all the processors have the same architecture is **homogeneous**.

There are a number of different interconnects for joining processors to memory in shared-memory systems and for interconnecting the processors in distributed-memory or hybrid systems. Two interconnects for shared memory are **buses** and **crossbars**. Distributed-memory systems sometimes use **direct** interconnects, such as **toroidal meshes** and **hypercubes**, and they sometimes use **indirect** interconnects, such as crossbars and **multistage** networks. Networks are often evaluated by examining the **bisection width** or the **bisection bandwidth** of the network. These give measures of how much simultaneous communication the network can support. For individual communications between nodes, authors often discuss the **latency** and **bandwidth** of the interconnect.

A potential problem with shared-memory systems is **cache coherence**. The same variable can be stored in the caches of two different cores, and if one core updates the value of the variable, the other core may be unaware of the change. There are two main methods for ensuring cache coherence: **snooping** and the use of **directories**. Snooping relies on the capability of the interconnect to broadcast information from each cache controller to every other cache controller. Directories are special distributed data structures, which store information on each cache line. Cache coherence introduces another problem for shared-memory programming: **false sharing**. When one core updates a variable in one cache line, and another core wants to access *another* variable in the same cache line, it will have to access main memory, since the unit of cache coherence is the cache line. That is, the second core only "knows" that the line it wants to access has been updated. It doesn't know that the variable it wants to access hasn't been changed.

2.10.3 Parallel software

In this text, we'll focus on developing software for homogeneous MIMD systems and heterogeneous systems with a CPU and a GPU. Most programs for such systems consist of a single program that obtains parallelism by partitioning the data among the threads/processes and/or branching. Such programs are often called **single program, multiple data** or **SPMD** programs. In shared-memory programs on CPUs and GPUs, we'll call the instances of running tasks threads; in distributed-memory programs, we'll call them processes.

Unless our problem is **embarrassingly parallel**, the development of a parallel program needs at a minimum to address the issues of **load balance**, **communication**, and **synchronization** among the processes or threads.

In shared-memory programs, the individual threads can have **private** and **shared** memory. Communication is usually done through **shared variables**. Anytime the processors execute asynchronously, there is the potential for **nondeterminism**, that is, for a given input, the behavior of the program can change from one run to the next. This can be a serious problem, especially in shared-memory programs. If the nondeterminism results from two threads' attempts to access the same resource, and it can result in an error, the program is said to have a **race condition**. One of the most common places for a race condition is a **critical section**, a block of code that can only be executed by one thread at a time. Two possible approaches to protecting a critical section are the use of **atomic operations** and **mutual exclusion**. When a thread carries out an atomic operation, other threads cannot modify the variables being updated. Some APIs provide special functions that operate on their data atomically. In most shared-memory APIs, **mutual exclusion** in a critical section can be enforced with an object called a **mutual exclusion lock** or **mutex**. Critical sections should be made as small as possible, since a mutex will allow only one thread at a time to execute the code in the critical section, effectively making the code serial.

A second potential problem with shared-memory programs is **thread safety**. A block of code that functions correctly when it is run by multiple threads is said to be **thread safe**. Functions that were written for use in serial programs can make unwitting use of shared data—for example, static variables—and this use in a multi-threaded program can cause errors. Such functions are not thread safe.

The most commonly used API for programming distributed-memory systems is **message-passing**. In message-passing, there are (at least) two distinct functions: a **send** function and a **receive** function. When processes need to communicate, one calls the send and the other calls the receive. There are a variety of possible behaviors for these functions. For example, the send can **block** or wait until the matching receive has started, or the message-passing software can copy the data for the message into its own storage, and the sending process can return before the matching receive has started. The most common behavior for receives is to block until the message has been received. The most widely used message-passing system is called the **Message-Passing Interface** or MPI. It provides a great deal of functionality beyond simple sends and receives.

Distributed-memory systems can also be programmed using **one-sided communications**, which provide functions for accessing memory belonging to another process, and **partitioned global address space** languages, which provide some shared-memory functionality in distributed-memory systems.

2.10.4 Input and output

In general parallel systems, multiple cores can access multiple secondary storage devices. We won't attempt to write programs that make use of this functionality. Rather, in our programs for homogeneous MIMD systems, we'll write programs, in which one process or thread can access `stdin`, and all processes can access `stdout` and `stderr`. However, because of nondeterminism—except for debug output—we'll usually have a single process or thread accessing `stdout`. For heterogeneous CPU-GPU systems, we'll use a single thread on the CPU for all I/O with the sole exception of debug output. For debug output, we'll use the threads on the GPU.

2.10.5 Performance

If we run a parallel program on a homogeneous MIMD system with p processes or threads and no more than one process/thread per core, then our ideal would be for our parallel program to run p times faster than the serial program. This is called **linear speedup**, but in practice it is usually not achieved. If we denote the run-time of the serial program by T_{serial} and the parallel program's run-time by $T_{parallel}$, then the **speedup** S and **efficiency** E of the parallel program are given by the formulas

$$S = \frac{T_{serial}}{T_{parallel}}, \text{ and } E = \frac{T_{serial}}{p T_{parallel}},$$

respectively. So linear speedup has $S = p$ and $E = 1$. In practice, we will almost always have $S < p$ and $E < 1$. If we fix the problem size, E usually decreases as we increase p, while if we fix the number of processes/threads, then S and E often increase as we increase the problem size. If the serial program is run on the same type of core as those on the parallel system, efficiency can also be viewed as the average fraction of the serial program executed on each parallel core.

The parallel cores on a GPU are fundamentally different from the cores on a CPU. So we usually don't talk about efficiency on a GPU: it doesn't make sense to look at the fraction of the serial program executed on the parallel cores. However, it *is* common to talk about the speedup of a GPU program over a serial program or a homogeneous MIMD program.

Amdahl's law provides an upper bound on the speedup that can be obtained by a parallel program: if a fraction r of the original, serial program isn't parallelized, then we can't possibly get a speedup better than $1/r$, regardless of how many processes/threads we use. In practice, many parallel programs obtain excellent speedups. One possible reason for this apparent contradiction is that Amdahl's law doesn't take into consideration the fact that the unparallelized part often decreases in size relative to the parallelized part as the problem size increases.

Scalability is a term that has many interpretations. One interpretation is that a program is scalable if it can obtain speedups when it's run on a larger system. Formally, a parallel program for a homogeneous MIMD system is scalable if there is a rate at which the problem size can be increased so that as the number of processes/threads is increased, the efficiency remains constant. A program is **strongly** scalable, if the problem size can remain fixed, and it's **weakly** scalable if the problem size needs to be increased at the same rate as the number of processes/threads. Scalability is only used in the informal sense for GPU programs: a GPU program is scalable if we can obtain speedups when the program is run on a larger system.

To determine T_{serial} and T_{parallel}, we usually need to include calls to a timer function in our source code. We want these timer functions to give **wall clock** time, not CPU time, since the program may be "active"—for example, waiting for a message—even when the core is idle. In homogeneous MIMD systems, we usually take parallel times by synchronizing the processes/threads before starting the timer, and, after stopping the timer, we find the maximum elapsed time among all the processes/threads. In GPUs, we usually take the time on the host by starting the timer before starting the GPU, and stopping the timer after the GPU is done. Because of system variability, we usually need to run a program several times with a given data set, and we usually take the minimum time from the multiple runs. To reduce variability and improve overall run-times on homogeneous MIMD systems, we usually run no more than one thread per core. On GPUs we let the system assign threads to cores.

2.10.6 Parallel program design

Foster's methodology provides a sequence of steps that can be used to design parallel programs. The steps are **partitioning** the problem to identify tasks, identifying

communication among the tasks, **agglomeration** or **aggregation** to group tasks, and **mapping** to assign aggregate tasks to processes/threads.

2.10.7 Assumptions

We'll be looking at the development of parallel programs for shared- and distributed-memory homogeneous MIMD systems and heterogeneous CPU-GPU systems. We'll be writing SPMD programs. On homogeneous systems we'll usually use **static** processes or threads—processes/threads that are created when the program begins execution, and are not shut down until the program terminates. On these systems, we'll also assume that we run at most one process or thread on each core of the system. On heterogeneous CPU-GPU systems, we'll start the program on the CPU, and the CPU will start threads on the GPU and wait for them to terminate.

2.11 Exercises

2.1 When we were discussing floating point addition, we made the simplifying assumption that each of the functional units took the same amount of time. Suppose that fetch and store each take 2 nanoseconds and the remaining operations each take 1 nanosecond.

   a. How long does a floating point addition take with these assumptions?

   b. How long will an unpipelined addition of 1000 pairs of floats take with these assumptions?

   c. How long will a pipelined addition of 1000 pairs of floats take with these assumptions?

   d. The time required for fetch and store may vary considerably if the operands/results are stored in different levels of the memory hierarchy. Suppose that a fetch from a level 1 cache takes 2 nanoseconds, while a fetch from a level 2 cache takes 5 nanoseconds, and a fetch from main memory takes 50 nanoseconds. What happens to the pipeline when there is a level 1 miss and a level 2 hit on a fetch of one of the operands? What happens when there is a level 1 and a level 2 miss?

2.2 Explain how a queue, implemented in hardware in the CPU, could be used to improve the performance of a write-through cache.

2.3 Recall the example involving cache reads of a two-dimensional array (p. 24). How do a larger matrix and a larger cache affect the performance of the two pairs of nested loops? What happens if MAX = 8 and the cache can store four lines? How many misses occur in the reads of A in the first pair of nested loops? How many misses occur in the second pair?

2.4 In Table 2.2, virtual addresses consist of a byte offset of 12 bits and a virtual page number of 20 bits. How many pages can a program have if it's run on a system with this page size and this virtual address size?

2.5 Does the addition of cache and virtual memory to a von Neumann system change its designation as a SISD system? What about the addition of pipelining? Multiple issue? Hardware multithreading?

2.6 Suppose that a vector processor has a memory system in which it takes 10 cycles to load a single 64-bit word from memory. How many memory banks are needed so that a stream of loads can, on average, require only one cycle per load?

2.7 Discuss the differences in how a GPU and a vector processor might execute the following code:

```
sum = 0.0;
for (i = 0; i < n; i++) {
    y[i] += a*x[i];
    sum += z[i]*z[i];
}
```

2.8 Consider the following pseudocode:

```
for (i = 0; i < n; i++) {
    if (my_rank == 0) {
        A(&args1, i);
        B(&args2, i);
    } else if (my_rank == 1) {
        C(&args3, i);
        D(&args4, i);
    } else if (my_rank == 2) {
        E(&args5, i);
        F(&args6, i);
    }
}
```

Suppose each of the tests my_rank == ... and each of the functions A, B, ..., F require 1 unit of time to execute.

a. Suppose the code is being executed on a SIMD system by three threads. Construct a table showing the execution of a single iteration of the body of the for statement. The table should show when each thread is executing and what it's executing, and when each thread is idle.

b. Now suppose the code is being executed on a MIMD system by three threads. Construct a table showing the execution of a single iteration of the body of the for statement.

2.9 Explain why the performance of a hardware multithreaded processing core might degrade if it had large caches and it ran many threads.

2.10 In our discussion of parallel hardware, we used Flynn's taxonomy to identify three types of parallel systems: SISD, SIMD, and MIMD. None of our systems were identified as multiple instruction, single data, or MISD. How would a MISD system work? Give an example.

2.11 Suppose a program must execute 10^{12} instructions to solve a particular problem. Suppose also that a single processor system can solve the problem in 10^6 seconds (about 11.6 days). So, on average, the single processor system executes 10^6 or a million instructions per second. Now suppose that the program has been parallelized for execution on a distributed-memory system. Suppose also that if the parallel program uses p processors, each processor will execute $10^{12}/p$ instructions and each processor must send $10^9(p-1)$ messages. Finally, suppose that there is no additional overhead in executing the parallel program. That is, the program will complete after each processor has executed all of its instructions and sent all of its messages, and there won't be any delays due to issues such as waiting for messages.

  a. Suppose it takes 10^{-9} seconds to send a message. How long will it take the program to run with 1000 processors if each processor is as fast as the single processor on which the serial program was run?

  b. Suppose it takes 10^{-3} seconds to send a message. How long will it take the program to run with 1000 processors?

2.12 Derive formulas for the total number of links in the various distributed-memory interconnects.

2.13 a. A planar mesh is just like a toroidal mesh, except that it doesn't have the "wraparound" links. What is the bisection width of a square planar mesh?

  b. A three-dimensional mesh is similar to a planar mesh, except that it also has depth. What is the bisection width of a three-dimensional mesh?

2.14 a. Sketch a four-dimensional hypercube.

  b. Use the inductive definition of a hypercube to explain why the bisection width of a hypercube is $p/2$.

2.15 a. Suppose the run-time of a serial program is given by $T_{serial} = n^2$, where the units of the run-time are in microseconds. Suppose that a parallelization of this program has run-time $T_{parallel} = n^2/p + \log_2(p)$. Write a program that finds the speedups and efficiencies of this program for various values of n and p. Run your program with $n = 10, 20, 40, \ldots, 320$, and $p = 1, 2, 4, \ldots, 128$. What happens to the speedups and efficiencies as p is increased and n is held fixed? What happens when p is fixed and n is increased?

  b. (Needs calculus.) Suppose that for a given program

$$T_{parallel} = T_{serial}/p + T_{overhead},$$

and consider the case when p is held fixed. In this setting suppose T'_{serial} and $T'_{overhead}$ denote the derivatives of T_{serial} and $T_{overhead}$ with respect to n for the fixed value of p. Also suppose that for the fixed value of p and for all n, the functions T_{serial}, $T_{overhead}$, T'_{serial}, and $T'_{overhead}$ are positive. Show that if

$$\frac{T'_{serial}}{T_{serial}} > \frac{T'_{overhead}}{T_{overhead}},$$

then for the fixed value of p, both the speedup and the efficiency are increasing functions of n. Hint: Work backward. Apply the quotient rule for derivatives to the definition of speedup, assume it's positive, and derive the formula above.

2.16 A parallel program that obtains a speedup greater than p—the number of processes or threads—is sometimes said to have **superlinear speedup**. However, many authors don't count programs that overcome "resource limitations" as having superlinear speedup. For example, a program that must use secondary storage for its data when it's run on a single processor system might be able to fit all its data into main memory when run on a large distributed-memory system. Give another example of how a program might overcome a resource limitation and obtain speedups greater than p.

2.17 Look at three programs you wrote in your Introduction to Computer Science class. Which parts of these programs (if any) are inherently serial? Does the inherently serial part of the work done by the program decrease as the problem size increases? Or does it remain roughly the same?

2.18 Suppose $T_{serial} = n$ and $T_{parallel} = n/p + \log_2(p)$, where times are in microseconds. If we increase p by a factor of k, find a formula for how much we'll need to increase n to maintain constant efficiency. How much should we increase n by if we double the number of processes from 8 to 16? Is the parallel program scalable?

2.19 Is a program that obtains linear speedup strongly scalable? Explain your answer.

2.20 Bob has a program that he wants to time with two sets of data, `input_data1` and `input_data2`. To get some idea of what to expect before adding timing functions to the code he's interested in, he runs the program with two sets of data and the Unix shell command `time`:

```
$ time ./bobs_prog < input_data1

real  0m0.001s
user  0m0.001s
sys   0m0.000s

$ time ./bobs_prog < input_data2

real  1m1.234s
user  1m0.001s
sys   0m0.111s
```

The timer function Bob is going to use in his program has millisecond resolution. Should Bob use it to time his program with the first set of data? What about the second set of data? Why or why not?

2.21 As we saw in the preceding problem, the Unix shell command `time` reports the user time, the system time, and the "real" time or total elapsed time. Suppose that Bob has defined the following functions that can be called in a C program:

```
double  utime ( void );
double  stime ( void );
double  rtime ( void );
```

The first returns the number of seconds of user time that have elapsed since the program started execution, the second returns the number of system seconds, and the third returns the total number of seconds. Roughly, user time is time spent in the user code, and library functions that don't need to use the operating system—for example, `sin` and `cos`. System time is time spent in functions that do need to use the operating system—for example, `printf` and `scanf`.

a. What is the mathematical relation among the three function values? That is, suppose the program contains the following code:

```
u = double  utime ( void );
s = double  stime ( void );
r = double  rtime ( void );
```

Write a formula that expresses the relation between `u`, `s`, and `r`. (You can assume that the time it takes to call the functions is negligible.)

b. On Bob's system, any time that an MPI process spends waiting for messages isn't counted by either `utime` or `stime`, but the time *is* counted by `rtime`. Explain how Bob can use these facts to determine whether an MPI process is spending too much time waiting for messages.

c. Bob has given Sally his timing functions. However, Sally has discovered that on her system, the time an MPI process spends waiting for messages is counted as user time. Furthermore, sending messages doesn't use any system time. Can Sally use Bob's functions to determine whether an MPI process is spending too much time waiting for messages? Explain your answer.

2.22 Apply Foster's method to an embarrassingly parallel problem—e.g., vector addition. What are the tasks? What are the communications among the tasks? How will you aggregate the tasks? How will you map composite tasks to processes/threads?

2.23 In our application of Foster's methodology to the construction of a histogram, we essentially identified aggregate tasks with elements of `data`. An apparent alternative would be to identify aggregate tasks with the elements of `bin_counts`. So an aggregate task would consist of all increments of `bin_counts[b]`. Each process/thread would be assigned a collection of bins, and for each of its bins b and for each element of `data` it would determine whether the element of data belongs to bin b. If the data structures have already been distributed and initialized, design a parallel program that uses this design. Which of the two designs is more scalable?

2.24 If you haven't already done so in Chapter 1, try to write pseudocode for our tree-structured global sum, which sums the elements of `loc_bin_cts`. First consider how this might be done in a shared-memory MIMD setting. Then consider how this might be done in a distributed-memory MIMD setting. In the shared-memory setting, which variables are shared and which are private?

Distributed memory programming with MPI

3

Recall that the world of parallel multiple instruction, multiple data, or MIMD, computers is, for the most part, divided into **distributed-memory** and **shared-memory** systems. From a programmer's point of view, a distributed-memory system consists of a collection of core-memory pairs connected by a network, and the memory associated with a core is directly accessible only to that core. (See Fig. 3.1.) On the other hand, from a programmer's point of view, a shared-memory system consists of a collection of cores connected to a globally accessible memory, in which each core can have access to any memory location. (See Fig. 3.2.) In this chapter, we're going to start looking at how to program distributed memory systems using **message-passing**.

Recall that in message-passing programs, a program running on one core-memory pair is usually called a **process**, and two processes can communicate by calling functions: one process calls a *send* function and the other calls a *receive* function. The implementation of message-passing that we'll be using is called **MPI**, which is an abbreviation of **Message-Passing Interface**. MPI is not a new programming language. It defines a *library* of functions that can be called from C and Fortran programs. We'll learn about some of MPI's different send and receive functions. We'll also

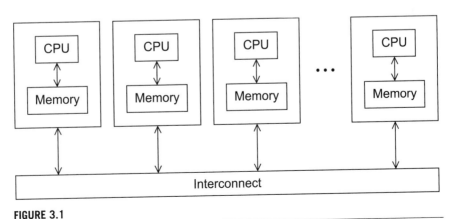

FIGURE 3.1

A distributed memory system.

An Introduction to Parallel Programming. https://doi.org/10.1016/B978-0-12-804605-0.00010-5

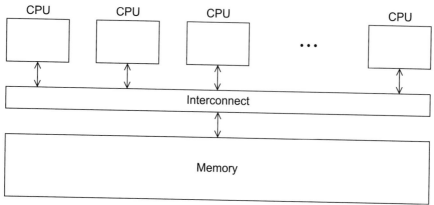

FIGURE 3.2

A shared memory system.

learn about some "global" communication functions that can involve more than two processes. These functions are called **collective** communications. In the process of learning about all of these MPI functions, we'll also learn about some of the fundamental issues involved in writing message-passing programs—issues such as data partitioning and I/O in distributed-memory systems. We'll also revisit the issue of parallel program performance.

3.1 Getting started

Perhaps the first program that many of us saw was some variant of the "hello, world" program in Kernighan and Ritchie's classic text [32]:

```
#include <stdio.h>

int main(void) {
   printf("hello, world\n");

   return 0;
}
```

Let's write a program similar to "hello, world" that makes some use of MPI. Instead of having each process simply print a message, we'll designate one process to do the output, and the other processes will send it messages, which it will print.

In parallel programming, it's common (one might say standard) for the processes to be identified by nonnegative integer *ranks*. So if there are p processes, the processes will have ranks $0, 1, 2, \ldots, p - 1$. For our parallel "hello, world," let's make process 0 the designated process, and the other processes will send it messages. See Program 3.1.

```
1  #include <stdio.h>
2  #include <string.h>   /* For strlen          */
3  #include <mpi.h>       /* For MPI functions, etc */
4
5  const int MAX_STRING = 100;
6
7  int main(void) {
8     char       greeting[MAX_STRING];
9     int        comm_sz;  /* Number of processes */
10    int        my_rank;  /* My process rank     */
11
12    MPI_Init(NULL, NULL);
13    MPI_Comm_size(MPI_COMM_WORLD, &comm_sz);
14    MPI_Comm_rank(MPI_COMM_WORLD, &my_rank);
15
16    if (my_rank != 0) {
17       sprintf(greeting, "Greetings from process %d of %d!",
18             my_rank, comm_sz);
19       MPI_Send(greeting, strlen(greeting)+1, MPI_CHAR, 0, 0,
20             MPI_COMM_WORLD);
21    } else {
22       printf("Greetings from process %d of %d!\n",
23             my_rank, comm_sz);
24       for (int q = 1; q < comm_sz; q++) {
25          MPI_Recv(greeting, MAX_STRING, MPI_CHAR, q,
26                0, MPI_COMM_WORLD, MPI_STATUS_IGNORE);
27          printf("%s\n", greeting);
28       }
29    }
30
31    MPI_Finalize();
32    return 0;
33 }  /* main */
```

Program 3.1: MPI program that prints greetings from the processes.

3.1.1 Compilation and execution

The details of compiling and running the program depend on your system, so you may need to check with a local expert. However, recall that when we need to be explicit, we'll assume that we're using a text editor to write the program source, and the command line to compile and run. Many systems use a command called mpicc for compilation[1]:

```
$ mpicc -g -Wall -o mpi_hello mpi_hello.c
```

Typically mpicc is a script that's a **wrapper** for the C compiler. A **wrapper script** is a script whose main purpose is to run some program. In this case, the program is the C compiler. However, the wrapper simplifies the running of the compiler by telling it where to find the necessary header files, and which libraries to link with the object file.

Many systems also support program startup with mpiexec:

```
$ mpiexec -n <number of processes> ./mpi_hello
```

So to run the program with one process, we'd type

```
$ mpiexec -n 1 ./mpi_hello
```

and to run the program with four processes, we'd type

```
$ mpiexec -n 4 ./mpi_hello
```

With one process, the program's output would be

```
Greetings from process 0 of 1!
```

and with four processes, the program's output would be

```
Greetings from process 0 of 4!
Greetings from process 1 of 4!
Greetings from process 2 of 4!
Greetings from process 3 of 4!
```

How do we get from invoking mpiexec to one or more lines of greetings? The mpiexec command tells the system to start <number of processes> instances of our <mpi_hello> program. It may also tell the system which core should run each instance of the program. After the processes are running, the MPI implementation takes care of making sure that the processes can communicate with each other.

3.1.2 MPI programs

Let's take a closer look at the program.

The first thing to observe is that this *is* a C program. For example, it includes the standard C header files stdio.h and string.h. It also has a main function, just like any other C program. However, there are many parts of the program that are new. Line 3 includes the mpi.h header file. This contains prototypes of MPI functions, macro definitions, type definitions, and so on; it contains all the definitions and declarations needed for compiling an MPI program.

---

[1] Recall that the dollar sign ($) is the shell prompt—so it shouldn't be typed in. Also recall that, for the sake of explicitness, we assume that we're using the Gnu C compiler, gcc, and we always use the options -g, -Wall, and -o. See Section 2.9 for further information.

The second thing to observe is that all of the identifiers defined by MPI start with the string `MPI_`. The first letter following the underscore is capitalized for function names and MPI-defined types. All of the letters in MPI-defined macros and constants are capitalized. So there's no question about what is defined by MPI and what's defined by the user program.

3.1.3 `MPI_Init` **and** `MPI_Finalize`

In Line 12, the call to `MPI_Init` tells the MPI system to do all of the necessary setup. For example, it might allocate storage for message buffers, and it might decide which process gets which rank. As a rule of thumb, no other MPI functions should be called before the program calls `MPI_Init`. Its syntax is

```
int MPI_Init(
        int*      argc_p   /* in/out */,
        char***   argv_p   /* in/out */);
```

The arguments, `argc_p` and `argv_p`, are pointers to the arguments to `main`, `argc` and `argv`. However, when our program doesn't use these arguments, we can just pass `NULL` for both. Like most MPI functions, `MPI_Init` returns an **int** error code. In most cases, we'll ignore the error codes, since checking them tends to clutter the code and make it more difficult to understand what it's doing.[2]

In Line 31, the call to `MPI_Finalize` tells the MPI system that we're done using MPI, and that any resources allocated for MPI can be freed. The syntax is quite simple:

```
int MPI_Finalize(void);
```

In general, no MPI functions should be called after the call to `MPI_Finalize`.

Thus a typical MPI program has the following basic outline:

```
. . .
#include <mpi.h>
. . .
int main(int argc, char* argv[]) {
    . . .
    /* No MPI calls before this */
    MPI_Init(&argc, &argv);
    . . .
    MPI_Finalize();
    /* No MPI calls after this */
    . . .
    return 0;
}
```

---

[2] Of course, when we're debugging our programs, we may make extensive use of the MPI error codes.

However, we've already seen that it's not necessary to pass pointers to `argc` and `argv` to `MPI_Init`. It's also not necessary that the calls to `MPI_Init` and `MPI_Finalize` be in `main`.

3.1.4 **Communicators,** `MPI_Comm_size,` **and** `MPI_Comm_rank`

In MPI a **communicator** is a collection of processes that can send messages to each other. One of the purposes of `MPI_Init` is to define a communicator that consists of all of the processes started by the user when she started the program. This communicator is called `MPI_COMM_WORLD`. The function calls in Lines 13 and 14 are getting information about `MPI_COMM_WORLD`. Their syntax is

```
int  MPI_Comm_size(
        MPI_Comm   comm          /* in   */,
        int*       comm_sz_p      /* out  */);

int  MPI_Comm_rank(
        MPI_Comm   comm          /* in   */,
        int*       my_rank_p      /* out  */);
```

For both functions, the first argument is a communicator and has the special type defined by MPI for communicators, `MPI_Comm`. `MPI_Comm_size` returns in its second argument the number of processes in the communicator, `MPI_Comm_rank` returns in its second argument the calling process's rank in the communicator. We'll often use the variable `comm_sz` for the number of processes in `MPI_COMM_WORLD`, and the variable `my_rank` for the process rank.

3.1.5 **SPMD programs**

Notice that we compiled a single program—we didn't compile a different program for each process—and we did this in spite of the fact that process 0 is doing something fundamentally different from the other processes: it's receiving a series of messages and printing them, while each of the other processes is creating and sending a message. This is quite common in parallel programming. In fact, *most* MPI programs are written in this way. That is, a single program is written so that different processes carry out different actions, and this can be achieved by simply having the processes branch on the basis of their process rank. Recall that this approach to parallel programming is called **single program, multiple data** or **SPMD**. The if –else statement in Lines 16–29 makes our program SPMD.

Also notice that our program will, in principle, run with any number of processes. We saw a little while ago that it can be run with one process or four processes, but if our system has sufficient resources, we could also run it with 1000 or even 100,000 processes. Although MPI doesn't require that programs have this property, it's almost always the case that we try to write programs that will run with any number of processes, because we usually don't know in advance the exact resources available to us. For example, we might have a 20-core system available today, but tomorrow we might have access to a 500-core system.

3.1.6 **Communication**

In Lines 17–18, each process, other than process 0, creates a message it will send to process 0. (The function sprintf is very similar to printf, except that instead of writing to stdout, it writes to a string.) Lines 19–20 actually send the message to process 0. Process 0, on the other hand, simply prints its message using printf, and then uses a **for** loop to receive and print the messages sent by processes $1, 2, \ldots$, comm_sz $- 1$. Lines 25–26 receive the message sent by process q, for $q = 1, 2, \ldots$, comm_sz $- 1$.

3.1.7 MPI_Send

The sends executed by processes $1, 2, \ldots$, comm_sz $- 1$ are fairly complex, so let's take a closer look at them. Each of the sends is carried out by a call to MPI_Send, whose syntax is

```
int MPI_Send(
        void*          msg_buf_p       /* in */,
        int            msg_size        /* in */,
        MPI_Datatype   msg_type        /* in */,
        int            dest            /* in */,
        int            tag             /* in */,
        MPI_Comm       communicator    /* in */);
```

The first three arguments, msg_buf_p, msg_size, and msg_type, determine the contents of the message. The remaining arguments, dest, tag, and communicator, determine the destination of the message.

The first argument, msg_buf_p, is a pointer to the block of memory containing the contents of the message. In our program, this is just the string containing the message, greeting. (Remember that in C an array, such as a string, is a pointer.) The second and third arguments, msg_size and msg_type, determine the amount of data to be sent. In our program, the msg_size argument is the number of characters in the message, plus one character for the '\0' character that terminates C strings. The msg_type argument is MPI_CHAR. These two arguments together tell the system that the message contains strlen(greeting)+1 **char**s.

Since C types (**int**, **char**, etc.) can't be passed as arguments to functions, MPI defines a special type, MPI_Datatype, that is used for the msg_type argument. MPI also defines a number of constant values for this type. The ones we'll use (and a few others) are listed in Table 3.1.

Notice that the size of the string greeting is not the same as the size of the message specified by the arguments msg_size and msg_type. For example, when we run the program with four processes, the length of each of the messages is 31 characters, while we've allocated storage for 100 characters in greetings. Of course, the size of the message sent should be less than or equal to the amount of storage in the buffer—in our case the string greeting.

The fourth argument, dest, specifies the rank of the process that should receive the message. The fifth argument, tag, is a nonnegative **int**. It can be used to distinguish

Table 3.1 Some predefined MPI datatypes.

MPI datatype	C datatype
MPI_CHAR	signed **char**
MPI_SHORT	signed **short int**
MPI_INT	signed **int**
MPI_LONG	signed **long int**
MPI_LONG_LONG	signed **long long int**
MPI_UNSIGNED_CHAR	**unsigned char**
MPI_UNSIGNED_SHORT	**unsigned short int**
MPI_UNSIGNED	**unsigned int**
MPI_UNSIGNED_LONG	**unsigned long int**
MPI_FLOAT	**float**
MPI_DOUBLE	**double**
MPI_LONG_DOUBLE	**long double**
MPI_BYTE	
MPI_PACKED	

messages that are otherwise identical. For example, suppose process 1 is sending **float**s to process 0. Some of the **float**s should be printed, while others should be used in a computation. Then the first four arguments to MPI_Send provide no information regarding which **float**s should be printed and which should be used in a computation. So process 1 can use (say) a tag of 0 for the messages that should be printed and a tag of 1 for the messages that should be used in a computation.

The final argument to MPI_Send is a communicator. All MPI functions that involve communication have a communicator argument. One of the most important purposes of communicators is to specify communication universes; recall that a communicator is a collection of processes that can send messages to each other. Conversely, a message sent by a process using one communicator cannot be received by a process that's using a different communicator. Since MPI provides functions for creating new communicators, this feature can be used in complex programs to ensure that messages aren't "accidentally received" in the wrong place.

An example will clarify this. Suppose we're studying global climate change, and we've been lucky enough to find two libraries of functions, one for modeling the earth's atmosphere and one for modeling the earth's oceans. Of course, both libraries use MPI. These models were built independently, so they don't communicate with each other, but they do communicate internally. It's our job to write the interface code. One problem we need to solve is ensuring that the messages sent by one library won't be accidentally received by the other. We might be able to work out some scheme with tags: the atmosphere library gets tags $0, 1, \ldots, n - 1$, and the ocean library gets tags $n, n + 1, \ldots, n + m$. Then each library can use the given range to figure out which tag it should use for which message. However, a much simpler solution is provided by communicators: we simply pass one communicator to the atmosphere library functions and a different communicator to the ocean library functions.

3.1.8 `MPI_Recv`

The first six arguments to `MPI_Recv` correspond to the first six arguments of `MPI_Send`:

```
int MPI_Recv(
        void*         msg_buf_p       /* out */,
        int           buf_size        /* in  */,
        MPI_Datatype  buf_type        /* in  */,
        int           source          /* in  */,
        int           tag             /* in  */,
        MPI_Comm      communicator    /* in  */,
        MPI_Status*   status_p        /* out */);
```

Thus the first three arguments specify the memory available for receiving the message: `msg_buf_p` points to the block of memory, `buf_size` determines the number of objects that can be stored in the block, and `buf_type` indicates the type of the objects. The next three arguments identify the message. The `source` argument specifies the process from which the message should be received. The `tag` argument should match the `tag` argument of the message being sent, and the `communicator` argument must match the communicator used by the sending process. We'll talk about the `status_p` argument shortly. In many cases, it won't be used by the calling function, and, as in our "greetings" program, the special MPI constant `MPI_STATUS_IGNORE` can be passed.

3.1.9 Message matching

Suppose process q calls `MPI_Send` with

```
MPI_Send(send_buf_p, send_buf_sz, send_type, dest, send_tag,
        send_comm);
```

Also suppose that process r calls `MPI_Recv` with

```
MPI_Recv(recv_buf_p, recv_buf_sz, recv_type, src, recv_tag,
        recv_comm, &status);
```

Then the message sent by q with the above call to `MPI_Send` can be received by r with the call to `MPI_Recv` if

- `recv_comm` = `send_comm`,
- `recv_tag` = `send_tag`,
- `dest` = r, and
- `src` = q.

These conditions aren't quite enough for the message to be *successfully* received, however. The parameters specified by the first three pairs of arguments, `send_buf_p`/`recv_buf_p`, `send_buf_sz`/`recv_buf_sz`, and `send_type`/`recv_type`, must specify compatible buffers. For detailed rules, see the MPI-3 specification [40]. Most of the time, the following rule will suffice:

- If `recv_type` = `send_type` and `recv_buf_sz` ≥ `send_buf_sz`, then the message sent by q can be successfully received by r.

Of course, it can happen that one process is receiving messages from multiple processes, *and* the receiving process doesn't know the order in which the other processes will send the messages. For example, suppose process 0 is doling out work to processes $1, 2, \ldots,$ comm_sz $- 1$, and processes $1, 2, \ldots,$ comm_sz $- 1$, send their results back to process 0 when they finish the work. If the work assigned to each process takes an unpredictable amount of time, then 0 has no way of knowing the order in which the processes will finish. If process 0 simply receives the results in process rank order—first the results from process 1, then the results from process 2, and so on—and if (say) process comm_sz-1 finishes first, it *could* happen that process comm_sz-1 could sit and wait for the other processes to finish. To avoid this problem MPI provides a special constant MPI_ANY_SOURCE that can be passed to MPI_Recv. Then, if process 0 executes the following code, it can receive the results in the order in which the processes finish:

```
for (i = 1; i < comm_sz; i++) {
    MPI_Recv(result, result_sz, result_type, MPI_ANY_SOURCE,
            result_tag, comm, MPI_STATUS_IGNORE);
    Process_result(result);
}
```

Similarly, it's possible that one process can be receiving multiple messages with different tags from another process, and the receiving process doesn't know the order in which the messages will be sent. For this circumstance, MPI provides the special constant MPI_ANY_TAG that can be passed to the tag argument of MPI_Recv.

A couple of points should be stressed in connection with these "wildcard" arguments:

1. Only a receiver can use a wildcard argument. Senders must specify a process rank and a nonnegative tag. So MPI uses a "push" communication mechanism rather than a "pull" mechanism.
2. There is no wildcard for communicator arguments; both senders and receivers must always specify communicators.

3.1.10 The status_p argument

If you think about these rules for a minute, you'll notice that a receiver can receive a message without knowing

1. the amount of data in the message,
2. the sender of the message, or
3. the tag of the message.

So how can the receiver find out these values? Recall that the last argument to MPI_Recv has type MPI_Status*. The MPI type MPI_Status is a struct with at least the three members MPI_SOURCE, MPI_TAG, and MPI_ERROR. Suppose our program contains the definition

```
MPI_Status status;
```

Then after a call to MPI_Recv, in which &status is passed as the last argument, we can determine the sender and tags by examining the two members:

```
status.MPI_SOURCE
status.MPI_TAG
```

The amount of data that's been received isn't stored in a field that's directly accessible to the application program. However, it can be retrieved with a call to MPI_Get_count. For example, suppose that in our call to MPI_Recv, the type of the receive buffer is recv_type and, once again, we passed in &status. Then the call

```
MPI_Get_count(&status, recv_type, &count)
```

will return the number of elements received in the count argument. In general, the syntax of MPI_Get_count is

```
int MPI_Get_count(
            MPI_Status*  status_p  /* in  */,
            MPI_Datatype type      /* in  */,
            int*         count_p   /* out */);
```

Note that the count isn't directly accessible as a member of the MPI_Status variable, simply because it depends on the type of the received data, and, consequently, determining it would probably require a calculation (e.g., (number of bytes received)/(bytes per object)). And if this information isn't needed, we shouldn't waste a calculation determining it.

3.1.11 **Semantics of** MPI_Send **and** MPI_Recv

What exactly happens when we send a message from one process to another? Many of the details depend on the particular system, but we can make a few generalizations. The sending process will assemble the message. For example, it will add the "envelope" information to the actual data being transmitted—the destination process rank, the sending process rank, the tag, the communicator, and some information on the size of the message. Once the message has been assembled, recall from Chapter 2 that there are essentially two possibilities: the sending process can **buffer** the message or it can **block**. If it buffers the message, the MPI system will place the message (data and envelope) into its own internal storage, and the call to MPI_Send will return.

Alternatively, if the system blocks, it will wait until it can begin transmitting the message, and the call to MPI_Send may not return immediately. Thus if we use MPI_Send, when the function returns, we don't actually know whether the message has been transmitted. We only know that the storage we used for the message, the send buffer, is available for reuse by our program. If we need to know that the message has been transmitted, or if we need for our call to MPI_Send to return immediately—regardless of whether the message has been sent—MPI provides alternative functions for sending. We'll learn about one of these alternative functions later.

The exact behavior of MPI_Send is determined by the MPI implementation. However, typical implementations have a default "cutoff" message size. If the size of a message is less than the cutoff, it will be buffered. If the size of the message is greater than the cutoff, MPI_Send will block.

Unlike MPI_Send, MPI_Recv always blocks until a matching message has been received. So when a call to MPI_Recv returns, we know that there is a message stored in the receive buffer (unless there's been an error). There is an alternate method for receiving a message, in which the system checks whether a matching message is available and returns, regardless of whether there is one.

MPI requires that messages be *non-overtaking*. This means that if process q sends two messages to process r, then the first message sent by q must be available to r before the second message. However, there is no restriction on the arrival of messages sent from different processes. That is, if q and t both send messages to r, then even if q sends its message before t sends its message, there is no requirement that q's message become available to r before t's message. This is essentially because MPI can't impose performance on a network. For example, if q happens to be running on a machine on Mars, while r and t are both running on the same machine in San Francisco, and if q sends its message a nanosecond before t sends its message, it would be extremely unreasonable to require that q's message arrive before t's.

3.1.12 Some potential pitfalls

Note that the semantics of MPI_Recv suggests a potential pitfall in MPI programming: If a process tries to receive a message and there's no matching send, then the process will block forever. That is, the process will **hang**. So when we design our programs, we need to be sure that every receive has a matching send. Perhaps even more important, we need to be very careful when we're coding that there are no inadvertent mistakes in our calls to MPI_Send and MPI_Recv. For example, if the tags don't match, or if the rank of the destination process is the same as the rank of the source process, the receive won't match the send, and either a process will hang, or, perhaps worse, the receive may match *another* send.

Similarly, if a call to MPI_Send blocks and there's no matching receive, then the sending process can hang. If, on the other hand, a call to MPI_Send is buffered and there's no matching receive, then the message will be lost.

3.2 The trapezoidal rule in MPI

Printing messages from processes is all well and good, but we're probably not taking the trouble to learn to write MPI programs just to print messages. So let's take a look at a somewhat more useful program—let's write a program that implements the trapezoidal rule for numerical integration.

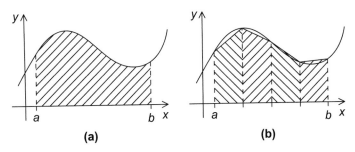

FIGURE 3.3

The trapezoidal rule: (a) area to be estimated and (b) approximate area using trapezoids.

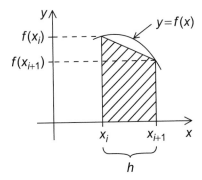

FIGURE 3.4

One trapezoid.

3.2.1 The trapezoidal rule

Recall that we can use the trapezoidal rule to approximate the area between the graph of a function, $y = f(x)$, two vertical lines, and the x-axis. (See Fig. 3.3.) The basic idea is to divide the interval on the x-axis into n equal subintervals. Then we approximate the area lying between the graph and each subinterval by a trapezoid, whose base is the subinterval, whose vertical sides are the vertical lines through the endpoints of the subinterval, and whose fourth side is the secant line joining the points where the vertical lines cross the graph. (See Fig. 3.4.) If the endpoints of the subinterval are x_i and x_{i+1}, then the length of the subinterval is $h = x_{i+1} - x_i$. Also, if the lengths of the two vertical segments are $f(x_i)$ and $f(x_{i+1})$, then the area of the trapezoid is

$$\text{Area of one trapezoid} = \frac{h}{2}[f(x_i) + f(x_{i+1})].$$

Since we chose the n subintervals so that they would all have the same length, we also know that if the vertical lines bounding the region are $x = a$ and $x = b$, then

$$h = \frac{b-a}{n}.$$

Thus if we call the leftmost endpoint x_0, and the rightmost endpoint x_n, we have that

$$x_0 = a, \; x_1 = a + h, \; x_2 = a + 2h, \; \ldots, \; x_{n-1} = a + (n-1)h, \; x_n = b,$$

and the sum of the areas of the trapezoids—our approximation to the total area—is

Sum of trapezoid areas $= h[f(x_0)/2 + f(x_1) + f(x_2) + \cdots + f(x_{n-1}) + f(x_n)/2]$.

Thus, pseudocode for a serial program might look something like this:

```
/* Input:  a, b, n */
h = (b-a)/n;
approx = (f(a) + f(b))/2.0;
for (i = 1; i <= n-1; i++) {
    x_i = a + i*h;
    approx += f(x_i);
}
approx = h*approx;
```

3.2.2 Parallelizing the trapezoidal rule

It is not the most attractive word, but, as we noted in Chapter 1, people who write parallel programs do use the verb "parallelize" to describe the process of converting a serial program or algorithm into a parallel program.

Recall that we can design a parallel program by using four basic steps:

1. Partition the problem solution into tasks.
2. Identify communication channels between the tasks.
3. Aggregate tasks into composite tasks.
4. Map composite tasks to cores.

In the partitioning phase, we usually try to identify as many tasks as possible. For the trapezoidal rule, we might identify two types of tasks: one type is finding the area of a single trapezoid, and the other is computing the sum of these areas. Then the communication channels will join each of the tasks of the first type to the single task of the second type. (See Fig. 3.5.)

So how can we aggregate the tasks and map them to the cores? Our intuition tells us that the more trapezoids we use, the more accurate our estimate will be. That is, we should use many trapezoids, and we will use many more trapezoids than cores. Thus we need to aggregate the computation of the areas of the trapezoids into groups. A natural way to do this is to split the interval $[a, b]$ up into comm_sz subintervals. If comm_sz evenly divides n, the number of trapezoids, we can simply apply the trapezoidal rule with $n/$comm_sz trapezoids to each of the comm_sz subintervals. To finish, we can have one of the processes, say process 0, add the estimates.

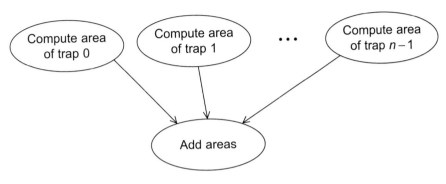

FIGURE 3.5

Tasks and communications for the trapezoidal rule.

Let's make the simplifying assumption that comm_sz evenly divides n. Then pseu-docode for the program might look something like the following:

```
1    Get  a,  b,  n;
2    h  =  (b−a)/n;
3    local_n  =  n/comm_sz;
4    local_a  =  a  +  my_rank*local_n*h;
5    local_b  =  local_a  +  local_n*h;
6    local_integral  =  Trap(local_a,  local_b,  local_n,  h);
7    if  (my_rank  !=  0)
8        Send  local_integral  to  process  0;
9    else  /*  my_rank  ==  0  */
10       total_integral  =  local_integral;
11       for  (proc  =  1;  proc  <  comm_sz;  proc++)  {
12           Receive  local_integral  from  proc;
13           total_integral  +=  local_integral;
14       }
15   }
16   if  (my_rank  ==  0)
17       print  result;
```

Let's defer, for the moment, the issue of input and just "hardwire" the values for a, b, and n. When we do this, we get the MPI program shown in Program 3.2. The Trap function is just an implementation of the serial trapezoidal rule. See Program 3.3.

Notice that in our choice of identifiers, we try to differentiate between *local* and *global* variables. **Local** variables are variables whose contents are significant only on the process that's using them. Some examples from the trapezoidal rule program are local_a, local_b, and local_n. Variables whose contents are significant to all the processes are sometimes called **global** variables. Some examples from the trapezoidal rule are a, b, and n. Note that this usage is different from the usage you learned in your introductory programming class, where local variables are private to a single function and global variables are accessible to all the functions. However, no confusion should arise, since the context should make the meaning clear.

```
1   int main(void) {
2      int my_rank, comm_sz, n = 1024, local_n;
3      double a = 0.0, b = 3.0, h, local_a, local_b;
4      double local_int, total_int;
5      int source;
6
7      MPI_Init(NULL, NULL);
8      MPI_Comm_rank(MPI_COMM_WORLD, &my_rank);
9      MPI_Comm_size(MPI_COMM_WORLD, &comm_sz);
10
11     h = (b-a)/n;            /* h is the same for all processes */
12     local_n = n/comm_sz;   /* So is the number of trapezoids  */
13
14     local_a = a + my_rank*local_n*h;
15     local_b = local_a + local_n*h;
16     local_int = Trap(local_a, local_b, local_n, h);
17
18     if (my_rank != 0) {
19        MPI_Send(&local_int, 1, MPI_DOUBLE, 0, 0,
20              MPI_COMM_WORLD);
21     } else {
22        total_int = local_int;
23        for (source = 1; source < comm_sz; source++) {
24           MPI_Recv(&local_int, 1, MPI_DOUBLE, source, 0,
25                 MPI_COMM_WORLD, MPI_STATUS_IGNORE);
26           total_int += local_int;
27        }
28     }
29
30     if (my_rank == 0) {
31        printf("With n = %d trapezoids, our estimate\n", n);
32        printf("of the integral from %f to %f = %.15e\n",
33              a, b, total_int);
34     }
35     MPI_Finalize();
36     return 0;
37  } /* main */
```

Program 3.2: First version of MPI trapezoidal rule.

3.3 Dealing with I/O

Of course, the current version of the parallel trapezoidal rule has a serious deficiency:
it will only compute the integral over the interval [0, 3] using 1024 trapezoids. We
can edit the code and recompile, but this is quite a bit of work compared to simply
typing in three new numbers. We need to address the problem of getting input from

```
1   double Trap(
2          double left_endpt   /* in */,
3          double right_endpt /* in */,
4          int    trap_count   /* in */,
5          double base_len     /* in */) {
6      double estimate, x;
7      int i;
8
9      estimate = (f(left_endpt) + f(right_endpt))/2.0;
10     for (i = 1; i <= trap_count -1; i++) {
11         x = left_endpt + i*base_len;
12         estimate += f(x);
13     }
14     estimate = estimate*base_len;
15
16     return estimate;
17  } /*  Trap  */
```

Program 3.3: Trap function in MPI trapezoidal rule.

the user. While we're talking about input to parallel programs, it might be a good idea to also take a look at output. We discussed these two issues in Chapter 2. So if you remember the discussion of nondeterminism and output, you can skip ahead to Section 3.3.2.

3.3.1 Output

In both the "greetings" program and the trapezoidal rule program, we've assumed that process 0 can write to stdout, i.e., its calls to printf behave as we might expect. Although the MPI standard doesn't specify which processes have access to which I/O devices, virtually all MPI implementations allow *all* the processes in MPI_COMM_WORLD full access to stdout and stderr. So most MPI implementations allow all processes to execute printf and fprintf(stderr, ...) .

However, most MPI implementations don't provide any automatic scheduling of access to these devices. That is, if multiple processes are attempting to write to, say, stdout, the order in which the processes' output appears will be unpredictable. Indeed, it can even happen that the output of one process will be interrupted by the output of another process.

For example, suppose we try to run an MPI program in which each process simply prints a message. (See Program 3.4.) On our cluster, if we run the program with five processes, it often produces the "expected" output:

```
Proc 0 of 5 > Does anyone have a toothpick?
Proc 1 of 5 > Does anyone have a toothpick?
Proc 2 of 5 > Does anyone have a toothpick?
```

```
#include <stdio.h>
#include <mpi.h>

int main(void) {
    int my_rank, comm_sz;

    MPI_Init(NULL, NULL);
    MPI_Comm_size(MPI_COMM_WORLD, &comm_sz);
    MPI_Comm_rank(MPI_COMM_WORLD, &my_rank);

    printf("Proc %d of %d > Does anyone have a toothpick?\n",
            my_rank, comm_sz);

    MPI_Finalize();
    return 0;
}   /* main */
```

Program 3.4: Each process just prints a message.

```
Proc 3 of 5 > Does anyone have a toothpick?
Proc 4 of 5 > Does anyone have a toothpick?
```

However, when we run it with six processes, the order of the output lines is unpredictable:

```
Proc 0 of 6 > Does anyone have a toothpick?
Proc 1 of 6 > Does anyone have a toothpick?
Proc 2 of 6 > Does anyone have a toothpick?
Proc 5 of 6 > Does anyone have a toothpick?
Proc 3 of 6 > Does anyone have a toothpick?
Proc 4 of 6 > Does anyone have a toothpick?
```

or

```
Proc 0 of 6 > Does anyone have a toothpick?
Proc 1 of 6 > Does anyone have a toothpick?
Proc 2 of 6 > Does anyone have a toothpick?
Proc 4 of 6 > Does anyone have a toothpick?
Proc 3 of 6 > Does anyone have a toothpick?
Proc 5 of 6 > Does anyone have a toothpick?
```

The reason this happens is that the MPI processes are "competing" for access to the shared output device, stdout, and it's impossible to predict the order in which the processes' output will be queued up. Such a competition results in **nondeterminism**. That is, the actual output will vary from one run to the next.

In any case, if we don't want output from different processes to appear in a random order, it's up to us to modify our program accordingly. For example, we can have each process other than 0 send its output to process 0, and process 0 can print the output in process rank order. This is exactly what we did in the "greetings" program.

3.3.2 Input

Unlike output, most MPI implementations only allow process 0 in MPI_COMM_WORLD access to stdin. This makes sense: If multiple processes have access to stdin, which process should get which parts of the input data? Should process 0 get the first line? process 1 get the second? Or should process 0 get the first character?

So, to write MPI programs that can use scanf, we need to branch on process rank, with process 0 reading in the data, and then sending it to the other processes. For example, we might write the Get_input function shown in Program 3.5 for our parallel trapezoidal rule program. In this function, process 0 simply reads in the values

```
1   void Get_input(
2         int       my_rank   /* in  */,
3         int       comm_sz   /* in  */,
4         double*   a_p       /* out */,
5         double*   b_p       /* out */,
6         int*      n_p       /* out */) {
7      int dest;
8
9      if (my_rank == 0) {
10        printf("Enter a, b, and n\n");
11        scanf("%lf %lf %d", a_p, b_p, n_p);
12        for (dest = 1; dest < comm_sz; dest++) {
13           MPI_Send(a_p, 1, MPI_DOUBLE, dest, 0, MPI_COMM_WORLD);
14           MPI_Send(b_p, 1, MPI_DOUBLE, dest, 0, MPI_COMM_WORLD);
15           MPI_Send(n_p, 1, MPI_INT, dest, 0, MPI_COMM_WORLD);
16        }
17     } else { /* my_rank != 0 */
18        MPI_Recv(a_p, 1, MPI_DOUBLE, 0, 0, MPI_COMM_WORLD,
19              MPI_STATUS_IGNORE);
20        MPI_Recv(b_p, 1, MPI_DOUBLE, 0, 0, MPI_COMM_WORLD,
21              MPI_STATUS_IGNORE);
22        MPI_Recv(n_p, 1, MPI_INT, 0, 0, MPI_COMM_WORLD,
23              MPI_STATUS_IGNORE);
24     }
25  }  /* Get_input */
```

Program 3.5: A function for reading user input.

for a, b, and n, and sends all three values to each process. So this function uses the same basic communication structure as the "greetings" program, except that now process 0 is sending to each process, while the other processes are receiving.

To use this function, we can simply insert a call to it inside our main function, being careful to put it after we've initialized `my_rank` and `comm_sz`:

```
. . .
MPI_Comm_rank(MPI_COMM_WORLD, &my_rank);
MPI_Comm_size(MPI_COMM_WORLD, &comm_sz);

Get_input(my_rank, comm_sz, &a, &b, &n);

h = (b-a)/n;
. . .
```

3.4 Collective communication

If we pause for a moment and think about our trapezoidal rule program, we can find several things that we might be able to improve on. One of the most obvious is the "global sum" after each process has computed its part of the integral. If we hire eight workers to, say, build a house, we might feel that we weren't getting our money's worth if seven of the workers told the first what to do, and then the seven collected their pay and went home. But this is very similar to what we're doing in our global sum. Each process with rank greater than 0 is "telling process 0 what to do" and then quitting. That is, each process with rank greater than 0 is, in effect, saying "add this number into the total." Process 0 is doing nearly all the work in computing the global sum, while the other processes are doing almost nothing. Sometimes it does happen that this is the best we can do in a parallel program, but if we imagine that we have eight students, each of whom has a number and we want to find the sum of all eight numbers, we can certainly come up with a more equitable distribution of the work than having seven of the eight give their numbers to one of the students and having the first do the addition.

3.4.1 Tree-structured communication

As we already saw in Chapter 1, we might use a "binary tree structure," like that illustrated in Fig. 3.6. In this diagram, initially students or processes 1, 3, 5, and 7 send their values to processes 0, 2, 4, and 6, respectively. Then processes 0, 2, 4, and 6 add the received values to their original values, and the process is repeated twice:

1. **a.** Processes 2 and 6 send their new values to processes 0 and 4, respectively.
   b. Processes 0 and 4 add the received values into their new values.
2. **a.** Process 4 sends its newest value to process 0.
   b. Process 0 adds the received value to its newest value.

This solution may not seem ideal, since half the processes (1, 3, 5, and 7) are doing the same amount of work that they did in the original scheme. However, if

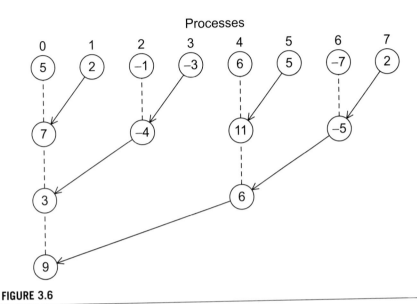

FIGURE 3.6

A tree-structured global sum.

you think about it, the original scheme required comm_sz − 1 = seven receives and seven adds by process 0, while the new scheme only requires three, and all the other processes do no more than two receives and adds. Furthermore, the new scheme has the property that a lot of the work is done concurrently by different processes. For example, in the first phase, the receives and adds by processes 0, 2, 4, and 6 can all take place simultaneously. So, if the processes start at roughly the same time, the total time required to compute the global sum will be the time required by process 0, i.e., three receives and three additions. So we've reduced the overall time by more than 50%. Furthermore, if we use more processes, we can do even better. For example, if comm_sz = 1024, then the original scheme requires process 0 to do 1023 receives and additions, while it can be shown (Exercise 3.5) that the new scheme requires process 0 to do only 10 receives and additions. This improves the original scheme by more than a factor of 100!

You may be thinking to yourself, this is all well and good, but coding this tree-structured global sum looks like it would take a quite a bit of work, and you'd be right. (See Programming Assignment 3.3.) In fact, the problem may be even harder. For example, it's perfectly feasible to construct a tree-structured global sum that uses different "process-pairings." For example, we might pair 0 and 4, 1 and 5, 2 and 6, and 3 and 7 in the first phase. Then we could pair 0 and 2, and 1 and 3 in the second, and 0 and 1 in the final. (See Fig. 3.7.) Of course, there are many other possibilities. How can we decide which is the best? Do we need to code each alternative and evaluate their performance? If we do, is it possible that one method works best for "small"

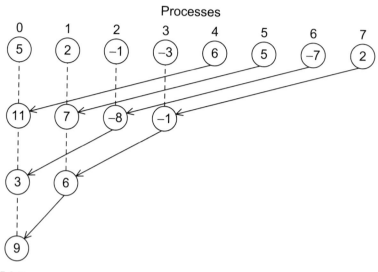

FIGURE 3.7

An alternative tree-structured global sum.

trees, while another works best for "large" trees? Even worse, one approach might work best on system A, while another might work best on system B.

3.4.2 MPI_Reduce

With virtually limitless possibilities, it's unreasonable to expect each MPI programmer to write an optimal global-sum function, so MPI specifically protects programmers against this trap of endless optimization by requiring that MPI implementations include implementations of global sums. This places the burden of optimization on the developer of the MPI implementation, rather than the application developer. The assumption here is that the developer of the MPI implementation should know enough about both the hardware and the system software so that she can make better decisions about implementation details.

Now a "global-sum function" will obviously require communication. However, unlike the MPI_Send-MPI_Recv pair, the global-sum function may involve more than two processes. In fact, in our trapezoidal rule program it will involve all the processes in MPI_COMM_WORLD. In MPI parlance, communication functions that involve all the processes in a communicator are called **collective communications**. To distinguish between collective communications and functions, such as MPI_Send and MPI_Recv, MPI_Send and MPI_Recv are often called **point-to-point** communications.

In fact, global-sum is just a special case of an entire class of collective communications. For example, it might happen that instead of finding the sum of a collection of comm_sz numbers distributed among the processes, we want to find the maximum or the minimum or the product, or any one of many other possibilities. So MPI general-

Table 3.2 Predefined Reduction Operators in MPI.

Operation Value	Meaning
MPI_MAX	Maximum
MPI_MIN	Minimum
MPI_SUM	Sum
MPI_PROD	Product
MPI_LAND	Logical and
MPI_BAND	Bitwise and
MPI_LOR	Logical or
MPI_BOR	Bitwise or
MPI_LXOR	Logical exclusive or
MPI_BXOR	Bitwise exclusive or
MPI_MAXLOC	Maximum and location of maximum
MPI_MINLOC	Minimum and location of minimum

ized the global-sum function so that any one of these possibilities can be implemented with a single function:

```
int MPI_Reduce(
        void*           input_data_p     /* in  */,
        void*           output_data_p    /* out */,
        int             count            /* in  */,
        MPI_Datatype    datatype         /* in  */,
        MPI_Op          operator         /* in  */,
        int             dest_process     /* in  */,
        MPI_Comm        comm             /* in  */);
```

The key to the generalization is the fifth argument, `operator`. It has type `MPI_Op`, which is a predefined MPI type—like `MPI_Datatype` and `MPI_Comm`. There are a number of predefined values in this type—see Table 3.2. It's also possible to define your own operators—for details, see the MPI-3 Standard [40].

The operator we want is `MPI_SUM`. Using this value for the `operator` argument, we can replace the code in Lines 18–28 of Program 3.2 with the single function call

```
MPI_Reduce(&local_int, &total_int, 1, MPI_DOUBLE, MPI_SUM,
        0, MPI_COMM_WORLD);
```

One point worth noting is that by using a `count` argument greater than 1, `MPI_Reduce` can operate on arrays instead of scalars. So the following code could be used to add a collection of N-dimensional vectors, one per process:

```
double local_x[N], sum[N];
. . .
MPI_Reduce(local_x, sum, N, MPI_DOUBLE, MPI_SUM, 0,
        MPI_COMM_WORLD);
```

Table 3.3 Multiple Calls to `MPI_Reduce`.

Time	Process 0	Process 1	Process 2
0	a = 1; c = 2	a = 1; c = 2	a = 1; c = 2
1	MPI_Reduce(&a, &b, ...)	MPI_Reduce(&c, &d, ...)	MPI_Reduce(&a, &b, ...)
2	MPI_Reduce(&c, &d, ...)	MPI_Reduce(&a, &b, ...)	MPI_Reduce(&c, &d, ...)

3.4.3 Collective vs. point-to-point communications

It's important to remember that collective communications differ in several ways from point-to-point communications:

1. *All* the processes in the communicator must call the same collective function. For example, a program that attempts to match a call to `MPI_Reduce` on one process with a call to `MPI_Recv` on another process is erroneous, and, in all likelihood, the program will hang or crash.
2. The arguments passed by each process to an MPI collective communication must be "compatible." For example, if one process passes in 0 as the `dest_process` and another passes in 1, then the outcome of a call to `MPI_Reduce` is erroneous, and, once again, the program is likely to hang or crash.
3. The `output_data_p` argument is only used on `dest_process`. However, all of the processes still need to pass in an actual argument corresponding to `output_data_p`, even if it's just `NULL`.
4. Point-to-point communications are matched on the basis of tags and communicators. Collective communications don't use tags. So they're matched solely on the basis of the communicator and the *order* in which they're called. As an example, consider the calls to `MPI_Reduce` shown in Table 3.3. Suppose that each process calls `MPI_Reduce` with operator `MPI_SUM`, and destination process 0. At first glance, it might seem that after the two calls to `MPI_Reduce`, the value of b will be 3, and the value of d will be 6. However, the names of the memory locations are irrelevant to the matching of the calls to `MPI_Reduce`. The *order* of the calls will determine the matching, so the value stored in b will be $1 + 2 + 1 = 4$, and the value stored in d will be $2 + 1 + 2 = 5$.

A final caveat: it might be tempting to call `MPI_Reduce` using the same buffer for both input and output. For example, if we wanted to form the global sum of x on each process and store the result in x on process 0, we might try calling

```
MPI_Reduce(&x, &x, 1, MPI_DOUBLE, MPI_SUM, 0, comm);
```

However, this call is illegal in MPI. So its result will be unpredictable: it might produce an incorrect result, it might cause the program to crash; it might even produce a correct result. It's illegal, because it involves **aliasing** of an output argument. Two arguments are aliased if they refer to the same block of memory, and MPI prohibits aliasing of arguments if one of them is an output or input/output argument. This is because the MPI Forum wanted to make the Fortran and C versions of MPI as similar as possible, and Fortran prohibits aliasing. In some instances MPI provides an alternative construction that effectively avoids this restriction. See Subsection 7.1.9 for an example.

3.4.4 MPI_Allreduce

In our trapezoidal rule program, we just print the result. So it's perfectly natural for only one process to get the result of the global sum. However, it's not difficult to imagine a situation in which *all* of the processes need the result of a global sum to complete some larger computation. In this situation, we encounter some of the same problems we encountered with our original global sum. For example, if we use a tree to compute a global sum, we might "reverse" the branches to distribute the global sum (see Fig. 3.8). Alternatively, we might have the processes *exchange* partial results instead of using one-way communications. Such a communication pattern is sometimes called a *butterfly*. (See Fig. 3.9.) Once again, we don't want to have to decide on which structure to use, or how to code it for optimal performance. Fortunately, MPI provides a variant of MPI_Reduce that will store the result on all the processes in the communicator:

```
int MPI_Allreduce(
        void*        input_data_p   /* in  */,
        void*        output_data_p  /* out */,
        int          count          /* in  */,
        MPI_Datatype datatype       /* in  */,
        MPI_Op       operator       /* in  */,
        MPI_Comm     comm           /* in  */);
```

The argument list is identical to that for MPI_Reduce, except that there is no dest_process since all the processes should get the result.

3.4.5 Broadcast

If we can improve the performance of the global sum in our trapezoidal rule program by replacing a loop of receives on process 0 with a tree structured communication, we ought to be able to do something similar with the distribution of the input data. In fact, if we simply "reverse" the communications in the tree-structured global sum in Fig. 3.6, we obtain the tree-structured communication shown in Fig. 3.10, and we can use this structure to distribute the input data. A collective communication in which data belonging to a single process is sent to all of the processes in the communicator is called a **broadcast**, and you've probably guessed that MPI provides a broadcast function:

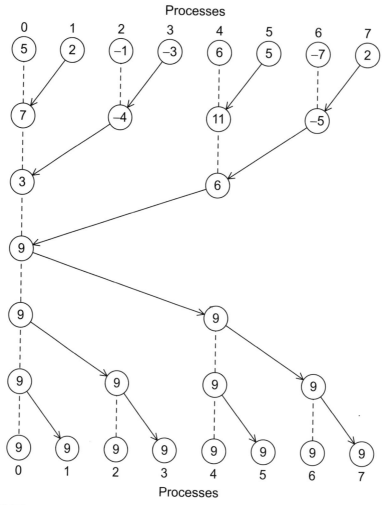

FIGURE 3.8

A global sum followed by distribution of the result.

```
int MPI_Bcast(
        void*           data_p          /* in/out */,
        int             count           /* in     */,
        MPI_Datatype    datatype        /* in     */,
        int             source_proc     /* in     */,
        MPI_Comm        comm            /* in     */);
```

The process with rank `source_proc` sends the contents of the memory referenced by `data_p` to all the processes in the communicator `comm`.

Processes

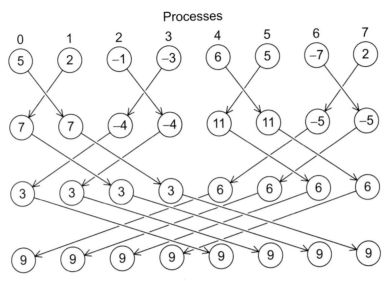

FIGURE 3.9

A butterfly-structured global sum.

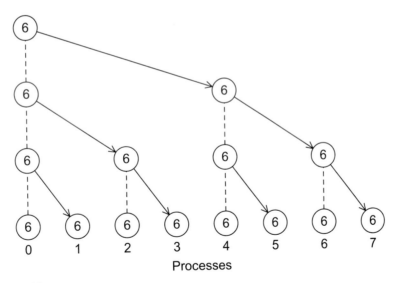

Processes

FIGURE 3.10

A tree-structured broadcast.

Program 3.6 shows how to modify the `Get_input` function shown in Program 3.5 so that it uses `MPI_Bcast`, instead of `MPI_Send` and `MPI_Recv`.

```
1   void Get_input(
2           int        my_rank   /* in  */,
3           int        comm_sz   /* in  */,
4           double*    a_p       /* out */,
5           double*    b_p       /* out */,
6           int*       n_p       /* out */) {
7
8       if (my_rank == 0) {
9           printf("Enter a, b, and n\n");
10          scanf("%lf %lf %d", a_p, b_p, n_p);
11      }
12      MPI_Bcast(a_p, 1, MPI_DOUBLE, 0, MPI_COMM_WORLD);
13      MPI_Bcast(b_p, 1, MPI_DOUBLE, 0, MPI_COMM_WORLD);
14      MPI_Bcast(n_p, 1, MPI_INT, 0, MPI_COMM_WORLD);
15  } /* Get_input */
```

Program 3.6: A version of Get_input that uses MPI_Bcast.

Recall that in serial programs an in/out argument is one whose value is both used and changed by the function. For MPI_Bcast, however, the data_p argument is an input argument on the process with rank source_proc and an output argument on the other processes. Thus when an argument to a collective communication is labeled in/out, it's possible that it's an input argument on some processes and an output argument on other processes.

3.4.6 Data distributions

Suppose we want to write a function that computes a vector sum:

$$\mathbf{x} + \mathbf{y} = (x_0, x_1, \ldots, x_{n-1}) + (y_0, y_1, \ldots, y_{n-1})$$
$$= (x_0 + y_0, x_1 + y_1, \ldots, x_{n-1} + y_{n-1})$$
$$= (z_0, z_1, \ldots, z_{n-1})$$
$$= \mathbf{z}$$

If we implement the vectors as arrays of, say, **double**s, we could implement serial vector addition with the code shown in Program 3.7.

How could we implement this using MPI? The work consists of adding the individual components of the vectors, so we might specify that the tasks are just the additions of corresponding components. Then there is no communication between the tasks, and the problem of parallelizing vector addition boils down to aggregating the tasks and assigning them to the cores. If the number of components is n and we have comm_sz cores or processes, let's assume that n is evenly divisible by comm_sz and define local_n $= n/$comm_sz. Then we can simply assign blocks of local_n consecutive components to each process. The four columns on the left of Table 3.4 show an

```
1   void Vector_sum(double x[], double y[], double z[], int n) {
2       int i;
3
4       for (i = 0; i < n; i++)
5           z[i] = x[i] + y[i];
6   }   /* Vector_sum */
```

Program 3.7: A serial implementation of vector addition.

Table 3.4 Different partitions of a 12-component vector among 3 processes.

	Components											
									Block-cyclic			
		Block				Cyclic			Blocksize = 2			
Process	0	1	2	3	0	3	6	9	0	1	6	7
0	4	5	6	7	1	4	7	10	2	3	8	9
1	8	9	10	11	2	5	8	11	4	5	10	11
2												

example when $n = 12$ and comm_sz $= 3$. This is often called a **block partition** of the vector.

An alternative to a block partition is a **cyclic partition**. In a cyclic partition, we assign the components in a round-robin fashion. The four columns in the middle of Table 3.4 show an example when $n = 12$ and comm_sz $= 3$. So process 0 gets component 0, process 1 gets component 1, process 2 gets component 2, process 0 gets component 3, and so on.

A third alternative is a **block-cyclic partition**. The idea here is that instead of using a cyclic distribution of individual components, we use a cyclic distribution of *blocks* of components. So a block-cyclic distribution isn't fully specified until we decide how large the blocks are. If comm_sz $= 3$, $n = 12$, and the blocksize $b = 2$, an example is shown in the four columns on the right of Table 3.4.

Once we've decided how to partition the vectors, it's easy to write a parallel vector addition function: each process simply adds its assigned components. Furthermore, regardless of the partition, each process will have local_n components of the vector, and, to save on storage, we can just store these on each process as an array of local_n elements. Thus each process will execute the function shown in Program 3.8. Although the names of the variables have been changed to emphasize the fact that the function is operating on only the process's portion of the vector, this function is virtually identical to the original serial function.

3.4.7 Scatter

Now suppose we want to test our vector addition function. It would be convenient to be able to read the dimension of the vectors and then read in the vectors **x** and **y**. We already know how to read in the dimension of the vectors: process 0 can prompt the

```
1    void  Parallel_vector_sum(
2             double   local_x []   /* in   */,
3             double   local_y []   /* in   */,
4             double   local_z []   /* out */,
5             int       local_n     /* in   */) {
6         int local_i;
7
8         for (local_i = 0; local_i < local_n; local_i++)
9             local_z[local_i] = local_x[local_i] + local_y[local_i];
10   }    /* Parallel_vector_sum */
```

Program 3.8: A parallel implementation of vector addition.

user, read in the value, and broadcast the value to the other processes. We might try something similar with the vectors: process 0 could read them in and broadcast them to the other processes. However, this could be very wasteful. If there are 10 processes and the vectors have 10,000 components, then each process will need to allocate storage for vectors with 10,000 components, when it is only operating on subvectors with 1000 components. If, for example, we use a block distribution, it would be better if process 0 sent only components 1000 to 1999 to process 1, components 2000 to 2999 to process 2, etc. Using this approach processes 1 to 9 would only need to allocate storage for the components they're actually using.

Thus we might try writing a function that reads in an entire vector on process 0, but only sends the needed components to each of the other processes. For the communication MPI provides just such a function:

```
int MPI_Scatter(
        void*           send_buf_p    /* in   */,
        int             send_count    /* in   */,
        MPI_Datatype    send_type     /* in   */,
        void*           recv_buf_p    /* out */,
        int             recv_count    /* in   */,
        MPI_Datatype    recv_type     /* in   */,
        int             src_proc      /* in   */,
        MPI_Comm        comm          /* in   */);
```

If the communicator comm contains comm_sz processes, then MPI_Scatter divides the data referenced by send_buf_p into comm_sz pieces—the first piece goes to process 0, the second to process 1, the third to process 2, and so on. For example, suppose we're using a block distribution and process 0 has read in all of an n-component vector into send_buf_p. Then process 0 will get the first local_n = n/comm_sz components, process 1 will get the next local_n components, and so on. Each process should pass its local vector as the recv_buf_p argument, and the recv_count argument should be local_n. Both send_type and recv_type should be MPI_DOUBLE, and src_proc should be 0. Perhaps surprisingly, send_count should also be local_n—send_count is the *amount of data going to each process*; it's *not* the amount of data in the memory referred to by send_buf_p. If we use a block distribution and MPI_Scatter, we can read in a vector using the function Read_vector shown in Program 3.9.

```
 1   void  Read_vector(
 2            double     local_a[]     /* out */,
 3            int        local_n       /* in  */,
 4            int        n             /* in  */,
 5            char       vec_name[]    /* in  */,
 6            int        my_rank       /* in  */,
 7            MPI_Comm   comm          /* in  */) {
 8
 9        double* a = NULL;
10        int  i;
11
12        if (my_rank == 0) {
13            a = malloc(n*sizeof(double));
14            printf("Enter the vector %s\n", vec_name);
15            for (i = 0; i < n; i++)
16                scanf("%lf", &a[i]);
17            MPI_Scatter(a, local_n, MPI_DOUBLE, local_a, local_n,
18                    MPI_DOUBLE, 0, comm);
19            free(a);
20        } else {
21            MPI_Scatter(a, local_n, MPI_DOUBLE, local_a, local_n,
22                    MPI_DOUBLE, 0, comm);
23        }
24   }  /* Read_vector */
```

Program 3.9: A function for reading and distributing a vector.

One point to note here is that MPI_Scatter sends the first block of send_count objects to process 0, the next block of send_count objects to process 1, and so on. So this approach to reading and distributing the input vectors will only be suitable if we're using a block distribution and n, the number of components in the vectors, is evenly divisible by comm_sz. We'll discuss how we might deal with a cyclic or block cyclic distribution in Section 3.5, and we'll look at dealing with the case in which n is not evenly divisible by comm_sz in Exercise 3.13.

3.4.8 Gather

Of course, our test program will be useless unless we can see the result of our vector addition. So we need to write a function for printing out a distributed vector. Our function can collect all of the components of the vector onto process 0, and then process 0 can print all of the components. The communication in this function can be carried out by MPI_Gather:

```
int MPI_Gather(
        void*           send_buf_p    /* in  */,
        int             send_count    /* in  */,
        MPI_Datatype    send_type     /* in  */,
        void*           recv_buf_p    /* out */,
        int             recv_count    /* in  */,
        MPI_Datatype    recv_type     /* in  */,
        int             dest_proc     /* in  */,
        MPI_Comm        comm          /* in  */);
```

The data stored in the memory referred to by send_buf_p on process 0 is stored in the first block in recv_buf_p, the data stored in the memory referred to by send_buf_p on process 1 is stored in the second block referred to by recv_buf_p, and so on. So, if we're using a block distribution, we can implement our distributed vector print function, as shown in Program 3.10. Note that recv_count is the number of data items received from *each* process, not the total number of data items received.

```
1   void Print_vector(
2           double     local_b[]   /* in */,
3           int        local_n     /* in */,
4           int        n           /* in */,
5           char       title[]     /* in */,
6           int        my_rank     /* in */,
7           MPI_Comm   comm        /* in */) {
8
9       double* b = NULL;
10      int i;
11
12      if (my_rank == 0) {
13          b = malloc(n*sizeof(double));
14          MPI_Gather(local_b, local_n, MPI_DOUBLE, b, local_n,
15                  MPI_DOUBLE, 0, comm);
16          printf("%s\n", title);
17          for (i = 0; i < n; i++)
18              printf("%f ", b[i]);
19          printf("\n");
20          free(b);
21      } else {
22          MPI_Gather(local_b, local_n, MPI_DOUBLE, b, local_n,
23                  MPI_DOUBLE, 0, comm);
24      }
25  } /* Print_vector */
```

Program 3.10: A function for printing a distributed vector.

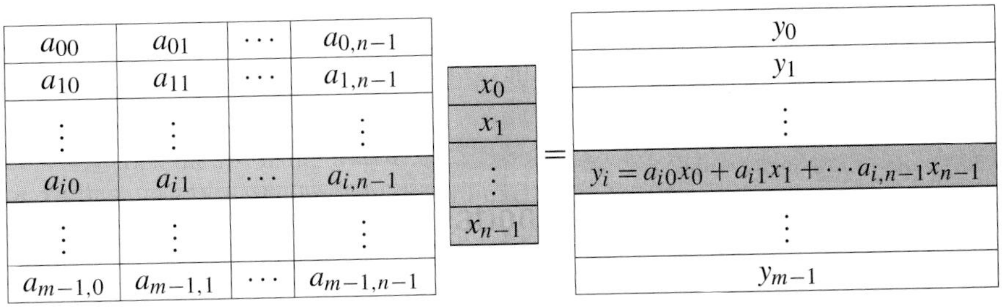

a_{00}	a_{01}	$\cdots$	$a_{0,n-1}$
a_{10}	a_{11}	$\cdots$	$a_{1,n-1}$
$\vdots$	$\vdots$		$\vdots$
a_{i0}	a_{i1}	$\cdots$	$a_{i,n-1}$
$\vdots$	$\vdots$		$\vdots$
$a_{m-1,0}$	$a_{m-1,1}$	$\cdots$	$a_{m-1,n-1}$

$$\begin{pmatrix} x_0 \\ x_1 \\ \vdots \\ x_{n-1} \end{pmatrix} = \begin{pmatrix} y_0 \\ y_1 \\ \vdots \\ y_i = a_{i0}x_0 + a_{i1}x_1 + \cdots a_{i,n-1}x_{n-1} \\ \vdots \\ y_{m-1} \end{pmatrix}$$

FIGURE 3.11

Matrix-vector multiplication.

The restrictions on the use of MPI_Gather are similar to those on the use of MPI_Scatter: our print function will only work correctly with vectors using a block distribution in which each block has the same size.

3.4.9 Allgather

As a final example, let's look at how we might write an MPI function that multiplies a matrix by a vector. Recall that if $A = (a_{ij})$ is an $m \times n$ matrix and $\mathbf{x}$ is a vector with n components, then $\mathbf{y} = A\mathbf{x}$ is a vector with m components, and we can find the ith component of y by forming the dot product of the ith row of A with $\mathbf{x}$:

$$y_i = a_{i0}x_0 + a_{i1}x_1 + a_{i2}x_2 + \cdots a_{i,n-1}x_{n-1}.$$

(See Fig. 3.11.)

So we might write pseudocode for serial matrix multiplication as follows:

```
/* For each row of A */
for (i = 0; i < m; i++) {
    /* Form dot product of ith row with x */
    y[i] = 0.0;
    for (j = 0; j < n; j++)
        y[i] += A[i][j]*x[j];
}
```

In fact, this could be actual C code. However, there are some peculiarities in the way that C programs deal with two-dimensional arrays (see Exercise 3.14). So C programmers frequently use one-dimensional arrays to "simulate" two-dimensional arrays. The most common way to do this is to list the rows one after another. For example, the two-dimensional array

$$\begin{pmatrix} 0 & 1 & 2 & 3 \\ 4 & 5 & 6 & 7 \\ 8 & 9 & 10 & 11 \end{pmatrix}$$

would be stored as the one-dimensional array

$$0\ 1\ 2\ 3\ 4\ 5\ 6\ 7\ 8\ 9\ 10\ 11.$$

In this example, if we start counting rows and columns from 0, then the element stored in row 2 and column 1—the 9—in the two-dimensional array is located in position $2 \times 4 + 1 = 9$ in the one-dimensional array. More generally, if our array has n columns, when we use this scheme, we see that the element stored in row i and column j is located in position $i \times n + j$ in the one-dimensional array.

Using this one-dimensional scheme, we get the C function shown in Program 3.11.

```
1   void  Mat_vect_mult(
2           double   A[]    /* in   */,
3           double   x[]    /* in   */,
4           double   y[]    /* out  */,
5           int      m      /* in   */,
6           int      n      /* in   */) {
7       int  i, j;
8
9       for (i = 0; i < m; i++) {
10          y[i] = 0.0;
11          for (j = 0; j < n; j++)
12              y[i] += A[i*n+j]*x[j];
13      }
14  }   /* Mat_vect_mult */
```

Program 3.11: Serial matrix-vector multiplication.

Now let's see how we might parallelize this function. An individual task can be the multiplication of an element of A by a component of **x** and the addition of this product into a component of **y**. That is, each execution of the statement

```
y[i] += A[i*n+j]*x[j];
```

is a task, so we see that if y[i] is assigned to process q, then it would be convenient to also assign row i of A to process q. This suggests that we partition A by *rows*. We could partition the rows using a block distribution, a cyclic distribution, or a block-cyclic distribution. In MPI it's easiest to use a block distribution. So let's use a block distribution of the rows of A, and, as usual, assume that comm_sz evenly divides m, the number of rows.

We are distributing A by rows so that the computation of y[i] will have all of the needed elements of A, so we should distribute y by blocks. That is, if the ith row of A is assigned to process q, then the ith component of y should also be assigned to process q.

Now the computation of y[i] involves all the elements in the *i*th row of A and *all* the components of x. So we could minimize the amount of communication by simply assigning all of x to each process. However, in actual applications—especially when the matrix is square—it's often the case that a program using matrix-vector multiplication will execute the multiplication many times, and the result vector **y** from one multiplication will be the input vector **x** for the next iteration. In practice, then, we usually assume that the distribution for x is the same as the distribution for y.

So if x has a block distribution, how can we arrange that each process has access to *all* the components of *x* before we execute the following loop?

```
for (j = 0; j < n; j++)
    y[i] += A[i*n+j]*x[j];
```

Using the collective communications we're already familiar with, we could execute a call to MPI_Gather, followed by a call to MPI_Bcast. This would, in all likelihood, involve two tree-structured communications, and we may be able to do better by using a butterfly. So, once again, MPI provides a single function:

```
int  MPI_Allgather(
        void*         send_buf_p   /* in  */,
        int           send_count   /* in  */,
        MPI_Datatype  send_type    /* in  */,
        void*         recv_buf_p   /* out */,
        int           recv_count   /* in  */,
        MPI_Datatype  recv_type    /* in  */,
        MPI_Comm      comm         /* in  */);
```

This function concatenates the contents of each process's send_buf_p and stores this in each process's recv_buf_p. As usual, recv_count is the amount of data being received from *each* process. So in most cases, recv_count will be the same as send_count.

We can now implement our parallel matrix-vector multiplication function as shown in Program 3.12. If this function is called many times, we can improve performance by allocating x once in the calling function and passing it as an additional argument.

3.5 MPI-derived datatypes

In virtually all distributed-memory systems, communication can be *much* more expensive than local computation. For example, sending a **double** from one node to another will take far longer than adding two **doubles** stored in the local memory of a node. Furthermore, the cost of sending a fixed amount of data in multiple messages is usually much greater than the cost of sending a single message with the same amount of data. For example, we would expect the following pair of **for** loops to be much slower than the single send/receive pair:

```
1   void Mat_vect_mult(
2         double      local_A[]   /* in  */,
3         double      local_x[]   /* in  */,
4         double      local_y[]   /* out */,
5         int         local_m     /* in  */,
6         int         n           /* in  */,
7         int         local_n     /* in  */,
8         MPI_Comm    comm        /* in  */) {
9      double* x;
10     int local_i, j;
11     int local_ok = 1;
12
13     x = malloc(n*sizeof(double));
14     MPI_Allgather(local_x, local_n, MPI_DOUBLE,
15           x, local_n, MPI_DOUBLE, comm);
16
17     for (local_i = 0; local_i < local_m; local_i++) {
18        local_y[local_i] = 0.0;
19        for (j = 0; j < n; j++)
20           local_y[local_i] += local_A[local_i*n+j]*x[j];
21     }
22     free(x);
23  }  /* Mat_vect_mult */
```

Program 3.12: An MPI matrix-vector multiplication function.

```
double x[1000];
.  .  .
if (my_rank == 0)
   for (i = 0; i < 1000; i++)
      MPI_Send(&x[i], 1, MPI_DOUBLE, 1, 0, comm);
else /* my_rank == 1 */
   for (i = 0; i < 1000; i++)
      MPI_Recv(&x[i], 1, MPI_DOUBLE, 0, 0, comm, &status);

if (my_rank == 0)
   MPI_Send(x, 1000, MPI_DOUBLE, 1, 0, comm);
else  /* my_rank == 1 */
   MPI_Recv(x, 1000, MPI_DOUBLE, 0, 0, comm, &status);
```

In fact, on one of our systems, the code with the loops of sends and receives takes nearly 50 times longer. On another system, the code with the loops takes more than 100 times longer. Thus if we can reduce the total number of messages we send, we're likely to improve the performance of our programs.

MPI provides three basic approaches to consolidating data that might otherwise require multiple messages: the count argument to the various communication functions, derived datatypes, and `MPI_Pack/Unpack`. We've already seen the `count` argument—it can be used to group contiguous array elements into a single message. In this section, we'll discuss one method for building derived datatypes. In the Exercises, we'll take a look at some other methods for building derived datatypes and `MPI_Pack/Unpack`.

In MPI, a **derived datatype** can be used to represent any collection of data items in memory by storing both the types of the items and their relative locations in memory. The idea here is that if a function that sends data knows the types and the relative locations in memory of a collection of data items, it can collect the items from memory before they are sent. Similarly, a function that receives data can distribute the items into their correct destinations in memory when they're received. As an example, in our trapezoidal rule program, we needed to call `MPI_Bcast` three times: once for the left endpoint a, once for the right endpoint b, and once for the number of trapezoids n. As an alternative, we could build a single derived datatype that consists of two **doubles** and one **int**. If we do this, we'll only need one call to `MPI_Bcast`. On process 0, a, b, and n will be sent with the one call, while on the other processes, the values will be received with the call.

Formally, a derived datatype consists of a sequence of basic MPI datatypes together with a *displacement* for each of the datatypes. In our trapezoidal rule example, suppose that on process 0 the variables a, b, and n are stored in memory locations with the following addresses:

Variable	Address
a	24
b	40
n	48

Then the following derived datatype could represent these data items:

$$\{(\texttt{MPI\_DOUBLE}, 0), (\texttt{MPI\_DOUBLE}, 16), (\texttt{MPI\_INT}, 24)\}.$$

The first element of each pair corresponds to the type of the data, and the second element of each pair is the displacement of the data element from the *beginning* of the type. We've assumed that the type begins with a, so it has displacement 0, and the other elements have displacements measured in bytes, from a: b is $40 - 24 = 16$ bytes beyond the start of a, and n is $48 - 24 = 24$ bytes beyond the start of a.

We can use `MPI_Type_create_struct` to build a derived datatype that consists of individual elements that have different basic types:

```
int MPI_Type_create_struct(
        int            count                        /* in  */,
        int            array_of_blocklengths[]      /* in  */,
        MPI_Aint       array_of_displacements[]     /* in  */,
        MPI_Datatype   array_of_types[]             /* in  */,
        MPI_Datatype*  new_type_p                   /* out */);
```

The argument `count` is the number of elements in the datatype, so for our example, it should be three. Each of the array arguments should have `count` elements. The first array, `array_of_block_lengths`, allows for the possibility that the individual data items might be arrays or subarrays. If, for example, the first element were an array containing five elements, we would have

```
array_of_blocklengths[0] = 5;
```

However, in our case, none of the elements is an array, so we can simply define

```
int array_of_blocklengths[3] = {1, 1, 1};
```

The third argument to `MPI_Type_create_struct`, `array_of_displacements` specifies the displacements in bytes, from the start of the message. So we want

```
array_of_displacements[] = {0, 16, 24};
```

To find these values, we can use the function `MPI_Get_address`:

```
int MPI_Get_address(
        void*        location_p   /* in  */,
        MPI_Aint*    address_p    /* out */);
```

It returns the address of the memory location referenced by `location_p`. The special type `MPI_Aint` is an integer type that is big enough to store an address on the system. Thus to get the values in `array_of_displacements`, we can use the following code:

```
MPI_Aint a_addr, b_addr, n_addr;

MPI_Get_address(&a, &a_addr);
array_of_displacements[0] = 0;
MPI_Get_address(&b, &b_addr);
array_of_displacements[1] = b_addr - a_addr;
MPI_Get_address(&n, &n_addr);
array_of_displacements[2] = n_addr - a_addr;
```

The `array_of_datatypes` should store the MPI datatypes of the elements. So we can just define

```
MPI_Datatype array_of_types[3] = {MPI_DOUBLE, MPI_DOUBLE,
        MPI_INT};
```

With these initializations, we can build the new datatype with the call

```
MPI_Datatype input_mpi_t;
. . .
MPI_Type_create_struct(3, array_of_blocklengths,
        array_of_displacements, array_of_types,
        &input_mpi_t);
```

Before we can use `input_mpi_t` in a communication function, we must first **commit** it with a call to

```
int MPI_Type_commit(
      MPI_Datatype* new_mpi_t_p   /* in/out */);
```

This allows the MPI implementation to optimize its internal representation of the datatype for use in communication functions.

Now, to use `input_mpi_t`, we make the following call to `MPI_Bcast` on each process:

```
MPI_Bcast(&a, 1, input_mpi_t, 0, comm);
```

So we can use `input_mpi_t`, just as we would use one of the basic MPI datatypes.

In constructing the new datatype, it's likely that the MPI implementation had to allocate additional storage internally. So when we're through using the new type, we can free any additional storage used with a call to

```
int MPI_Type_free(MPI_Datatype* old_mpi_t_p   /* in/out */);
```

We used the steps outlined here to define a `Build_mpi_type` function that our `Get_input` function can call. The new function and the updated `Get_input` function are shown in Program 3.13.

3.6 Performance evaluation of MPI programs

Let's take a look at the performance of the matrix-vector multiplication program. For the most part, we write parallel programs because we expect that they'll be faster than a serial program that solves the same problem. How can we verify this? We spent some time discussing this in Section 2.6, so we'll start by recalling some of the material we learned there.

3.6.1 Taking timings

We're usually not interested in the time taken from the start of program execution to the end of program execution. For example, in the matrix-vector multiplication, we're not interested in the time it takes to type in the matrix or print out the product. We're only interested in the time it takes to do the actual multiplication, so we need to modify our source code by adding in calls to a function that will tell us the amount of time that elapses from the beginning to the end of the actual matrix-vector multiplication. MPI provides a function, `MPI_Wtime`, that returns the number of seconds that have elapsed since some time in the past:

```
double MPI_Wtime(void);
```

We can time a block of MPI code as follows:

```
double start, finish;
. . .
start = MPI_Wtime();
/* Code to be timed */
. . .
```

```
void Build_mpi_type(
      double*          a_p              /* in  */,
      double*          b_p              /* in  */,
      int*             n_p              /* in  */,
      MPI_Datatype*    input_mpi_t_p    /* out */) {

   int array_of_blocklengths[3] = {1, 1, 1};
   MPI_Datatype array_of_types[3] = {MPI_DOUBLE, MPI_DOUBLE,
      MPI_INT};
   MPI_Aint a_addr, b_addr, n_addr;
   MPI_Aint array_of_displacements[3] = {0};

   MPI_Get_address(a_p, &a_addr);
   MPI_Get_address(b_p, &b_addr);
   MPI_Get_address(n_p, &n_addr);
   array_of_displacements[1] = b_addr-a_addr;
   array_of_displacements[2] = n_addr-a_addr;
   MPI_Type_create_struct(3, array_of_blocklengths,
         array_of_displacements, array_of_types,
         input_mpi_t_p);
   MPI_Type_commit(input_mpi_t_p);
}  /* Build_mpi_type */

void Get_input(int my_rank, int comm_sz, double* a_p,
      double* b_p, int* n_p) {
   MPI_Datatype input_mpi_t;

   Build_mpi_type(a_p, b_p, n_p, &input_mpi_t);

   if (my_rank == 0) {
      printf("Enter a, b, and n\n");
      scanf("%lf %lf %d", a_p, b_p, n_p);
   }
   MPI_Bcast(a_p, 1, input_mpi_t, 0, MPI_COMM_WORLD);

   MPI_Type_free(&input_mpi_t);
}  /* Get_input */
```

Program 3.13: The Get_input function with a derived datatype.

```
finish = MPI_Wtime();
printf("Proc %d > Elapsed time = %e seconds\n"
      my_rank, finish-start);
```

To time serial code, it's not necessary to link in the MPI libraries. There is a POSIX library function called gettimeofday that returns the number of microseconds that have elapsed since some point in the past. The syntax details aren't too important: there's a C macro GET_TIME defined in the header file timer.h that can be downloaded from the book's website. This macro should be called with a double argument:

```
#include "timer.h"
. . .
double now;
. . .
GET_TIME(now);
```

After executing this macro, now will store the number of seconds since some time in the past. So we can get the elapsed time of serial code with microsecond resolution by executing

```
#include "timer.h"
. . .
double start, finish;
. . .
GET_TIME(start);
/* Code to be timed */
. . .
GET_TIME(finish);
printf("Elapsed time = %e seconds\n", finish-start);
```

One point to stress here: GET_TIME is a macro, so the code that defines it is inserted directly into your source code by the preprocessor. Hence, it can operate directly on its argument, and the argument is a **double**, *not* a pointer to a **double**. A final note in this connection: Since timer.h is not in the system include file directory, it's necessary to tell the compiler where to find it if it's not in the directory where you're compiling. For example, if it's in the directory /home/peter/my_include, the following command can be used to compile a serial program that uses GET_TIME:

```
$ gcc -g -Wall -I/home/peter/my_include -o <executable>
      <source_code.c>
```

Both MPI_Wtime and GET_TIME return *wall clock* time. Recall that a timer, such as the C clock function, returns CPU time: the time spent in user code, library functions, and operating system code. It doesn't include idle time, which can be a significant part of parallel run time. For example, a call to MPI_Recv may spend a significant amount of time waiting for the arrival of a message. Wall clock time, on the other hand, gives total elapsed time. So it includes idle time.

There are still a few remaining issues. First, as we've described it, our parallel program will report comm_sz times: one for each process. We would like to have it report a single time. Ideally, all of the processes would start execution of the matrix-vector multiplication at the same time, and then we would report the time that elapsed when the last process finished. In other words, the parallel execution time would be the time it took the "slowest" process to finish. We can't get exactly this time because we can't ensure that all the processes start at the same instant. However, we can come reasonably close. The MPI collective communication function MPI_Barrier ensures that no process will return from calling it until every process in the communicator has started calling it. Its syntax is

```
int MPI_Barrier(MPI_Comm   comm   /* in */);
```

So the following code can be used to time a block of MPI code and report a single elapsed time:

```
double local_start, local_finish, local_elapsed, elapsed;
. . .
MPI_Barrier(comm);
local_start = MPI_Wtime();
/* Code to be timed */
. . .
local_finish = MPI_Wtime();
local_elapsed = local_finish - local_start;
MPI_Reduce(&local_elapsed, &elapsed, 1, MPI_DOUBLE,
   MPI_MAX, 0, comm);

if (my_rank == 0)
   printf("Elapsed time = %e seconds\n", elapsed);
```

Note that the call to MPI_Reduce is using the MPI_MAX operator: it finds the largest of the input arguments local_elapsed.

As we noted in Chapter 2, we also need to be aware of variability in timings: when we run a program several times, we're likely to see a substantial variation in the times. This will be true, even if for each run we use the same input, the same number of processes, and the same system. This is because the interaction of the program with the rest of the system, especially the operating system, is unpredictable. Since this interaction will almost certainly not make the program run faster than it would run on a "quiet" system, we usually report the *minimum* run-time, rather than the mean or median. (For further discussion of this, see [6].)

Finally, when we run an MPI program on a hybrid system in which the nodes are multicore processors, we'll only run one MPI process on each node. This may reduce contention for the interconnect and result in somewhat better run-times. It may also reduce variability in run-times.

3.6.2 Results

The results of timing the matrix-vector multiplication program are shown in Table 3.5. The input matrices were square. The times shown are in milliseconds, and we've rounded each time to two significant digits. The times for comm_sz = 1 are the run-times of the serial program running on a single core of the distributed memory system. Not surprisingly, if we fix comm_sz, and increase n, the order of the matrix, the run-times increase. For relatively small numbers of processes, doubling n results in roughly a four-fold increase in the run-time. However, for large numbers of processes, this formula breaks down.

Table 3.5 Run-times of serial and parallel matrix-vector multiplication (times are in milliseconds).

comm_sz	Order of Matrix				
	1024	2048	4096	8192	16,384
1	4.1	16.0	64.0	270	1100
2	2.3	8.5	33.0	140	560
4	2.0	5.1	18.0	70	280
8	1.7	3.3	9.8	36	140
16	1.7	2.6	5.9	19	71

If we fix n and increase comm_sz, the run-times usually decrease. In fact, for large values of n, doubling the number of processes roughly halves the overall run-time. However, for small n, there is very little benefit in increasing comm_sz. In fact, in going from 8 to 16 processes when $n = 1024$, the overall run time is unchanged.

These timings are fairly typical of parallel run-times—as we increase the problem size, the run-times increase, and this is true regardless of the number of processes. The rate of increase can be fairly constant (e.g., the one process times) or it can vary wildly (e.g., the sixteen process times). As we increase the number of processes, the run-times typically decrease for a while. However, at some point, the run-times can actually start to get worse. The closest we came to this behavior was going from 8 to 16 processes when the matrix had order 1024.

The explanation for this is that there is a fairly common relation between the run-times of serial programs and the run-times of corresponding parallel programs. Recall that we denote the serial run-time by T_{serial}. Since it typically depends on the size of the input, n, we'll frequently denote it as $T_{\text{serial}}(n)$. Also recall that we denote the parallel run-time by T_{parallel}. Since it depends on both the input size, n, and the number of processes, comm_sz $= p$, we'll frequently denote it as $T_{\text{parallel}}(n, p)$. As we noted in Chapter 2, it's often the case that the parallel program will divide the work of the serial program among the processes, and add in some overhead time, which we denoted T_{overhead}:

$$T_{\text{parallel}}(n, p) = T_{\text{serial}}(n)/p + T_{\text{overhead}}.$$

In MPI programs, the parallel overhead typically comes from communication, and it can depend on both the problem size and the number of processes.

It's not to hard to see that this formula applies to our matrix-vector multiplication program. The heart of the serial program is the pair of nested for loops:

```
for (i = 0; i < m; i++) {
    y[i] = 0.0;
    for (j = 0; j < n; j++)
        y[i] += A[i*n+j]*x[j];
}
```

If we only count floating point operations, the inner loop carries out n multiplications and n additions, for a total of $2n$ floating point operations. Since we execute the inner loop m times, the pair of loops executes a total of $2mn$ floating point operations. So when $m = n$,

$$T_{serial}(n) \approx an^2$$

for some constant a. (The symbol $\approx$ means "is approximately equal to.")

If the serial program multiplies an $n \times n$ matrix by an n-dimensional vector, then each process in the parallel program multiplies an $n/p \times n$ matrix by an n-dimensional vector. The *local* matrix-vector multiplication part of the parallel program therefore executes n^2/p floating point operations. Thus it appears that this *local* matrix-vector multiplication reduces the work per process by a factor of p.

However, the parallel program also needs to complete a call to `MPI_Allgather` before it can carry out the local matrix-vector multiplication. In our example, it appears that

$$T_{parallel}(n, p) = T_{serial}(n)/p + T_{allgather}.$$

Furthermore, in light of our timing data, it appears that for smaller values of p and larger values of n, the dominant term in our formula is $T_{serial}(n)/p$. To see this, observe first that for small p (e.g., $p = 2, 4$), doubling p roughly halves the overall run time. For example,

$$T_{serial}(4096) = 1.9 \times T_{parallel}(4096, 2)$$
$$T_{serial}(8192) = 1.9 \times T_{parallel}(8192, 2)$$
$$T_{parallel}(8192, 2) = 2.0 \times T_{parallel}(8192, 4)$$
$$T_{serial}(16{,}384) = 2.0 \times T_{parallel}(16{,}384, 2)$$
$$T_{parallel}(16{,}384, 2) = 2.0 \times T_{parallel}(16{,}384, 4)$$

Also, if we fix p at a small value (e.g., $p = 2, 4$), then increasing n seems to have approximately the same effect as increasing n for the serial program. For example,

$$T_{serial}(4096) = 4.0 \times T_{serial}(2048)$$
$$T_{parallel}(4096, 2) = 3.9 \times T_{serial}(2048, 2)$$
$$T_{parallel}(4096, 4) = 3.5 \times T_{serial}(2048, 4)$$
$$T_{serial}(8192) = 4.2 \times T_{serial}(4096)$$
$$T_{serial}(8192, 2) = 4.2 \times T_{parallel}(4096, 2)$$
$$T_{parallel}(8192, 4) = 3.9 \times T_{parallel}(8192, 4)$$

These observations suggest that the parallel run-times are behaving much as the run-times of the serial program, i.e., $T_{serial}(n)/p$. In other words, the overhead $T_{allgather}$ has little effect on the performance.

Table 3.6 Speedups of parallel matrix-vector multiplication.

comm_sz	Order of Matrix				
	1024	2048	4096	8192	16,384
1	1.0	1.0	1.0	1.0	1.0
2	1.8	1.9	1.9	1.9	2.0
4	2.1	3.1	3.6	3.9	3.9
8	2.4	4.8	6.5	7.5	7.9
16	2.4	6.2	10.8	14.2	15.5

On the other hand, for small n and large p, these patterns break down. For example,

$$T_{\text{parallel}}(1024, 8) = 1.0 \times T_{\text{parallel}}(1024, 16)$$
$$T_{\text{parallel}}(2048, 16) = 1.5 \times T_{\text{parallel}}(1024, 16)$$

So it appears that for small n and large p, the dominant term in our formula for T_{parallel} is $T_{\text{allgather}}$.

3.6.3 Speedup and efficiency

Recall that the most widely used measure of the relation between the serial and the parallel run-times is the **speedup**. It's just the ratio of the serial run-time to the parallel run-time:

$$S(n, p) = \frac{T_{\text{serial}}(n)}{T_{\text{parallel}}(n, p)}.$$

The ideal value for $S(n, p)$ is p. If $S(n, p) = p$, then our parallel program with comm_sz $= p$ processes is running p times faster than the serial program. In practice, this speedup, sometimes called **linear speedup**, is not often achieved. Our matrix-vector multiplication program got the speedups shown in Table 3.6. For small p and large n, our program obtained nearly linear speedup. On the other hand, for large p and small n, the speedup was considerably less than p. The worst case being $n = 1024$ and $p = 16$, when we only managed a speedup of 2.4.

Also recall that another widely used measure of parallel performance is **parallel efficiency**. This is "per process" speedup:

$$E(n, p) = \frac{S(n, p)}{p} = \frac{T_{\text{serial}}(n)}{p \times T_{\text{parallel}}(n, p)}.$$

So linear speedup corresponds to a parallel efficiency of $p/p = 1.0$, and, in general, we expect that our efficiencies will usually be less than 1.

Table 3.7 Efficiencies of parallel matrix-vector multiplication.

comm_sz	Order of Matrix				
	1024	2048	4096	8192	16,384
1	1.00	1.00	1.00	1.00	1.00
2	0.89	0.94	0.97	0.96	0.98
4	0.51	0.78	0.89	0.96	0.98
8	0.30	0.61	0.82	0.94	0.98
16	0.15	0.39	0.68	0.89	0.97

The efficiencies for the matrix-vector multiplication program are shown in Table 3.7. Once again, for small p and large n, our parallel efficiencies are near linear, and for large p and small n, they are very far from linear.

3.6.4 Scalability

Our parallel matrix-vector multiplication program doesn't come close to obtaining linear speedup for small n and large p. Does this mean that it's not a good program? Many computer scientists answer this question by looking at the "scalability" of the program. Recall that very roughly speaking a program is **scalable** if the problem size can be increased at a rate so that the efficiency doesn't decrease as the number of processes increase.

The problem with this definition is the phrase "the problem size can be increased at a rate ...." Consider two parallel programs: program A and program B. Suppose that if $p \geq 2$, the efficiency of program A is 0.75, regardless of problem size. Also suppose that the efficiency of program B is $n/(625p)$, provided $p \geq 2$ and $1000 \leq n \leq 625p$. Then according to our "definition," both programs are scalable. For program A, the rate of increase needed to maintain constant efficiency is 0, while for program B if we increase n at the same rate as we increase p, we'll maintain a constant efficiency. For example, if $n = 1000$ and $p = 2$, the efficiency of B is 0.80. If we then double p to 4 and we leave the problem size at $n = 1000$, the efficiency will drop to 0.4, but if we also double the problem size to $n = 2000$, the efficiency will remain constant at 0.8. Program A is thus *more* scalable than B, but both satisfy our definition of scalability.

Looking at our table of parallel efficiencies (Table 3.7), we see that our matrix-vector multiplication program definitely doesn't have the same scalability as program A: in almost every case when p is increased, the efficiency decreases. On the other hand, the program is somewhat like program B: if $p \geq 2$ and we increase both p and n by a factor of 2, the parallel efficiency, for the most part, actually increases. Furthermore, the only exceptions occur when we increase p from 2 to 4, and when computer scientists discuss scalability, they're usually interested in large values of p. When p is increased from 4 to 8 or from 8 to 16, our efficiency always increases when we increase n by a factor of 2.

Recall that programs that can maintain a constant efficiency without increasing the problem size are sometimes said to be **strongly scalable**. Programs that can maintain a constant efficiency if the problem size increases at the same rate as the number of processes are sometimes said to be **weakly scalable**. Program A is strongly scalable, and program B is weakly scalable. Furthermore, our matrix-vector multiplication program is also apparently weakly scalable.

3.7 A parallel sorting algorithm

What do we mean by a parallel sorting algorithm in a distributed memory environment? What would its "input" be and what would its "output" be? The answers depend on where the keys are stored. We can start or finish with the keys distributed among the processes or assigned to a single process. In this section, we'll look at an algorithm that starts and finishes with the keys distributed among the processes. In programming assignment 3.8, we'll look at an algorithm that finishes with the keys assigned to a single process.

If we have a total of n keys and $p = $ comm_sz processes, our algorithm will start and finish with n/p keys assigned to each process. (As usual, we'll assume n is evenly divisible by p.) At the start, there are no restrictions on which keys are assigned to which processes. However, when the algorithm terminates,

- the keys assigned to each process should be sorted in (say) increasing order, and
- if $0 \leq q < r < p$, then each key assigned to process q should be less than or equal to every key assigned to process r.

So if we lined up the keys according to process rank—keys from process 0 first, then keys from process 1, and so on—then the keys would be sorted in increasing order. For the sake of explicitness, we'll assume our keys are ordinary **ints**.

3.7.1 Some simple serial sorting algorithms

Before starting, let's look at a couple of simple serial sorting algorithms. Perhaps the best-known serial sorting algorithm is bubble sort. (See Program 3.14.) The array a stores the unsorted keys when the function is called, and the sorted keys when the function returns. The number of keys in a is n. The algorithm proceeds by comparing the elements of the list a pairwise: a[0] is compared to a[1], a[1] is compared to a[2], and so on. Whenever a pair is out of order, the entries are swapped, so in the first pass through the outer loop, when list_length = n, the largest value in the list will be moved into a[n−1]. The next pass will ignore this last element, and it will move the next-to-the-largest element into a[n−2]. Thus as list_length decreases, successively more elements get assigned to their final positions in the sorted list.

There isn't much point in trying to parallelize this algorithm because of the inherently sequential ordering of the comparisons. To see this, suppose that a[i−1] = 9, a[i] = 5, and a[i+1] = 7. The algorithm will first compare 9 and 5 and swap them;

```
void Bubble_sort(
        int  a[]  /* in/out */,
        int  n    /* in     */) {
    int list_length, i, temp;

    for (list_length = n; list_length >= 2; list_length--)
        for (i = 0; i < list_length-1; i++)
            if (a[i] > a[i+1]) {
                temp = a[i];
                a[i] = a[i+1];
                a[i+1] = temp;
            }
}  /* Bubble_sort */
```

Program 3.14: Serial bubble sort.

it will then compare 9 and 7 and swap them; and we'll have the sequence $5, 7, 9$. If we try to do the comparisons out of order, that is, if we compare the 5 and 7 first and then compare the 9 and 5, we'll wind up with the sequence $5, 9, 7$. Therefore the order in which the "compare-swaps" take place is essential to the correctness of the algorithm.

A variant of bubble sort, known as *odd-even transposition sort*, has considerably more opportunities for parallelism. The key idea is to "decouple" the compare-swaps. The algorithm consists of a sequence of *phases*, of two different types. During *even* phases, compare-swaps are executed on the pairs

$$(a[0], a[1]), (a[2], a[3]), (a[4], a[5]), \ldots,$$

and during *odd* phases, compare-swaps are executed on the pairs

$$(a[1], a[2]), (a[3], a[4]), (a[5], a[6]), \ldots.$$

Here's a small example:

> *Start*: $5, 9, 4, 3$
> *Even phase*: Compare-swap $(5, 9)$ and $(4, 3)$, getting the list $5, 9, 3, 4$.
> *Odd phase*: Compare-swap $(9, 3)$, getting the list $5, 3, 9, 4$.
> *Even phase*: Compare-swap $(5, 3)$, and $(9, 4)$ getting the list $3, 5, 4, 9$.
> *Odd phase*: Compare-swap $(5, 4)$ getting the list $3, 4, 5, 9$.

This example required four phases to sort a four-element list. In general, it may require fewer phases, but the following theorem guarantees that we can sort a list of n elements in at most n phases.

Theorem. *Suppose A is a list with n keys, and A is the input to the odd-even transposition sort algorithm. Then after n phases, A will be sorted.*

Program 3.15 shows code for a serial odd-even transposition sort function.

```
void Odd_even_sort(
      int   a[]   /* in/out */,
      int   n     /* in     */) {
   int phase, i, temp;

   for (phase = 0; phase < n; phase++)
      if (phase % 2 == 0) { /* Even phase */
         for (i = 1; i < n; i += 2)
            if (a[i-1] > a[i]) {
               temp = a[i];
               a[i] = a[i-1];
               a[i-1] = temp;
            }
      } else { /* Odd phase */
         for (i = 1; i < n-1; i += 2)
            if (a[i] > a[i+1]) {
               temp = a[i];
               a[i] = a[i+1];
               a[i+1] = temp;
            }
      }
} /* Odd_even_sort */
```

Program 3.15: Serial odd-even transposition sort.

3.7.2 Parallel odd-even transposition sort

It should be clear that odd-even transposition sort has considerably more opportunities for parallelism than bubble sort, because all of the compare-swaps in a single phase can happen simultaneously. Let's try to exploit this.

There are a number of possible ways to apply Foster's methodology. Here's one:

- *Tasks*: Determine the value of a[i] at the end of phase j.
- *Communications*: The task that's determining the value of a[i] needs to communicate with either the task determining the value of a[i−1] or a[i+1]. Also the value of a[i] at the end of phase j needs to be available for determining the value of a[i] at the end of phase $j + 1$.

This is illustrated in Fig. 3.12, where we've labeled the task determining the value of a[k] with a[k].

Now recall that when our sorting algorithm starts and finishes execution, each process is assigned n/p keys. In this case, our aggregation and mapping are at least partially specified by the description of the problem. Let's look at two cases.

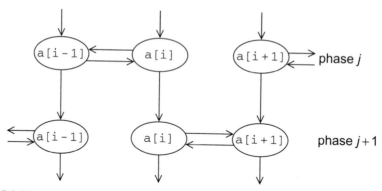

FIGURE 3.12

Communications among tasks in odd-even sort. Tasks determining a[k] are labeled with a[k].

When $n = p$, Fig. 3.12 makes it fairly clear how the algorithm should proceed. Depending on the phase, process i can send its current value, a[i], either to process $i - 1$ or process $i + 1$. At the same time, it should receive the value stored on process $i - 1$ or process $i + 1$, respectively, and then decide which of the two values it should store as a[i] for the next phase.

However, it's unlikely that we'll actually want to apply the algorithm when $n = p$, since we're unlikely to have more than a few hundred or a few thousand processors at our disposal, and sorting a few thousand values is usually a fairly trivial matter for a single processor. Furthermore, even if we do have access to thousands or even millions of processors, the added cost of sending and receiving a message for each compare-exchange will slow the program down so much that it will be useless. Remember that the cost of communication is usually much greater than the cost of "local" computation—for example, a compare-swap.

How should this be modified when each process is storing $n/p > 1$ elements? (Recall that we're assuming that n is evenly divisible by p.) Let's look at an example. Suppose we have $p = 4$ processes and $n = 16$ keys, as shown in Table 3.8. In the first place, we can apply a fast serial sorting algorithm to the keys assigned to each process. For example, we can use the C library function qsort on each process to sort the local keys. Now if we had one element per process, 0 and 1 would exchange elements, and 2 and 3 would exchange. So let's try this: Let's have 0 and 1 exchange *all* their elements, and 2 and 3 exchange all of theirs. Then it would seem natural for 0 to keep the four smaller elements and 1 to keep the larger. Similarly, 2 should keep the smaller and 3 the larger. This gives us the situation shown in the third row of the table. Once again, looking at the one element per process case, in phase 1, processes 1 and 2 exchange their elements and processes 0 and 4 are idle. If process 1 keeps the smaller and 2 the larger elements, we get the distribution shown in the fourth row. Continuing this process for two more phases results in a sorted list. That is, each

Table 3.8 Parallel odd-even transposition sort.

	Process			
Time	**0**	**1**	**2**	**3**
Start	15, 11, 9, 16	3, 14, 8, 7	4, 6, 12, 10	5, 2, 13, 1
After Local Sort	9, 11, 15, 16	3, 7, 8, 14	4, 6, 10, 12	1, 2, 5, 13
After Phase 0	3, 7, 8, 9	11, 14, 15, 16	1, 2, 4, 5	6, 10, 12, 13
After Phase 1	3, 7, 8, 9	1, 2, 4, 5	11, 14, 15, 16	6, 10, 12, 13
After Phase 2	1, 2, 3, 4	5, 7, 8, 9	6, 10, 11, 12	13, 14, 15, 16
After Phase 3	1, 2, 3, 4	5, 6, 7, 8	9, 10, 11, 12	13, 14, 15, 16

process's keys are stored in increasing order, and if $q < r$, then the keys assigned to process q are less than or equal to the keys assigned to process r.

In fact, our example illustrates the worst-case performance of this algorithm:

Theorem. *If parallel odd-even transposition sort is run with p processes, then after p phases, the input list will be sorted.*

The parallel algorithm is clear to a human computer:

```
Sort local keys;
for (phase = 0; phase < comm_sz; phase++) {
    partner = Compute_partner(phase, my_rank);
    if (I'm not idle) {
        Send my keys to partner;
        Receive keys from partner;
        if (my_rank < partner)
            Keep smaller keys;
        else
            Keep larger keys;
    }
}
```

However, there are some details that we need to clear up before we can convert the algorithm into an MPI program.

First, how do we compute the partner rank? And what is the partner rank when a process is idle? If the phase is even, then odd-ranked partners exchange with my_rank − 1 and even-ranked partners exchange with my_rank+1. In odd phases, the calculations are reversed. However, these calculations can return some invalid ranks: if my_rank = 0 or my_rank = comm_sz−1, the partner rank can be -1 or comm_sz. But when either partner = −1 or partner = comm_sz, the process should be idle. So we can use the rank computed by Compute_partner to determine whether a process is idle:

```
if (phase % 2 == 0)          /* Even phase */
    if (my_rank % 2 != 0)    /* Odd rank */
        partner = my_rank - 1;
    else                     /* Even rank */
        partner = my_rank + 1;
```

```
    else                    /* Odd phase */
        if (my_rank % 2 != 0)       /* Odd rank */
            partner = my_rank + 1;
        else                    /* Even rank */
            partner = my_rank - 1;
    if (partner == -1 || partner == comm_sz)
        partner = MPI_PROC_NULL;
```

MPI_PROC_NULL is a constant defined by MPI. When it's used as the source or destination rank in a point-to-point communication, no communication will take place, and the call to the communication will simply return.

3.7.3 Safety in MPI programs

If a process is not idle, we might try to implement the communication with a call to MPI_Send and a call to MPI_Recv:

```
MPI_Send(my_keys, n/comm_sz, MPI_INT, partner, 0, comm);
MPI_Recv(temp_keys, n/comm_sz, MPI_INT, partner, 0, comm,
    MPI_STATUS_IGNORE);
```

This, however, might result in the program's hanging or crashing. Recall that the MPI standard allows MPI_Send to behave in two different ways: it can simply copy the message into an MPI-managed buffer and return, or it can block until the matching call to MPI_Recv starts. Furthermore, many implementations of MPI set a threshold at which the system switches from buffering to blocking. That is, messages that are relatively small will be buffered by MPI_Send, but for larger messages, it will block. If the MPI_Send executed by each process blocks, no process will be able to start executing a call to MPI_Recv, and the program will hang or **deadlock**: that is, each process is blocked waiting for an event that will never happen.

A program that relies on MPI-provided buffering is said to be **unsafe**. Such a program may run without problems for various sets of input, but it may hang or crash with other sets. If we use MPI_Send and MPI_Recv in this way, our program will be unsafe, and it's likely that for small values of n the program will run without problems, while for larger values of n, it's likely that it will hang or crash.

There are a couple of questions that arise here:

1. In general, how can we tell if a program is safe?
2. How can we modify the communication in the parallel odd-even sort program so that it is safe?

To answer the first question, we can use an alternative to MPI_Send defined by the MPI standard. It's called MPI_Ssend. The extra "s" stands for *synchronous*, and MPI_Ssend is guaranteed to block until the matching receive starts. So, we can check whether a program is safe by replacing the calls to MPI_Send with calls to MPI_Ssend. If the program doesn't hang or crash when it's run with appropriate input and comm_sz, then the original program was safe. The arguments to MPI_Ssend are the same as the arguments to MPI_Send:

```
int MPI_Ssend(
        void*           msg_buf_p       /* in */,
        int             msg_size        /* in */,
        MPI_Datatype    msg_type        /* in */,
        int             dest            /* in */,
        int             tag             /* in */,
        MPI_Comm        communicator    /* in */);
```

The answer to the second question is that the communication must be restructured. The most common cause of an unsafe program is multiple processes simultaneously first sending to each other and then receiving. Our exchanges with partners is one example. Another example is a "ring pass," in which each process q sends to the process with rank $q + 1$, except that process comm_sz $- 1$ sends to 0:

```
MPI_Send(msg, size, MPI_INT, (my_rank+1) % comm_sz, 0,
        comm);
MPI_Recv(new_msg, size, MPI_INT,
        (my_rank+comm_sz -1) % comm_sz, 0, comm,
        MPI_STATUS_IGNORE);
```

In both settings, we need to restructure the communications so that some of the processes receive before sending. For example, the preceding communications could be restructured as follows:

```
if (my_rank % 2 == 0) {
    MPI_Send(msg, size, MPI_INT, (my_rank+1) % comm_sz, 0,
            comm);
    MPI_Recv(new_msg, size, MPI_INT,
            (my_rank+comm_sz -1) % comm_sz, 0, comm,
            MPI_STATUS_IGNORE);
} else {
    MPI_Recv(new_msg, size, MPI_INT,
            (my_rank+comm_sz -1) % comm_sz, 0, comm,
            MPI_STATUS_IGNORE);
    MPI_Send(msg, size, MPI_INT, (my_rank+1) % comm_sz, 0,
            comm);
}
```

It's fairly clear that this will work if comm_sz is even. If, say, comm_sz $= 4$, then processes 0 and 2 will first send to 1 and 3, respectively, while processes 1 and 3 will receive from 0 and 2, respectively. The roles are reversed for the next send-receive pairs: processes 1 and 3 will send to 2 and 0, respectively, while 2 and 0 will receive from 1 and 3.

However, it may not be clear that this scheme is also safe if comm_sz is odd (and greater than 1). Suppose, for example, that comm_sz $= 5$. Then Fig. 3.13 shows a possible sequence of events. The solid arrows show a completed communication, and the dashed arrows show a communication waiting to complete.

MPI provides an alternative to scheduling the communications ourselves—we can call the function MPI_Sendrecv:

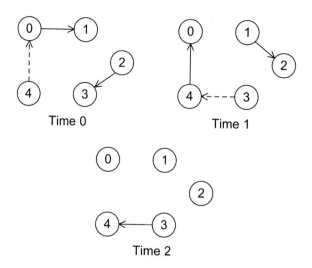

FIGURE 3.13

Safe communication with five processes.

```
int MPI_Sendrecv(
        void*          send_buf_p        /* in  */,
        int            send_buf_size     /* in  */,
        MPI_Datatype   send_buf_type     /* in  */,
        int            dest              /* in  */,
        int            send_tag          /* in  */,
        void*          recv_buf_p        /* out */,
        int            recv_buf_size     /* in  */,
        MPI_Datatype   recv_buf_type     /* in  */,
        int            source            /* in  */,
        int            recv_tag          /* in  */,
        MPI_Comm       communicator      /* in  */,
        MPI_Status*    status_p          /* in  */);
```

This function carries out a blocking send and a receive in a single call. The dest and the source can be the same or different. What makes it especially useful is that the MPI implementation schedules the communications so that the program won't hang or crash. The complex code we used above—the code that checks whether the process rank is odd or even—can be replaced with a single call to MPI_Sendrecv. If it happens that the send and the receive buffers should be the same, MPI provides the alternative:

```
int MPI_Sendrecv_replace(
        void*          buf_p        /* in/out */,
        int            buf_size     /* in     */,
        MPI_Datatype   buf_type     /* in     */,
        int            dest         /* in     */,
```

```
int              send_tag        /* in      */,
int              source          /* in      */,
int              recv_tag        /* in      */,
MPI_Comm         communicator    /* in      */,
MPI_Status*      status_p        /* in      */);
```

3.7.4 Final details of parallel odd-even sort

Recall that we had developed the following parallel odd-even transposition sort algorithm:

```
Sort local keys;
for (phase = 0; phase < comm_sz; phase++) {
    partner = Compute_partner(phase, my_rank);
    if (I'm not idle) {
        Send my keys to partner;
        Receive keys from partner;
        if (my_rank < partner)
            Keep smaller keys;
        else
            Keep larger keys;
    }
}
```

In light of our discussion of safety in MPI, it probably makes sense to implement the send and the receive with a single call to MPI_Sendrecv:

```
MPI_Sendrecv(my_keys, n/comm_sz, MPI_INT, partner, 0,
        recv_keys, n/comm_sz, MPI_INT, partner, 0, comm,
        MPI_Status_ignore);
```

So it only remains to identify which keys we keep. Suppose for the moment that we want to keep the smaller keys. Then we want to keep the smallest n/p keys in a collection of $2n/p$ keys. An obvious approach to doing this is to sort (using a serial sorting algorithm) the list of $2n/p$ keys, and keep the first half of the list. However, sorting is a relatively expensive operation, and we can exploit the fact that we already have two sorted lists of n/p keys to reduce the cost by *merging* the two lists into a single list. In fact, we can do even better: we don't need a fully general merge; once we've found the smallest n/p keys, we can quit. See Program 3.16.

To get the largest n/p keys, we simply reverse the order of the merge. That is, we start with local_n−1 and work backward through the arrays. A final improvement avoids copying the arrays; it simply swaps pointers. (See Exercise 3.28.)

Run-times for the version of parallel odd-even sort with the "final improvement" are shown in Table 3.9. Note that if parallel odd-even sort is run on a single processor, it will use whatever serial sorting algorithm we use to sort the local keys. So the times for a single process use serial quicksort, not serial odd-even sort, which would be *much* slower. We'll take a closer look at these times in Exercise 3.27.

```
void Merge_low(
        int   my_keys[],      /* in/out   */
        int   recv_keys[],    /* in       */
        int   temp_keys[],    /* scratch  */
        int   local_n         /* = n/p, in */) {
    int m_i, r_i, t_i;

    m_i = r_i = t_i = 0;
    while (t_i < local_n) {
        if (my_keys[m_i] <= recv_keys[r_i]) {
            temp_keys[t_i] = my_keys[m_i];
            t_i++; m_i++;
        } else {
            temp_keys[t_i] = recv_keys[r_i];
            t_i++; r_i++;
        }
    }

    for (m_i = 0; m_i < local_n; m_i++)
        my_keys[m_i] = temp_keys[m_i];
}  /* Merge_low */
```

Program 3.16: The Merge_low function in parallel odd-even transposition sort.

Table 3.9 Run-times of parallel odd-even sort (times are in milliseconds).

Processes	Number of Keys (in thousands)				
	200	400	800	1600	3200
1	88	190	390	830	1800
2	43	91	190	410	860
4	22	46	96	200	430
8	12	24	51	110	220
16	7.5	14	29	60	130

3.8 Summary

MPI, or the Message-Passing Interface, is a library of functions that can be called from C or Fortran programs. Many systems use mpicc to compile MPI programs and mpiexec to run them. C MPI programs should include the mpi.h header file to get function prototypes, types, macros, and so on defined by MPI.

MPI_Init does the setup needed to run MPI. It should be called before other MPI functions are called. When your program doesn't use argc and argv, NULL can be passed for both arguments.

In MPI a **communicator** is a collection of processes that can send messages to each other. After an MPI program is started, MPI always creates a communicator consisting of all the processes. It's called MPI_COMM_WORLD.

Many parallel programs use the **single-program multiple data** or **SPMD** approach: running a single program obtains the effect of running multiple different programs by including branches on data, such as the process rank.

When you're done using MPI, you should call MPI_Finalize.

To send a message from one MPI process to another, you can use MPI_Send. To receive a message, you can use MPI_Recv. The arguments to MPI_Send describe the contents of the message and its destination. The arguments to MPI_Recv describe the storage that the message can be received into, and where the message should be received from. MPI_Recv is **blocking**. That is, a call to MPI_Recv won't return until the message has been received (or an error has occurred). The behavior of MPI_Send is defined by the MPI implementation. It can either block or **buffer** the message. When it blocks, it won't return until the matching receive has started. If the message is buffered, MPI will copy the message into its own private storage, and MPI_Send will return as soon as the message is copied.

When you're writing MPI programs, it's important to differentiate between **local** and **global** variables. Local variables have values that are specific to the process on which they're defined, while global variables are the same on all the processes. In the trapezoidal rule program, the total number of trapezoids, n, was a global variable, while the left and right endpoints of each process's subinterval were local variables.

Most serial programs are **deterministic**. This means that if we run the same program with the same input, we'll get the same output. Recall that parallel programs often don't possess this property—if multiple processes are operating more or less independently, the processes may reach various points at different times, depending on events outside the control of the process. Thus parallel programs can be **nondeterministic**. That is, the same input can result in different outputs. If all the processes in an MPI program are printing output, the order in which the output appears may be different each time the program is run. For this reason, it's common in MPI programs to have a single process (e.g., process 0) handle all the output. This rule of thumb is usually ignored during debugging, when we allow each process to print debug information.

Most MPI implementations allow all the processes to print to stdout and stderr. However, every implementation we've encountered only allows at most one process (usually process 0 in MPI_COMM_WORLD) to read from stdin.

Collective communications involve all the processes in a communicator, so they're different from MPI_Send and MPI_Recv, which only involve two processes. To distinguish between the two types of communications, functions such as MPI_Send and MPI_Recv are often called **point-to-point** communications.

Two of the most commonly used collective communication functions are MPI_Reduce and MPI_Allreduce. MPI_Reduce stores the result of a global operation (e.g., a global sum) on a single designated process, while MPI_Allreduce stores the result on all the processes in the communicator.

In MPI functions, such as `MPI_Reduce`, it may be tempting to pass the same actual argument to both the input and output buffers. This is called **argument aliasing**, and MPI explicitly prohibits aliasing an output argument with another argument.

We learned about a number of other important MPI collective communications:

- `MPI_Bcast` sends data from a single process to all the processes in a communicator. This is very useful if, for example, process 0 reads data from `stdin` and the data needs to be sent to all the processes.
- `MPI_Scatter` distributes the elements of an array among the processes. If the array to be distributed contains n elements, and there are p processes, then the first n/p are sent to process 0, the next n/p to process 1, and so on.
- `MPI_Gather` is the "inverse operation" to `MPI_Scatter`. If each process stores a sub-array containing m elements, `MPI_Gather` will collect all of the elements on a designated process, putting the elements from process 0 first, then the elements from process 1, and so on.
- `MPI_Allgather` is like `MPI_Gather`, except that it collects all of the elements on *all* the processes.
- `MPI_Barrier` approximately synchronizes the processes; no process can return from a call to `MPI_Barrier` until all the processes in the communicator have started the call.

In distributed-memory systems there is no globally shared memory, so partition-ing global data structures among the processes is a key issue in writing MPI programs. For ordinary vectors and arrays, we saw that we could use block partitioning, cyclic partitioning, or block-cyclic partitioning. If the global vector or array has n compo-nents and there are p processes, a **block partition** assigns the first n/p to process 0, the next n/p to process 1, and so on. A **cyclic partition** assigns the elements in a "round-robin" fashion: the first element goes to 0, the next to 1, ... the pth to $p-1$. After assigning the first p elements, we return to process 0, so the $(p+1)$st goes to process 0, the $(p+2)$nd to process 1, and so on. A **block-cyclic partition** assigns blocks of elements to the processes in a cyclic fashion.

Compared to operations involving only the CPU and main memory, sending mes-sages is expensive. Furthermore, sending a given volume of data in fewer messages is usually less expensive than sending the same volume in more messages. Thus it often makes sense to reduce the number of messages sent by combining the con-tents of multiple messages into a single message. MPI provides three methods for doing this: the `count` argument to communication functions, derived datatypes, and `MPI_Pack/Unpack`. Derived datatypes describe arbitrary collections of data by spec-ifying the types of the data items and their relative positions in memory. In this chapter, we took a brief look at the use of `MPI_Type_create_struct` to build a de-rived datatype. In the exercises, we'll explore some other methods, and we'll take a look at `MPI_Pack/Unpack`

When we time parallel programs, we're usually interested in elapsed time or "wall clock time," which is the total time taken by a block of code. It includes time in user code, time in library functions, time in operating system functions started by the user code, and idle time. We learned about two methods for finding wall clock time. GET_TIME and MPI_Wtime. GET_TIME is a macro defined in the file timer.h that can be downloaded from the book's website. It can be used in serial code as follows:

```
#include "timer.h"    // From the book's website
. . .
double start, finish, elapsed;
. . .
GET_TIME(start);
/* Code to be timed */
. . .
GET_TIME(finish);
elapsed = finish - start;
printf("Elapsed time = %e seconds\n", elapsed);
```

MPI provides a function, MPI_Wtime, that can be used instead of GET_TIME. In spite of this, timing parallel code is more complex, since—ideally—we'd like to synchronize the processes at the start of the code, and then report the time it took for the "slowest" process to complete the code. MPI_Barrier does a fairly good job of synchronizing the processes. A process that calls it will block until all the processes in the communicator have called it. We can use the following template for finding the run-time of MPI code:

```
double start, finish, loc_elapsed, elapsed;
. . .
MPI_Barrier(comm);
start = MPI_Wtime();
/* Code to be timed */
. . .
finish = MPI_Wtime();
loc_elapsed = finish - start;
MPI_Reduce(&loc_elapsed, &elapsed, 1, MPI_DOUBLE,
      MPI_MAX, 0, comm);
if (my_rank == 0)
    printf("Elapsed time = %e seconds\n", elapsed);
```

A further problem with taking timings lies in the fact that there is ordinarily considerable variation if the same code is timed repeatedly. For example, the operating system may idle one or more of our processes so that other processes can run. Therefore we typically take several timings and report their minimum.

After taking timings, we can use the **speedup** or the **efficiency** to evaluate the program performance. The speedup is the ratio of the serial run-time to the parallel run-time, and the efficiency is the speedup divided by the number of parallel processes. The ideal value for speedup is p, the number of processes, and the ideal value for the efficiency is 1.0. We often fail to achieve these ideals, but it's not uncommon to see programs that get close to these values, especially when p is small and n, the problem size, is large. **Parallel overhead** is the part of the parallel run-time that's due to any additional work that isn't done by the serial program. In MPI programs, parallel overhead will come from communication. When p is large and n is small, it's not unusual for parallel overhead to dominate the total run-time, and speedups and efficiencies can be quite low.

If it's possible to increase the problem size (n) so that the efficiency doesn't decrease as p is increased, a parallel program is said to be **scalable**.

Recall that MPI_Send can either block or buffer its input. An MPI program is **unsafe** if its correct behavior depends on the fact that MPI_Send is buffering its input. This typically happens when multiple processes first call MPI_Send and then call MPI_Recv. If the calls to MPI_Send don't buffer the messages, then they'll block until the matching calls to MPI_Recv have started. However, this will never happen. For example, if process 0 and process 1 are trying to exchange data, and both call MPI_Send before calling MPI_Recv, then each process will wait forever for the other process to call MPI_Recv, since both will block in their calls to MPI_Send. That is, the processes will hang or **deadlock**—they'll block forever waiting for events that will never happen.

An MPI program can be checked for safety by replacing each call to MPI_Send with a call to MPI_Ssend. MPI_Ssend takes the same arguments as MPI_Send, but it always blocks until the matching receive has started. The extra "s" stands for *synchronous*. If the program completes correctly with MPI_Ssend for the desired inputs and communicator sizes, then the program is safe.

An unsafe MPI program can be made safe in several ways. The programmer can schedule the calls to MPI_Send and MPI_Recv so that some processes (e.g., even-ranked processes) first call MPI_Send, while others (e.g., odd-ranked processes) first call MPI_Recv. Alternatively, we can use MPI_Sendrecv or MPI_Sendrecv_replace. These functions execute both a send and a receive, but they're guaranteed to schedule them so that the program won't hang or crash. MPI_Sendrecv uses different arguments for the send and the receive buffers, while MPI_Sendrecv_replace uses the same buffer for both.

3.9 Exercises

3.1 What happens in the greetings program if, instead of strlen(greeting) + 1, we use strlen(greeting) for the length of the message being sent by processes 1, 2, ..., comm_sz−1? What happens if we use MAX_STRING instead of strlen(greeting) + 1? Can you explain these results?

3.2 Modify the trapezoidal rule so that it will correctly estimate the integral, even if comm_sz doesn't evenly divide n. (You can still assume that $n \geq$ comm_sz.)

3.3 Determine which of the variables in the trapezoidal rule program are local and which are global.

3.4 Modify the program that just prints a line of output from each process (mpi_output.c) so that the output is printed in process rank order: process 0's output first, then process 1's, and so on.

3.5 In a binary tree, there is a unique shortest path from each node to the root. The length of this path is often called the **depth** of the node. A binary tree, in which

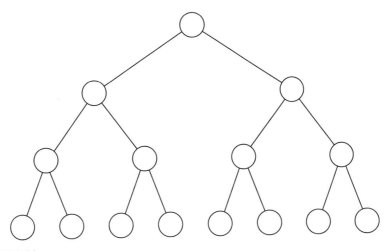

FIGURE 3.14

A complete binary tree.

every non-leaf has two children, is called a **full** binary tree; and a full binary tree, in which every leaf has the same depth, is sometimes called a **complete** binary tree. (See Fig. 3.14.) Use the principle of mathematical induction to prove that if T is a complete binary tree with n leaves, then the depth of the leaves is $\log_2(n)$.

3.6 Suppose `comm_sz` $= 4$, and suppose that $\mathbf{x}$ is a vector with $n = 14$ components.

  a. How would the components of $\mathbf{x}$ be distributed among the processes in a program that used a block distribution?

  b. How would the components of $\mathbf{x}$ be distributed among the processes in a program that used a cyclic distribution?

  c. How would the components of $\mathbf{x}$ be distributed among the processes in a program that used a block-cyclic distribution with blocksize $b = 2$?

  You should try to make your distributions general so that they could be used, regardless of what `comm_sz` and n are. You should also try to make your distributions "fair" so that if q and r are any two processes, the difference between the number of components assigned to q and the number of components assigned to r is as small as possible.

3.7 What do the various MPI collective functions do if the communicator contains a single process?

3.8 Suppose `comm_sz` $= 8$ and $n = 16$.

  a. Draw a diagram that shows how `MPI_Scatter` can be implemented using tree-structured communication with `comm_sz` processes when process 0 needs to distribute an array containing n elements.

  b. Draw a diagram that shows how `MPI_Gather` can be implemented using tree-structured communication when an n-element array that has been distributed among `comm_sz` processes needs to be gathered onto process 0.

3.9 Write an MPI program that implements multiplication of a vector by a scalar and dot product. The user should enter two vectors and a scalar, all of which are read in by process 0 and distributed among the processes. The results are calculated and collected onto process 0, which prints them. You can assume that n, the order of the vectors, is evenly divisible by `comm_sz`.

3.10 In the `Read_vector` function shown in Program 3.9, we use `local_n` as the actual argument for two of the formal arguments to `MPI_Scatter`: `send_count` and `recv_count`. Why is it OK to alias these arguments?

3.11 Finding **prefix sums** is a generalization of global sum. Rather than simply finding the sum of n values,

$$x_0 + x_1 + \cdots + x_{n-1},$$

the prefix sums are the n partial sums

$$x_0, \quad x_0 + x_1, \quad x_0 + x_1 + x_2, \quad \ldots, \quad x_0 + x_1 + \cdots + x_{n-1}.$$

a. Devise a serial algorithm for computing the n prefix sums of an array with n elements.

b. Parallelize your serial algorithm for a system with n processes, each of which is storing one of the x_i's.

c. Suppose $n = 2^k$ for some positive integer k. Can you devise a serial algorithm and a parallelization of the serial algorithm so that the parallel algorithm requires only k communication phases? (You might want to look for this online.)

d. MPI provides a collective communication function, `MPI_Scan`, that can be used to compute prefix sums:

```
int MPI_Scan(
        void*           sendbuf_p    /* in  */,
        void*           recvbuf_p    /* out */,
        int             count        /* in  */,
        MPI_Datatype    datatype     /* in  */,
        MPI_Op          op           /* in  */,
        MPI_Comm        comm         /* in  */);
```

It operates on arrays with `count` elements; both `sendbuf_p` and `recvbuf_p` should refer to blocks of `count` elements of type `datatype`. The `op` argument is the same as `op` for `MPI_Reduce`.

Write an MPI program that generates a random array of `count` elements on each MPI process, finds the prefix sums, and prints the results.

3.12 An alternative to a butterfly-structured allreduce is a "ring-pass" structure. In a ring-pass, if there are p processes, each process q sends data to process $q + 1$, except that process $p - 1$ sends data to process 0. This is repeated until each process has the desired result. Thus we can implement allreduce with the following code:

```
sum = temp_val = my_val;
for (i = 1; i < p; i++) {
    MPI_Sendrecv_replace(&temp_val, 1, MPI_INT, dest,
        sendtag, source, recvtag, comm, &status);
    sum += temp_val;
}
```

  a. Write an MPI program that implements this algorithm for allreduce. How does its performance compare to the butterfly-structured allreduce?

  b. Modify the MPI program you wrote in the first part so that it implements prefix sums.

3.13 MPI_Scatter and MPI_Gather have the limitation that each process must send or receive the same number of data items. When this is not the case, we must use the MPI functions MPI_Gatherv and MPI_Scatterv. Look at the man pages for these functions, and modify your vector sum, dot product program so that it can correctly handle the case when n isn't evenly divisible by comm_sz.

3.14 a. Write a serial C program that defines a two-dimensional array in the main function. Just use numeric constants for the dimensions:

  int two_d[3][4];

  Initialize the array in the main function. After the array is initialized, call a function that attempts to print the array. The prototype for the function should look something like this.

  void Print_two_d(**int** two_d[][], **int** rows,
      int cols);

  After writing the function, try to compile the program. Can you explain why it won't compile?

  b. After consulting a C reference (e.g., Kernighan and Ritchie [32]), modify the program so that it will compile and run, but so that it still uses a two-dimensional C array.

3.15 What is the relationship between the "row-major" storage for two-dimensional arrays that we discussed in Section 2.2.3 and the one-dimensional storage we use in Section 3.4.9?

3.16 Suppose comm_sz = 8 and the vector $\mathbf{x} = (0, 1, 2, \ldots, 15)$ has been distributed among the processes using a block distribution. Draw a diagram illustrating the steps in a butterfly implementation of allgather of $\mathbf{x}$.

3.17 MPI_Type_contiguous can be used to build a derived datatype from a collection of contiguous elements in an array. Its syntax is

```
int MPI_Type_contiguous(
    int              count          /* in  */,
    MPI_Datatype     old_mpi_t      /* in  */,
    MPI_Datatype*    new_mpi_t_p    /* out */);
```

  Modify the Read_vector and Print_vector functions so that they use an MPI datatype created by a call to MPI_Type_contiguous and a count argument of 1 in the calls to MPI_Scatter and MPI_Gather.

3.18 MPI_Type_vector can be used to build a derived datatype from a collection of blocks of elements in an array, as long as the blocks all have the same size and they're equally spaced. Its syntax is

```
int MPI_Type_vector (
    int                count          /* in  */,
    int                blocklength    /* in  */,
    int                stride         /* in  */,
    MPI_Datatype       old_mpi_t      /* in  */,
    MPI_Datatype*      new_mpi_t_p    /* out */);
```

For example, if we had an array x of 18 **doubles** and we wanted to build a type corresponding to the elements in positions 0, 1, 6, 7, 12, 13, we could call

```
int MPI_Type_vector (3, 2, 6, MPI_DOUBLE, &vect_mpi_t );
```

since the type consists of 3 blocks, each of which has 2 elements, and the spacing between the starts of the blocks is 6 **doubles**.

Write Read_vector and Print_vector functions that will allow process 0 to read and print, respectively, a vector with a block-cyclic distribution. But beware! Do *not* use MPI_Scatter or MPI_Gather. There is a technical issue involved in using these functions with types created with MPI_Type_vector. (See, for example, [23].) Just use a loop of sends on process 0 in Read_vector and a loop of receives on process 0 in Print_vector. The other processes should be able to complete their calls to Read_vector and Print_vector with a single call to MPI_Recv and MPI_Send, respectively. The communication on process 0 should use a derived datatype created by MPI_Type_vector. The calls on the other processes should just use the count argument to the communication function, since they're receiving/sending elements that they will store in contiguous array locations.

3.19 MPI_Type_indexed can be used to build a derived datatype from arbitrary array elements. Its syntax is

```
int MPI_Type_indexed (
    int                count                       /* in  */,
    int                array_of_blocklengths []     /* in  */,
    int                array_of_displacements []    /* in  */,
    MPI_Datatype       old_mpi_t                   /* in  */,
    MPI_Datatype*      new_mpi_t_p                 /* out */);
```

Unlike MPI_Type_create_struct, the displacements are measured in units of old_mpi_t—not bytes. Use MPI_Type_indexed to create a derived datatype that corresponds to the upper triangular part of a square matrix. For example, in the 4 × 4 matrix

$$\begin{pmatrix} 0 & 1 & 2 & 3 \\ 4 & 5 & 6 & 7 \\ 8 & 9 & 10 & 11 \\ 12 & 13 & 14 & 15 \end{pmatrix}$$

the upper triangular part is the elements 0, 1, 2, 3, 5, 6, 7, 10, 11, 15. Process 0 should read in an $n \times n$ matrix as a one-dimensional array, create the derived datatype, and send the upper triangular part with a single call to MPI_Send. Process 1 should receive the upper triangular part with a single call to MPI_Recv, and then print the data it received.

3.20 The functions MPI_Pack and MPI_Unpack provide an alternative to derived datatypes for grouping data. MPI_Pack copies the data to be sent, one block at a time, into a user-provided buffer. The buffer can then be sent and received. After the data is received, MPI_Unpack can be used to unpack it from the receive buffer. The syntax of MPI_Pack is

```
int MPI_Pack(
    void*          in_buf             /* in     */,
    int            in_buf_count       /* in     */,
    MPI_Datatype   datatype           /* in     */,
    void*          pack_buf           /* out    */,
    int            pack_buf_sz        /* in     */,
    int*           position_p         /* in/out */,
    MPI_Comm       comm               /* in     */);
```

We could therefore pack the input data to the trapezoidal rule program with the following code:

```
char pack_buf[100];
int position = 0;

MPI_Pack(&a, 1, MPI_DOUBLE, pack_buf, 100, &position,
    comm);
MPI_Pack(&b, 1, MPI_DOUBLE, pack_buf, 100, &position,
    comm);
MPI_Pack(&n, 1, MPI_INT, pack_buf, 100, &position,
    comm);
```

The key is the position argument. When MPI_Pack is called, position should refer to the first available slot in pack_buf. When MPI_Pack returns, it refers to the first available slot *after* the data that was just packed, so after process 0 executes this code, all the processes can call MPI_Bcast:

```
MPI_Bcast(pack_buf, 100, MPI_PACKED, 0, comm);
```

Note that the MPI datatype for a packed buffer is MPI_PACKED. Now the other processes can unpack the data using MPI_Unpack:

```
int MPI_Unpack(
    void*          pack_buf           /* in     */,
    int            pack_buf_sz        /* in     */,
    int*           position_p         /* in/out */,
    void*          out_buf            /* out    */,
    int            out_buf_count      /* in     */,
    MPI_Datatype   datatype           /* in     */,
    MPI_Comm       comm               /* in     */);
```

This can be used by "reversing" the steps in MPI_Pack. That is, the data is unpacked one block at a time starting with position = 0.

Write another Get_input function for the trapezoidal rule program. This one should use MPI_Pack on process 0, and MPI_Unpack on the other processes.

3.21 How does your system compare to ours? What run-times does your system get for matrix-vector multiplication? What kind of variability do you see in the times for a given value of comm_sz and n? Do the results tend to cluster around the minimum, the mean, or the median?

3.22 Time our implementation of the trapezoidal rule that uses MPI_Reduce. How will you choose n, the number of trapezoids? How do the minimum times compare to the mean and median times? What are the speedups? What are the efficiencies? On the basis of the data you collected, would you say that the trapezoidal rule is scalable?

3.23 Although we don't know the internals of the implementation of MPI_Reduce, we might guess that it uses a structure similar to the binary tree we discussed. If this is the case, we would expect that its run-time would grow roughly at the rate of $\log_2(p)$, since there are roughly $\log_2(p)$ levels in the tree. (Here $p = $ comm_sz.) Since the run-time of the serial trapezoidal rule is roughly proportional to n, the number of trapezoids, and the parallel trapezoidal rule simply applies the serial rule to n/p trapezoids on each process, with our assumption about MPI_Reduce, we get a formula for the overall run-time of the parallel trapezoidal rule that looks like

$$T_{\text{parallel}}(n, p) \approx a \times \frac{n}{p} + b \log_2(p)$$

for some constants a and b.

a. Use the formula, the times you've taken in Exercise 3.22, and your favorite program for doing mathematical calculations (e.g., MATLAB®) to get a least-squares estimate of the values of a and b.

b. Comment on the quality of the predicted run-times using the formula and the values for a and b computed in part (a).

3.24 Take a look at Programming Assignment 3.7. The code that we outlined for timing the cost of sending messages should work, even if the count argument is zero. What happens on your system when the count argument is 0? Can you explain why you get a nonzero elapsed time when you send a zero-byte message?

3.25 If comm_sz = p, we mentioned that the "ideal" speedup is p. Is it possible to do better?

a. Consider a parallel program that computes a vector sum. If we only time the vector sum—that is, we ignore input and output of the vectors—how might this program achieve speedup greater than p?

b. A program that achieves speedup greater than p is said to have **superlinear** speedup. Our vector sum example only achieved superlinear speedup by overcoming certain "resource limitations." What were these resource limitations? Is it possible for a program to obtain superlinear speedup without overcoming resource limitations?

3.26 Serial odd-even transposition sort of an n-element list can sort the list in considerably fewer than n phases. As an extreme example, if the input list is already sorted, the algorithm requires 0 phases.

    a. Write a serial Is_sorted function that determines whether a list is sorted.

    b. Modify the serial odd-even transposition sort program so that it checks whether the list is sorted after each phase.

    c. If this program is tested on a random collection of n-element lists, roughly what fraction gets improved performance by checking whether the list is sorted?

3.27 Find the speedups and efficiencies of the parallel odd-even sort. Does the program obtain linear speedups? Is it scalable? Is it strongly scalable? Is it weakly scalable?

3.28 Modify the parallel odd-even transposition sort so that the Merge functions simply swap array pointers after finding the smallest or largest elements. What effect does this change have on the overall run-time?

3.10 Programming assignments

3.1 Use MPI to implement the histogram program discussed in Section 2.7.1. Have process 0 read in the input data and distribute it among the processes. Also have process 0 print out the histogram.

3.2 Suppose we toss darts randomly at a square dartboard, whose bullseye is at the origin, and whose sides are 2 feet in length. Suppose also that there's a circle inscribed in the square dartboard. The radius of the circle is 1 foot, and it's area is π square feet. If the points that are hit by the darts are uniformly distributed (and we always hit the square), then the number of darts that hit inside the circle should approximately satisfy the equation

$$\frac{\text{number in circle}}{\text{total number of tosses}} = \frac{\pi}{4},$$

since the ratio of the area of the circle to the area of the square is $\pi/4$.
We can use this formula to estimate the value of π with a random number generator:

```
number_in_circle = 0;
for (toss = 0; toss < number_of_tosses; toss++) {
    x = random double between -1 and 1;
    y = random double between -1 and 1;
    distance_squared = x*x + y*y;
    if (distance_squared <= 1) number_in_circle++;
}
```

```
pi_estimate =
          4*number_in_circle/((double) number_of_tosses);
```

This is called a "Monte Carlo" method, since it uses randomness (the dart tosses).

Write an MPI program that uses a Monte Carlo method to estimate π. Process 0 should read in the total number of tosses and broadcast it to the other processes. Use `MPI_Reduce` to find the global sum of the local variable `number_in_circle`, and have process 0 print the result. You may want to use **long long int**s for the number of hits in the circle and the number of tosses, since both may have to be very large to get a reasonable estimate of π.

3.3 Write an MPI program that computes a tree-structured global sum. First write your program for the special case in which `comm_sz` is a power of two. Then, after you've gotten this version working, modify your program so that it can handle any `comm_sz`.

3.4 Write an MPI program that computes a global sum using a butterfly. First write your program for the special case in which `comm_sz` is a power of two. Can you modify your program so that it will handle any number of processes?

3.5 Implement matrix-vector multiplication using a block-column distribution of the matrix. You can have process 0 read in the matrix and simply use a loop of sends to distribute it among the processes. Assume the matrix is square of order n and that n is evenly divisible by `comm_sz`. You may want to look at the MPI function `MPI_Reduce_scatter`.

3.6 Implement matrix-vector multiplication using a block-submatrix distribution of the matrix. Assume that the vectors are distributed among the diagonal processes. Once again, you can have process 0 read in the matrix and aggregate the submatrices before sending them to the processes. Assume `comm_sz` is a perfect square and that $\sqrt{\text{comm\_sz}}$ evenly divides the order of the matrix.

3.7 A **ping-pong** is a communication in which two messages are sent, first from process A to process B (ping), and then from process B back to process A (pong). Timing blocks of repeated ping-pongs is a common way to estimate the cost of sending messages. Time a ping-pong program using the C `clock` function on your system. How long does the code have to run before `clock` gives a nonzero run-time? How do the times you got with the `clock` function compare to times taken with `MPI_Wtime`?

3.8 Parallel merge sort starts with $n/\text{comm\_sz}$ keys assigned to each process. It ends with all the keys stored on process 0 in sorted order. To achieve this, it uses the same tree-structured communication that we used to implement a global sum. However, when a process receives another process's keys, it merges the new keys into its already sorted list of keys. Write a program that implements parallel mergesort. Process 0 should read in n and broadcast it to the other processes. Each process should use a random number generator to create a local list of $n/\text{comm\_sz}$ **int**s. Each process should then sort its local list, and process 0 should gather and print the local lists. Then the processes should use tree-structured communication to merge the global list onto process 0, which prints the result.

3.9 Write a program that can be used to determine the cost of changing the distribution of a distributed data structure. How long does it take to change from a block distribution of a vector to a cyclic distribution? How long does the reverse redistribution take?

Shared-memory programming with Pthreads

4

Recall that from a programmer's point of view, a shared-memory system is one in which all the cores can access all the memory locations (see Fig. 4.1). Thus an obvious approach to the problem of coordinating the work of the cores is to specify that certain memory locations are "shared." This is a very natural approach to parallel programming. Indeed, we might well wonder why all parallel programs don't use this shared-memory approach. However, we'll see in this chapter that there are problems that arise with programming shared-memory systems; problems that are often different from the problems encountered in distributed memory programming.

For example, in Chapter 2 we saw that if different cores attempt to update a single shared-memory location, then the contents of the shared location can be unpredictable. The code that updates the shared location is an example of a *critical section*. We'll see some other examples of critical sections, and we'll learn several methods for controlling access to a critical section.

We'll also learn about other issues and techniques in shared-memory programming. In shared-memory programming, an instance of a program running on a processor is usually called a **thread** (unlike MPI, where it's called a process). We'll learn

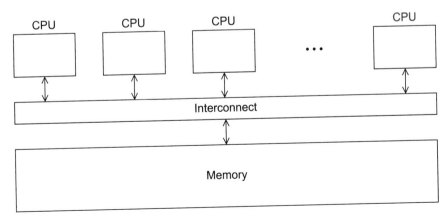

FIGURE 4.1

A shared-memory system.

An Introduction to Parallel Programming. https://doi.org/10.1016/B978-0-12-804605-0.00011-7

how to synchronize threads so that each thread will wait to execute a block of statements until another thread has completed some work. We'll learn how to put a thread "to sleep" until a condition has occurred. We'll see that there are some circumstances in which it may at first seem that a critical section must be quite large. However, we'll also see that there are tools that can allow us to "fine-tune" access to these large blocks of code so that more of the program can truly be executed in parallel. We'll see that the use of cache memories can actually cause a shared-memory program to run more slowly. Finally, we'll observe that functions that "maintain state" between successive calls can cause inconsistent or even incorrect results.

In this chapter we'll be using POSIX threads for most of our shared-memory functions. In the next chapter we'll look at an alternative approach to shared-memory programming called OpenMP.

4.1 Processes, threads, and Pthreads

Recall from Chapter 2 that in shared-memory programming, a thread is somewhat analogous to a process in MPI programming. However, it can, in principle, be "lighter-weight." A process is an instance of a running (or suspended) program. In addition to its executable, it consists of the following:

- A block of memory for the stack
- A block of memory for the heap
- Descriptors of resources that the system has allocated for the process—for example, file descriptors (including `stdout`, `stdin`, and `stderr`)
- Security information—for example, information about which hardware and software resources the process can access
- Information about the state of the process, such as whether the process is ready to run or is waiting on a resource, the content of the registers, including the program counter, and so on

In most systems, by default, a process's memory blocks are private: another process can't directly access the memory of a process unless the operating system intervenes. This makes sense. If you're using a text editor to write a program (one process—the running text editor), you don't want your browser (another process) overwriting your text editor's memory. This is even more crucial in a multiuser environment. Ordinarily, one user's processes shouldn't be allowed access to the memory of another user's processes.

However, this isn't desirable when we're running shared-memory programs. At a minimum, we'd like certain variables to be available to multiple processes, allowing much easier memory access. It is also convenient for the processes to share access to things like `stdout` and all other process-specific resources, except for their stacks and program counters. This can be arranged by starting a single process and then having the process start these additional "lighter-weight" processes. For this reason, they're often called **light-weight processes**.

The more commonly used term, **thread**, comes from the concept of "thread of control." A thread of control is just a sequence of statements in a program. The term suggests a stream of control in a single process, and in a shared-memory program a single *process* may have multiple *threads* of control.

As we noted earlier, in this chapter the particular implementation of threads that we'll be using is called POSIX threads or, more often, **Pthreads**. POSIX [46] is a standard for Unix-like operating systems—for example, Linux and macOS. It specifies a variety of facilities that should be available in such systems. In particular, it specifies an application programming interface (API) for *multithreaded* programming.

Pthreads is not a programming language (like C or Java). Rather, like MPI, Pthreads specifies a *library* that can be linked with C programs. Unlike MPI, the Pthreads API is only available on POSIX systems — Linux, macOS, Solaris, HPUX, and so on. Also unlike MPI, there are a number of other widely used specifications for multithreaded programming: Java threads, Windows threads, Solaris threads. However, all of the thread specifications support the same basic ideas, so once you've learned how to program in Pthreads, it won't be difficult to learn how to program with another thread API.

Since Pthreads is a C library, it can, in principle, be used in C++ programs. However, the recent C++11 standard includes its own shared-memory programming model with support for threads (std::thread), so it may make sense to use it instead if you're writing C++ programs.

4.2 Hello, world

Let's take a look at a Pthreads program. In Program 4.1, the main function starts up several threads. Each thread prints a message and then quits.

4.2.1 Execution

The program is compiled like an ordinary C program, with the possible exception that we may need to link in the Pthreads library[1]:

```
$ gcc -g -Wall -o pth_hello pth_hello.c -lpthread
```

The -lpthread tells the compiler that we want to link in the Pthreads library. Note that it's -lpthread, *not* -lpthreads. On some systems the compiler will automatically link in the library, and -lpthread won't be needed.

---

[1] Recall that the dollar sign ($) is the shell prompt, so it shouldn't be typed in. Also recall that for the sake of explicitness, we assume that we're using the Gnu C compiler, gcc, and we always use the options -g, -Wall, and -o. See page 77 for further information.

```
1   #include <stdio.h>
2   #include <stdlib.h>
3   #include <pthread.h>
4
5   /* Global variable:  accessible to all threads */
6   int thread_count;
7
8   void *Hello(void* rank);   /* Thread function */
9
10  int main(int argc, char* argv[]) {
11     long thread;  /* Use long in case of a 64-bit system */
12     pthread_t* thread_handles;
13
14     /* Get number of threads from command line */
15     thread_count = strtol(argv[1], NULL, 10);
16
17     thread_handles = malloc (thread_count*sizeof(pthread_t));
18
19     for (thread = 0; thread < thread_count; thread++)
20        pthread_create(&thread_handles[thread], NULL,
21            Hello, (void*) thread);
22
23     printf("Hello from the main thread\n");
24
25     for (thread = 0; thread < thread_count; thread++)
26        pthread_join(thread_handles[thread], NULL);
27
28     free(thread_handles);
29     return 0;
30  }  /* main */
31
32  void *Hello(void* rank) {
33     /* Use long in case of 64-bit system */
34     long my_rank = (long) rank;
35
36     printf("Hello from thread %ld of %d\n",
37            my_rank, thread_count);
38
39     return NULL;
40  }  /* Hello */
```

Program 4.1: A Pthreads "hello, world" program.

To run the program, we just type

```
$ ./pth_hello <number of threads>
```

For example, to run the program with 1 thread, we type

```
$ ./pth_hello 1
```

and the output will look something like this:

```
Hello  from  the  main  thread
Hello  from  thread  0  of  1
```

To run the program with four threads, we type

```
$ ./pth_hello 4
```

and the output will look something like this:

```
Hello  from  the  main  thread
Hello  from  thread  0  of  4
Hello  from  thread  1  of  4
Hello  from  thread  2  of  4
Hello  from  thread  3  of  4
```

If your output appears out of order, don't worry. As we will discuss later, we usually do not have direct control of the order in which threads execute.

4.2.2 Preliminaries

Let's take a closer look at the source code in Program 4.1. First notice that this *is* just a C program with a main function and one other function. The program includes the familiar stdio.h and stdlib.h header files. However, there's a lot that's new and different.

In Line 3 we include pthread.h, the Pthreads header file, which declares the various Pthreads functions, constants, types, and so on.

In Line 6 we define a *global* variable thread_count. In Pthreads programs, global variables are shared by all the threads. Local variables and function arguments—that is, variables declared in functions—are (ordinarily) private to the thread executing the function. If several threads are executing the same function, each thread will have its own private copies of the local variables and function arguments. This makes sense if you recall that each thread has its own stack.

We should keep in mind that global variables can introduce subtle and confusing bugs. For example, suppose we write a program in which we declare a global variable **int** x. Then we write a function f, in which we intend to use a local variable called x, but we forget to declare it. The program will compile with no warnings, since f has access to the global x. But when we run the program, it produces very strange output, which we eventually determine to have been caused by the fact that the global variable x has a strange value. Days later, we finally discover that the strange value came from f. As a rule of thumb, we should try to limit our use of global variables to situations in which they're really needed—for example, for a shared variable.

In Line 15 the program gets the number of threads it should start from the command line. Unlike MPI programs, Pthreads programs are typically compiled and run just like serial programs, and one relatively simple way to specify the number of threads that should be started is to use a command-line argument. This isn't a requirement, it's simply a convenient convention we'll be using.

The strtol function converts a string into a **long int**. It's declared in stdlib.h, and its syntax is

```
long strtol(
        const char*    number_p    /* in   */,
        char**         end_p       /* out  */,
        int            base        /* in   */);
```

It returns a **long int** corresponding to the string referred to by number_p. The base of the representation of the number is given by the base argument. If end_p isn't NULL, it will point to the first invalid (that is, nonnumeric) character in number_p.

4.2.3 Starting the threads

As we already noted, unlike MPI programs, in which the processes are usually started by a script, in Pthreads the threads are started by the program executable. This introduces a bit of additional complexity, as we need to include code in our program to explicitly start the threads, and we need data structures to store information on the threads.

In Line 17 we allocate storage for one pthread_t object for each thread. The pthread_t data structure is used for storing thread-specific information. It's declared in pthread.h.

The pthread_t objects are examples of **opaque** objects. The actual data that they store is system specific, and their data members aren't directly accessible to user code. However, the Pthreads standard guarantees that a pthread_t object does store enough information to uniquely identify the thread with which it's associated. So, for example, there is a Pthreads function that a thread can use to retrieve its associated pthread_t object, and there is a Pthreads function that can determine whether two threads are in fact the same by examining their associated pthread_t objects.

In Lines 19–21, we use the pthread_create function to start the threads. Like most Pthreads functions, its name starts with the string pthread_. The syntax of pthread_create is

```
int pthread_create(
    pthread_t*              thread_p              /* out */,
    const pthread_attr_t*   attr_p                /* in  */,
    void*                   (*start_routine)(void*) /* in  */,
    void*                   args_p                /* in  */);
```

The first argument is a pointer to the appropriate `pthread_t` object. Note that the object is not allocated by the call to `pthread_create`; it must be allocated *before* the call. We won't be using the second argument, so we just pass `NULL` in our function call.[2] The third argument is the function that the thread is to run, and the last argument is a pointer to the argument that should be passed to the function `start_routine`. The return value for most Pthreads functions indicates if there's been an error in the function call. To reduce the clutter in our examples, in this chapter (as in most of the rest of the book) we'll generally ignore the return values of Pthreads functions.

Let's take a closer look at the last two arguments. The function that's started by `pthread_create` should have a prototype that looks something like this:

```
void* thread_function(void* args_p);
```

Recall that the type **void**∗ can be cast to any pointer type in C, so `args_p` can point to a list containing one or more values needed by `thread_function`. Similarly, the return value of `thread_function` can point to a list of one or more values.

In our call to `pthread_create`, the final argument is a fairly common kluge: we're effectively assigning each thread a unique integer *rank*. Let's first look at why we are doing this; then we'll worry about the details of how to do it.

Consider the following problem: We start a Pthreads program that uses two threads, but one of the threads encounters an error. How do we, the users, know which thread encountered the error? We can't just print out the `pthread_t` object, since it's opaque. However, if when we start the threads, we assign the first thread rank 0, and the second thread rank 1, we can easily determine which thread ran into trouble by just including the thread's rank in the error message.

Since the thread function takes a **void**∗ argument, we could allocate one **int** in `main` for each thread and assign each allocated **int** a unique value. When we start a thread, we could then pass a pointer to the appropriate **int** in the call to `pthread_create`. However, most programmers resort to some trickery with casts. Instead of creating an **int** in `main` for the "rank," we cast the loop variable `thread` to have type **void**∗. Then in the thread function, `hello`, we cast the argument back to a **long** (Line 34).

The result of carrying out these casts is "system-defined," but most C compilers do allow this. However, if the size of pointer types is different from the size of the integer type you use for the rank, you may get a warning. On the machines we used, pointers are 64 bits, and **int**s are only 32 bits, so we use **long** instead of **int**.

Note that our method of assigning thread ranks and, indeed, the thread ranks themselves are just a convenient convention that we'll use. There is no requirement that a thread rank be passed in the call to `pthread_create`, nor a requirement that a thread be assigned a rank. The following thread procedure expects a pointer to a **struct** to be passed in for `args_p`. The **struct** contains both a rank and the name of the task. (Imagine distinguishing between different requests in a web server, for instance.)

---

[2] Passing `NULL` here uses the default set of Pthread *attributes*—settings that specify a variety of properties, including operating system scheduling parameters and the stack size of the new thread.

```
struct thread_args {
    long my_rank;
    char *task_name;
};

void *Hello(void *args) {
    struct thread_args* t_args
        = (struct thread_args *) args;
    printf("Thread %ld is working on task '%s'\n",
        t_args->my_rank, t_args->task_name);
    return NULL;
}
```

When we create the thread, a pointer to the appropriate **struct** is passed to `pthread_create`. We can add the logic to do this at Line 19 (in this case, each thread has the same "task name"):

```
struct thread_args *t_args
    = malloc(sizeof(struct thread_args));

t_args->my_rank = thread;
t_args->task_name = "Hello task";

pthread_create(&thread_handles[thread],
    NULL;          .
    Hello,
    (void *) t_args);
```

Also note that there is no technical reason for each thread to run the same function; we could have one thread run `hello`, another run `goodbye`, and so on. However, as with the MPI programs, we'll typically use "single program, multiple data" style parallelism with our Pthreads programs. That is, each thread will run the same thread function, but we'll obtain the effect of different thread functions by branching within a thread.

4.2.4 Running the threads

The thread that's running the `main` function is sometimes called the **main thread**. Hence, after starting the threads, it prints the message

```
Hello from the main thread
```

In the meantime, the threads started by the calls to `pthread_create` are also running. They get their ranks by casting in Line 34, and then print their messages. Note that when a thread is done, since the type of its function has a return value, the thread should return something. In this example, the threads don't actually need to return anything, so they return `NULL`.

As we hinted earlier, in Pthreads the programmer doesn't directly control where the threads are run.[3] There's no argument in `pthread_create` saying which core should run which thread; thread placement is controlled by the operating system. Indeed, on a heavily loaded system, the threads may all be run on the same core. In fact, if a program starts more threads than cores, we should expect multiple threads to be run on a single core. However, if there is a core that isn't being used, operating systems will typically place a new thread on such a core.

4.2.5 Stopping the threads

In Lines 25 and 26, we call the function `pthread_join` once for each thread. A single call to `pthread_join` will wait for the thread associated with the `pthread_t` object to complete. The syntax of `pthread_join` is

```
int pthread_join(
      pthread_t    thread      /* in  */,
      void**       ret_val_p    /* out */);
```

The second argument can be used to receive any return value computed by the thread. In the example, each thread returns `NULL`, and eventually the main thread will call `pthread_join` on that thread to complete its termination.

This function is called `pthread_join` because of a diagramming style that is often used to describe the threads in a multithreaded process. If we think of the main thread as a single line in our diagram, then, when we call `pthread_create`, we can create a *branch* or *fork* off the main thread. Multiple calls to `pthread_create` will result in multiple branches or forks. Then, when the threads started by `pthread_create` terminate, the diagram shows the branches *joining* the main thread. See Fig. 4.2.

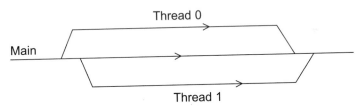

FIGURE 4.2

Main thread forks and joins two threads.

As noted previously, every thread requires a variety of resources to be allocated, including stacks and local variables. The `pthread_join` function not only allows us to wait for a particular thread to finish its execution but also frees the resources associated with the thread. In fact, not joining threads that have finished execution produces

---

[3] Some systems (for example, some implementations of Linux) do allow the programmer to specify where a thread is run. However, these constructions will not be portable.

zombie threads that waste resources and may even prevent the creation of new threads if left unchecked. If your program does not need to wait for a particular thread to finish, it can be *detached* with the `pthread_detach` function to indicate that its resources should be freed automatically upon termination. See Exercise 4.7 for an example of using `pthread_detach`.

4.2.6 Error checking

In the interest of keeping the program compact and easy to read, we have resisted the temptation to include many details that would be important in a "real" program. The most likely source of problems in this example (and in many programs) is the user input (or lack thereof). Therefore it would be a very good idea to check that the program was started with command line arguments, and, if it was, to check the actual value of the number of threads to see if it's reasonable. If you visit the book's website, you can download a version of the program that includes this basic error checking.

In general, it is good practice to always check the error codes returned by the Pthreads functions. This can be especially useful when you're just starting to use Pthreads and some of the details of function use aren't completely clear. We'd suggest getting in the habit of consulting the "RETURN VALUE" sections of the man pages for Pthreads functions (for instance, see `man pthread_create`; you will note several return values that indicate a variety of errors).

4.2.7 Other approaches to thread startup

In our example, the user specifies the number of threads to start by typing in a command-line argument. The main thread then creates all of the "subsidiary" threads. While the threads are running, the main thread prints a message, and then waits for the other threads to terminate. This approach to threaded programming is very similar to our approach to MPI programming, in which the MPI system starts a collection of processes and waits for them to complete.

There is, however, a very different approach to the design of multithreaded programs. In this approach, subsidiary threads are only started as the need arises. As an example, imagine a Web server that handles requests for information about highway traffic in the San Francisco Bay Area. Suppose that the main thread receives the requests and subsidiary threads fulfill the requests. At 1 o'clock on a typical Tuesday morning, there will probably be very few requests, while at 5 o'clock on a typical Tuesday evening, there will probably be thousands. Thus a natural approach to the design of this Web server is to have the main thread start subsidiary threads when it receives requests.

Intuitively, thread startup involves some overhead. The time required to start a thread will be much greater than, for instance, a floating point arithmetic operation, so in applications that need maximum performance the "start threads as needed" approach may not be ideal. In such a case, it is usually more performant to employ a scheme that leverages the strengths of both approaches: our main thread will start all the threads it anticipates needing at the beginning of the program, but the threads will sit idle instead of terminating when they finish their work. Once another request arrives, an idle thread can fulfill it without incurring thread creation overhead. This approach is called a *thread pool*, which we'll cover in Programming Assignment 4.5.

4.3 Matrix-vector multiplication

Let's take a look at writing a Pthreads matrix-vector multiplication program. Recall that if $A = (a_{ij})$ is an $m \times n$ matrix and $\mathbf{x} = (x_0, x_1, \ldots, x_{n-1})^T$ is an n-dimensional column vector,[4] then the matrix-vector product $A\mathbf{x} = \mathbf{y}$ is an m-dimensional column vector, $\mathbf{y} = (y_0, y_1, \ldots, y_{m-1})^T$, in which the ith component y_i is obtained by finding the dot product of the ith row of A with $\mathbf{x}$:

$$y_i = \sum_{j=0}^{n-1} a_{ij} x_j.$$

(See Fig. 4.3.)

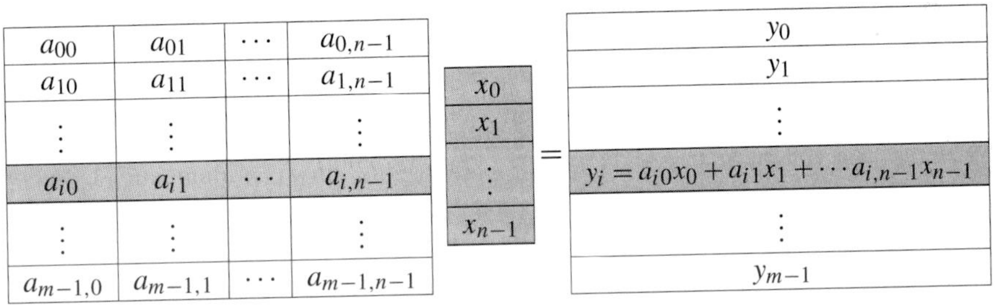

FIGURE 4.3

Matrix-vector multiplication.

Thus pseudocode for a *serial* program for matrix-vector multiplication might look like this:

```
/* For each row of A */
for (i = 0; i < m; i++) {
    y[i] = 0.0;
    /* For each element of the row and each element of x */
    for (j = 0; j < n; j++)
        y[i] += A[i][j]* x[j];
}
```

---

[4] Recall that we use the convention that matrix and vector subscripts start with 0. Also recall that if $\mathbf{b}$ is a matrix or a vector, then $\mathbf{b}^T$ denotes its transpose.

We want to parallelize this by dividing the work among the threads. One possibility is to divide the iterations of the outer loop among the threads. If we do this, each thread will compute some of the components of y. For example, suppose that $m = n = 6$ and the number of threads, `thread_count` or t, is three. Then the computation could be divided among the threads as follows:

Thread	Components of y
0	y[0], y[1]
1	y[2], y[3]
2	y[4], y[5]

To compute y[0], thread 0 will need to execute the code

```
y[0] = 0.0;
for (j = 0; j < n; j++)
    y[0] += A[0][j]* x[j];
```

Therefore thread 0 will need to access every element of row 0 of A and every element of x. More generally, the thread that has been assigned y[i] will need to execute the code

```
y[i] = 0.0;
for (j = 0; j < n; j++)
    y[i] += A[i][j]*x[j];
```

Thus this thread will need to access every element of row i of A and every element of x. We see that each thread needs to access every component of x, while each thread only needs to access its assigned rows of A and assigned components of y. This suggests that, at a minimum, x should be shared. Let's also make A and y shared. This might seem to violate our principle that we should only make variables global that need to be global. However, in the exercises, we'll take a closer look at some of the issues involved in making the A and y variables local to the thread function, and we'll see that making them global can make good sense. At this point, we'll just observe that if they are global, the main thread can easily initialize all of A by just reading its entries from `stdin`, and the product vector y can be easily printed by the main thread.

Having made these decisions, we only need to write the code that each thread will use for deciding which components of y it will compute. To simplify the code, let's assume that both m and n are evenly divisible by t. Our example with $m = 6$ and $t = 3$ suggests that each thread gets m/t components. Furthermore, thread 0 gets the first m/t, thread 1 gets the next m/t, and so on. Thus the formulas for the components assigned to thread q might be

$$\text{first component: } q \times \frac{m}{t}$$

and

$$\text{last component: } (q + 1) \times \frac{m}{t} - 1.$$

With these formulas, we can write the thread function that carries out matrix-vector multiplication. (See Program 4.2.) Note that in this code, we're assuming that A, x, y, m, and n are all global and shared.

```
void *Pth_mat_vect(void* rank) {
   long my_rank = (long) rank;
   int i, j;
   int local_m = m/thread_count;
   int my_first_row = my_rank*local_m;
   int my_last_row = (my_rank+1)*local_m - 1;

   for (i = my_first_row; i <= my_last_row; i++) {
      y[i] = 0.0;
      for (j = 0; j < n; j++)
         y[i] += A[i][j]*x[j];
   }

   return NULL;
}  /* Pth_mat_vect */
```

Program 4.2: Pthreads matrix-vector multiplication.

If you have already read the MPI chapter, you may recall that it took more work to write a matrix-vector multiplication program using MPI. This was because of the fact that the data structures were necessarily distributed, that is, each MPI process only has direct access to its own local memory. Thus for the MPI code, we need to explicitly *gather* all of x into each process's memory. We see from this example that there are instances in which writing shared-memory programs is easier than writing distributed-memory programs. However, we'll shortly see that there are situations in which shared-memory programs can be more complex.

4.4 Critical sections

Matrix-vector multiplication was very easy to code, because the shared-memory locations were accessed in a highly desirable way. After initialization, all of the variables—except y—are only *read* by the threads. That is, except for y, none of the shared variables are changed after they've been initialized by the main thread. Furthermore, although the threads do make changes to y, only one thread makes changes to any individual component, so there are no attempts by two (or more) threads to modify any single component. What happens if this isn't the case? That is, what happens when multiple threads update a single memory location? We also discuss this in Chapters 2 and 5, so if you've read one of these chapters, you already know the answer. But let's look at an example.

Let's try to estimate the value of π. There are lots of different formulas we could use. One of the simplest is

$$\pi = 4\left(1 - \frac{1}{3} + \frac{1}{5} - \frac{1}{7} + \cdots + (-1)^n \frac{1}{2n+1} + \cdots\right).$$

This isn't the best formula for computing π, because it takes *a lot* of terms on the right-hand side before it is very accurate. However, for our purposes, lots of terms will be better to demonstrate the effects of parallelism.

The following *serial* code uses this formula:

```
double factor = 1.0;
double sum = 0.0;
for (i = 0; i < n; i++, factor = -factor) {
    sum += factor/(2*i+1);
}
pi = 4.0*sum;
```

We can try to parallelize this in the same way we parallelized the matrix-vector multiplication program: divide up the iterations in the **for** loop among the threads and make sum a shared variable. To simplify the computations, let's assume that the number of threads, thread_count or t, evenly divides the number of terms in the sum, n. Then, if $\bar{n} = n/t$, thread 0 can add the first $\bar{n}$ terms. Therefore for thread 0, the loop variable i will range from 0 to $\bar{n} - 1$. Thread 1 will add the next $\bar{n}$ terms, so for thread 1, the loop variable will range from $\bar{n}$ to $2\bar{n} - 1$. More generally, for thread q the loop variable will range over

$$q\bar{n}, q\bar{n} + 1, q\bar{n} + 2, \ldots, (q+1)\bar{n} - 1.$$

Furthermore, the sign of the first term, term $q\bar{n}$, will be positive if $q\bar{n}$ is even and negative if $q\bar{n}$ is odd. The thread function might use the code shown in Program 4.3.

If we run the Pthreads program with two threads and n is relatively small, we find that the results of the Pthreads program are in agreement with the serial sum program. However, as n gets larger, we start getting some peculiar results. For example, with a dual-core processor we get the following results:

	10^5	10^6	10^7	10^8
π	3.14159	3.141593	3.1415927	3.14159265
1 Thread	3.14158	3.141592	3.1415926	3.14159264
2 Threads	3.14158	3.141480	3.1413692	3.14164686

Notice that as we increase n, the estimate with one thread gets better and better. In fact, with each factor of 10 increase in n, we get another correct digit. With $n = 10^5$, the result as computed by a single thread has five correct digits. With $n = 10^6$, it has six correct digits, and so on. The result computed by two threads agrees with the

```
1   void* Thread_sum(void* rank) {
2      long my_rank = (long) rank;
3      double factor;
4      long long i;
5      long long my_n = n/thread_count;
6      long long my_first_i = my_n*my_rank;
7      long long my_last_i = my_first_i + my_n;
8
9      if (my_first_i % 2 == 0)   /* my_first_i is even */
10        factor = 1.0;
11     else   /* my_first_i is odd */
12        factor = -1.0;
13
14     for (i = my_first_i; i < my_last_i; i++, factor = -factor) {
15        sum += factor/(2*i+1);
16     }
17
18     return NULL;
19  }   /* Thread_sum */
```

Program 4.3: An attempt at a thread function for computing π.

result computed by one thread when $n = 10^5$. However, for larger values of n, the result computed by two threads actually gets worse. In fact, if we ran the program several times with two threads and the same value of n, we would see that the result computed by two threads *changes* from run to run. The answer to our original question must clearly be, "Yes, it matters if multiple threads try to update a single shared variable."

Let's recall why this is the case. Remember that the addition of two values is typically *not* a single machine instruction. For example, although we can add the contents of a memory location y to a memory location x with a single C statement,

```
x = x + y;
```

what the machine does is typically more complicated. The current values stored in x and y will, in general, be stored in the computer's main memory, which has no circuitry for carrying out arithmetic operations. Before the addition can be carried out, the values stored in x and y may therefore have to be transferred from main memory to registers in the CPU. Once the values are in registers, the addition can be carried out. After the addition is completed, the result may have to be transferred from a register back to memory.

Suppose that we have two threads, and each computes a value that is stored in its private variable y. Also suppose that we want to add these private values together into a shared variable x that has been initialized to 0 by the main thread. Each thread will execute the following code:

```
y = Compute(my_rank);
x = x + y;
```

Let's also suppose that thread 0 computes $y = 1$ and thread 1 computes $y = 2$. The "correct" result should then be $x = 3$. Here's one possible scenario:

Time	Thread 0	Thread 1
1	Started by main thread	
2	Call Compute()	Started by main thread
3	Assign y = 1	Call Compute()
4	Put x=0 and y=1 into registers	Assign y = 2
5	Add 0 and 1	Put x=0 and y=2 into registers
6	Store 1 in memory location x	Add 0 and 2
7		Store 2 in memory location x

So we see that if thread 1 copies x from memory to a register *before* thread 0 stores its result, the computation carried out by thread 0 will be *overwritten* by thread 1. The problem could be reversed: if thread 1 *races* ahead of thread 0, then its result may be overwritten by thread 0. In fact, unless one of the threads stores its result *before* the other thread starts reading x from memory, the "winner's" result will be overwritten by the "loser."

This example illustrates a fundamental problem in shared-memory programming: when multiple threads attempt to update a shared resource—in our case a shared variable—the result may be unpredictable. Recall that more generally, when multiple threads attempt to access a shared resource, such as a shared variable or a shared file, at least one of the accesses is an update, and the accesses can result in an error, we have a **race condition**. In our example, in order for our code to produce the correct result, we need to make sure that once one of the threads starts executing the statement $x = x + y$, it finishes executing the statement *before* the other thread starts executing the statement. Therefore the code $x = x + y$ is a **critical section**. That is, it's a block of code that updates a shared resource that can only be updated by one thread at a time.

To further illustrate the concept of a race condition, imagine a bank wants to improve the performance of its checking account system. An obvious first step would be to make the system multithreaded; rather than processing a single transaction at a time, banking operations should be spread across multiple threads to take advantage of parallelism. This works well—until multiple transactions modify an account at the same time. Consider two pending transactions on a checking account with an initial balance of $1000:

- A $100 utility bill payment
- A $500 salary deposit

After the transactions complete, the new account balance should be $1400. The salary deposit will require an addition operation and the utility payment will require a subtraction. However, as mentioned previously, these simple math operations will be broken into more than one machine instruction. One possible outcome is:

Time	Thread 0 (Bill Payment)	Thread 1 (Salary Deposit)
1		Read Balance ($1000)
2	Read Balance ($1000)	Calculate Balance + $500
3	Calculate Balance - $100	Write Balance ($1500)
4	Write Balance ($900)	

Rather than the expected ending balance of $1400, we get $900 instead, because the transaction processed by thread 1 was overwritten by thread 0.

These types of issues are particularly difficult to debug, because the outcome is non-deterministic. It is entirely possible that the error shown above occurs less than 1% of the time and could be influenced by external factors, including the hardware, operating system, or process scheduling algorithm. Even worse, attaching a debugger or adding `printf` statements to the code may change the relative timing of the threads and seemingly "correct" the issue temporarily. Such bugs that disappear when inspected are known as *Heisenbugs* (the act of observing the system alters its state).

4.5 Busy-waiting

To avoid race conditions, threads need exclusive access to shared memory regions. When, say, thread 0 wants to execute the statement $x = x + y$, it needs to first make sure that thread 1 is not already executing the statement. Once thread 0 makes sure of this, it needs to provide some way for thread 1 to determine that it, thread 0, is executing the statement, so that thread 1 won't attempt to start executing the statement until thread 0 is done. Finally, after thread 0 has completed execution of the statement, it needs to provide some way for thread 1 to determine that it is done, so that thread 1 can safely start executing the statement.

A simple approach that doesn't involve any new concepts is the use of a flag variable. Suppose `flag` is a shared **int** that is set to 0 by the main thread. Further, suppose we add the following code to our example:

```
1   y = Compute(my_rank);
2   while (flag != my_rank);
3   x = x + y;
4   flag++;
```

Let's suppose that thread 1 finishes the assignment in Line 1 before thread 0. What happens when it reaches the **while** statement in Line 2? If you look at the **while** statement for a minute, you'll see that it has the somewhat peculiar property that its body is empty. So if the test `flag != my_rank` is true, then thread 1 will just execute the test a second time. In fact, it will keep re-executing the test until the test is false. When the test is false, thread 1 will go on to execute the code in the critical section $x = x + y$.

Since we're assuming that the main thread has initialized `flag` to 0, thread 1 won't proceed to the critical section in Line 3 until thread 0 executes the statement `flag++`. In fact, we see that unless some catastrophe befalls thread 0, it will eventually catch up to thread 1. However, when thread 0 executes its first test of `flag != my_rank`, the condition is false, and it will go on to execute the code in the critical section `x = x + y`. When it's done with this, we see that it will execute `flag++`, and thread 1 can finally enter the critical section.

The key here is that thread 1 *cannot enter the critical section until thread 0 has completed the execution of* `flag++`. And, provided the statements are executed exactly as they're written, this means that thread 1 cannot enter the critical section until thread 0 has completed it.

The **while** loop is an example of **busy-waiting**. In busy-waiting, a thread repeatedly tests a condition, but, effectively, does no useful work until the condition has the appropriate value (false in our example).

Note that we said that the busy-wait solution would work "provided the statements are executed exactly as they're written." If compiler optimization is turned on, it *is* possible that the compiler will make changes that will affect the correctness of busy-waiting. The reason for this is that the compiler is unaware that the program is multithreaded, so it doesn't "know" that the variables `x` and `flag` can be modified by another thread. For example, if our code

```
y = Compute(my_rank);
while (flag != my_rank);
x = x + y;
flag++;
```

is run by just one thread, the order of the statements **while** `(flag != my_rank)` and `x = x + y` is unimportant. An optimizing compiler might therefore determine that the program would make better use of registers if the order of the statements were switched. Of course, this will result in the code

```
y = Compute(my_rank);
x = x + y;
while (flag != my_rank);
flag++;
```

which defeats the purpose of the busy-wait loop. The simplest solution to this problem is to turn compiler optimizations off when we use busy-waiting. For an alternative to completely turning off optimizations, see Exercise 4.3.

We can immediately see that busy-waiting is not an ideal solution to the problem of controlling access to a critical section. Since thread 1 will execute the test over and over until thread 0 executes `flag++`, if thread 0 is delayed (for example, if the operating system preempts it to run something else), thread 1 will simply "spin" on the test, eating up CPU cycles. This approach—often called a *spinlock*—can be positively disastrous for performance. Turning off compiler optimizations can also seriously degrade performance.

Before going on, though, let's return to our π calculation program in Program 4.3 and correct it by using busy-waiting. The critical section in this function is Line 15. We can therefore precede this with a busy-wait loop. However, when a thread is done with the critical section, if it simply increments flag, eventually flag will be greater than t, the number of threads, and none of the threads will be able to return to the critical section. That is, after executing the critical section once, all the threads will be stuck forever in the busy-wait loop. Thus, in this instance, we don't want to simply increment flag. Rather, the last thread, thread $t - 1$, should reset flag to zero. This can be accomplished by replacing flag++ with

```
flag = (flag + 1) % thread_count;
```

With this change, we get the thread function shown in Program 4.4. If we compile the program and run it with two threads, we see that it is computing the correct results.

```
1   void* Thread_sum(void* rank) {
2      long my_rank = (long) rank;
3      double factor;
4      long long i;
5      long long my_n = n/thread_count;
6      long long my_first_i = my_n*my_rank;
7      long long my_last_i = my_first_i + my_n;
8
9      if (my_first_i % 2 == 0)
10         factor = 1.0;
11     else
12         factor = -1.0;
13
14     for (i = my_first_i; i < my_last_i; i++, factor = -factor) {
15         while (flag != my_rank);
16         sum += factor/(2*i+1);
17         flag = (flag+1) % thread_count;
18     }
19
20     return NULL;
21  }  /* Thread_sum */
```

Program 4.4: Pthreads global sum with busy-waiting.

However, if we add in code for computing elapsed time, we see that when $n = 10^8$, the serial sum is consistently faster than the parallel sum. For example, on the dual-core system, the elapsed time for the sum as computed by two threads is about 19.5 seconds, while the elapsed time for the serial sum is about 2.8 seconds!

Why is this? Of course, there's overhead associated with starting up and joining the threads. However, we can estimate this overhead by writing a Pthreads program in which the thread function simply returns:

```
void* Thread_function(void* ignore) {
    return NULL;
}   /* Thread_function */
```

When we find the time that's elapsed between starting the first thread and joining the second thread, we see that on this particular system, the overhead is less than 0.3 milliseconds, so the slowdown isn't due to thread overhead. If we look closely at the thread function that uses busy-waiting, we see that the threads alternate between executing the critical section code in Line 16. Initially `flag` is 0, so thread 1 must wait until thread 0 executes the critical section and increments `flag`. Then, thread 0 must wait until thread 1 executes and increments. The threads will alternate between waiting and executing, and evidently the waiting and the incrementing increase the overall run-time by a factor of seven.

As we'll see, busy-waiting isn't the only solution to protecting a critical section. In fact, there are much better solutions. However, since the code in a critical section can only be executed by one thread at a time, no matter how we limit access to the critical section, we'll effectively serialize the code in the critical section. Therefore, if it's at all possible, we should minimize the number of times we execute critical section code. One way to greatly improve the performance of the sum function is to have each thread use a *private* variable to store its total contribution to the sum. Then, each thread can add in its contribution to the global sum once, *after* the **for** loop. See Program 4.5. When we run this on the dual core system with $n = 10^8$, the elapsed time is reduced to 1.5 seconds for two threads, a *substantial* improvement.

4.6 Mutexes

Since a thread that is busy-waiting may continually use the CPU, busy-waiting is generally not an ideal solution to the problem of limiting access to a critical section. Two better solutions are mutexes and semaphores. **Mutex** is an abbreviation of *mutual exclusion*, and a mutex is a special type of variable that, together with a couple of special functions, can be used to restrict access to a critical section to a single thread at a time. Thus a mutex can be used to guarantee that one thread "excludes" all other threads while it executes the critical section. Hence, the mutex guarantees mutually exclusive access to the critical section.

The Pthreads standard includes a special type for mutexes: `pthread_mutex_t`. A variable of type `pthread_mutex_t` needs to be initialized by the system before it's used. This can be done with a call to

```
int pthread_mutex_init(
        pthread_mutex_t*             mutex_p    /* out */,
        const pthread_mutexattr_t*   attr_p     /* in  */);
```

We won't make use of the second argument, so we'll just pass in `NULL` to use the default attributes. You may also occasionally encounter the following *static* mutex initialization that declares a mutex and initializes it in a single line of code:

```
void* Thread_sum(void* rank) {
   long my_rank = (long) rank;
   double factor, my_sum = 0.0;
   long long i;
   long long my_n = n/thread_count;
   long long my_first_i = my_n*my_rank;
   long long my_last_i = my_first_i + my_n;

   if (my_first_i % 2 == 0)
      factor = 1.0;
   else
      factor = -1.0;

   for (i = my_first_i; i < my_last_i; i++, factor = -factor)
      my_sum += factor/(2*i+1);

   while (flag != my_rank);
   sum += my_sum;
   flag = (flag+1) % thread_count;

   return NULL;
}  /* Thread_sum */
```

Program 4.5: Global sum function with critical section after loop.

```
pthread_mutex_t mutex = PTHREAD_MUTEX_INITIALIZER;
```

Although in general `pthread_mutex_init` is more flexible, this initialization is fine in many, if not most, cases.

When a Pthreads program finishes using a mutex (regardless of how they are initialized), it should call

```
int pthread_mutex_destroy(
      pthread_mutex_t* mutex_p  /* in/out */);
```

The point of a mutex is to protect a critical section from being entered by more than one thread at a time. To gain access to a critical section, a thread will lock the mutex, do its work, and then unlock the mutex to let other threads execute the critical section. To lock the mutex and gain exclusive access to the critical section, a thread calls

```
int pthread_mutex_lock(
      pthread_mutex_t* mutex_p  /* in/out */);
```

When a thread is finished executing the code in a critical section, it should call

```
int pthread_mutex_unlock(
      pthread_mutex_t* mutex_p  /* in/out */);
```

The call to `pthread_mutex_lock` will cause the thread to wait until no other thread is in the critical section, and the call to `pthread_mutex_unlock` notifies the system that the calling thread has completed execution of the code in the critical section.

We can use mutexes instead of busy-waiting in our global sum program by declaring a global mutex variable, having the main thread initialize it, and then, instead of busy-waiting and incrementing a flag, the threads call `pthread_mutex_lock` before entering the critical section, and they call `pthread_mutex_unlock` when they're done with the critical section. (See Program 4.6.) The first thread to call `pthread_mutex_lock`

```
1   void* Thread_sum(void* rank) {
2      long my_rank = (long) rank;
3      double factor;
4      long long i;
5      long long my_n = n/thread_count;
6      long long my_first_i = my_n*my_rank;
7      long long my_last_i = my_first_i + my_n;
8      double my_sum = 0.0;
9
10     if (my_first_i % 2 == 0)
11        factor = 1.0;
12     else
13        factor = -1.0;
14
15     for (i = my_first_i; i < my_last_i; i++, factor = -factor) {
16        my_sum += factor/(2*i+1);
17     }
18     pthread_mutex_lock(&mutex);
19     sum += my_sum;
20     pthread_mutex_unlock(&mutex);
21
22     return NULL;
23   }  /* Thread_sum */
```

Program 4.6: Global sum function that uses a mutex.

will, effectively, "lock the door" to the critical section: any other thread that attempts to execute the critical section code must first also call `pthread_mutex_lock`, and until the first thread calls `pthread_mutex_unlock`, all the threads that have called `pthread_mutex_lock` will **block** in their calls—they'll just wait until the first thread is done. After the first thread calls `pthread_mutex_unlock`, the system will choose one of the blocked threads and allow it to execute the code in the critical section. This process will be repeated until all the threads have completed executing the critical section.

"Locking" and "unlocking" the door to the critical section isn't the only metaphor that's used in connection with mutexes. Programmers often say that the thread that has returned from a call to `pthread_mutex_lock` has "obtained the mutex" or "obtained the lock." When this terminology is used, a thread that calls `pthread_mutex_unlock` relinquishes the mutex or lock. (You may also encounter terminology referring to this as "acquiring" and "releasing" the lock.)

Notice that with mutexes (unlike our busy-waiting solution), the order in which the threads execute the code in the critical section is more or less random: the first thread to call `pthread_mutex_lock` will be the first to execute the code in the critical section. Subsequent accesses will be scheduled by the system. Pthreads doesn't guarantee that the threads will obtain the lock in the order in which they called `Pthread_mutex_lock`. However, in our setting, a finite number of threads will try to acquire the lock and they are guaranteed to eventually obtain it.

If we look at the (unoptimized) performance of the busy-wait π program (with the critical section after the loop) and the mutex program, we see that for both versions the ratio of the run-time of the single-threaded program with the multithreaded program is equal to the number of threads, as long as the number of threads is no greater than the number of cores. That is,

$$\frac{T_{\text{serial}}}{T_{\text{parallel}}} \approx \texttt{thread\_count},$$

provided `thread_count` is less than or equal to the number of cores. Recall that $T_{\text{serial}}/T_{\text{parallel}}$ is called the *speedup*, and when the speedup is equal to the number of threads, we have achieved more or less "ideal" performance or *linear speedup*.

If we compare the performance of the version that uses busy-waiting with the version that uses mutexes, we don't see much difference in the overall run-time when the programs are run with fewer threads than cores. This shouldn't be surprising, as each thread only enters the critical section once; unless the critical section is very long, or the Pthreads functions are very slow, we wouldn't expect the threads to be delayed very much by waiting to enter the critical section. However, if we start increasing the number of threads beyond the number of cores, the performance of the version that uses mutexes remains largely unchanged, while the performance of the busy-wait version degrades. (See Table 4.1.)

We see that when we use busy-waiting, performance can degrade if there are more threads than cores.[5] This should make sense. For example, suppose we have two cores and five threads. Also suppose that thread 0 is in the critical section, thread 1 is in the busy-wait loop, and threads 2, 3, and 4 have been descheduled by the operating system. After thread 0 completes the critical section and sets `flag = 1`, it will be terminated, and thread 1 can enter the critical section so the operating system can schedule

---

[5] These are typical run-times. When using busy-waiting and the number of threads is greater than the number of cores, the run-times vary considerably.

Table 4.1 Run-times (in seconds) of π programs using $n = 10^8$ terms on a system with two four-core processors.

Threads	Busy-Wait	Mutex
1	2.90	2.90
2	1.45	1.45
4	0.73	0.73
8	0.38	0.38
16	0.50	0.38
32	0.80	0.40
64	3.56	0.38

Table 4.2 Possible sequence of events with busy-waiting and more threads than cores.

Time	`flag`	Thread				
		0	**1**	**2**	**3**	**4**
0	0	crit sect	busy wait	susp	susp	susp
1	1	terminate	crit sect	susp	busy wait	susp
2	2	—	terminate	susp	busy wait	busy wait
⋮	⋮			⋮	⋮	⋮
?	2	—	—	crit sect	susp	busy wait

thread 2, thread 3, or thread 4. Suppose it schedules thread 3, which will spin in the **while** loop. When thread 1 finishes the critical section and sets `flag = 2`, the operating system can schedule thread 2 or thread 4. If it schedules thread 4, then both thread 3 and thread 4 will be busily spinning in the busy-wait loop until the operating system deschedules one of them and schedules thread 2. (See Table 4.2.)

4.7 Producer–consumer synchronization and semaphores

Although busy-waiting is generally wasteful of CPU resources, it does have the property that we know, in advance, the order in which the threads will execute the code in the critical section: thread 0 is first, then thread 1, then thread 2, and so on. With mutexes, the order in which the threads execute the critical section is left to chance and the system. Since addition is commutative, this doesn't matter in our program for estimating π. However, it's not difficult to think of situations in which we also want to control the order in which the threads execute the code in the critical section. For example, suppose each thread generates an $n \times n$ matrix, and we want to multiply the matrices together in thread-rank order. Since matrix multiplication isn't commutative, our mutex solution would have problems:

```
/* n and product_matrix are shared and initialized by
 * the main thread. product_matrix is initialized
 * to be the identity matrix. */
void* Thread_work(void* rank) {
   long my_rank = (long) rank;
   matrix_t my_mat = Allocate_matrix(n);
   Generate_matrix(my_mat);
   pthread_mutex_lock(&mutex);
   Multiply_matrix(product_mat, my_mat);
   pthread_mutex_unlock(&mutex);
   Free_matrix(&my_mat);
   return NULL;
}  /* Thread_work */
```

A somewhat more complicated example involves having each thread "send a message" to another thread. For example, suppose we have thread_count or *t* threads and we want thread 0 to send a message to thread 1, thread 1 to send a message to thread 2, ..., thread *t* − 2 to send a message to thread *t* − 1 and thread *t* − 1 to send a message to thread 0. After a thread "receives" a message, it can print the message and terminate. To implement the message transfer, we can allocate a shared array of **char**∗. Then each thread can allocate storage for the message it's sending, and, after it has initialized the message, set a pointer in the shared array to refer to it. To avoid dereferencing undefined pointers, the main thread can set the individual entries in the shared array to NULL. (See Program 4.7.) When we run the program with more

```
1    /* 'messages' has type char**. It's allocated in main. */
2    /* Each entry is set to NULL in main.                   */
3    void *Send_msg(void* rank) {
4       long my_rank = (long) rank;
5       long dest = (my_rank + 1) % thread_count;
6       long source = (my_rank + thread_count - 1) % thread_count;
7       char* my_msg = malloc(MSG_MAX*sizeof(char));
8
9       sprintf(my_msg, "Hello to %ld from %ld", dest, my_rank);
10      messages[dest] = my_msg;
11
12      if (messages[my_rank] != NULL)
13         printf("Thread %ld > %s\n", my_rank, messages[my_rank]);
14      else
15         printf("Thread %ld > No message from %ld\n",
16                 my_rank, source);
17
18      return NULL;
19   }  /* Send_msg */
```

Program 4.7: A first attempt at sending messages using pthreads.

than a couple of threads on a dual core system, we see that some of the messages are never received. For example, thread 0, which is started first, will typically finish before thread $t - 1$ has copied the message into the messages array.

This isn't surprising, and we could fix the problem by replacing the **if** statement in Line 12 with a busy-wait **while** statement:

```
while (messages[my_rank] == NULL);
printf("Thread %ld > %s\n", my_rank, messages[my_rank]);
```

Of course, this solution would have the same problems that any busy-waiting solution has, so we'd prefer a different approach.

After executing the assignment in Line 10, we'd like to "notify" the thread with rank dest that it can proceed to print the message. We'd like to do something like this:

```
. . .
messages[dest] = my_msg;
Notify thread dest that it can proceed;

Await notification from thread source
printf("Thread %ld > %s\n", my_rank, messages[my_rank]);
. . .
```

It's not at all clear how mutexes can help here. We might try calling pthread_mutex_unlock to "notify" the thread dest. However, mutexes are initialized to be *unlocked*, so we'd need to add a call *before* initializing messages[dest] to lock the mutex. This will be a problem, since we don't know when the threads will reach the calls to pthread_mutex_lock.

To make this a little clearer, suppose that the main thread creates and initializes an array of mutexes, one for each thread. Then we're trying to do something like this:

```
1    . . .
2    pthread_mutex_lock(&mutex[dest]);
3    . . .
4    messages[dest] = my_msg;
5    pthread_mutex_unlock(&mutex[dest]);
6    . . .
7    pthread_mutex_lock(&mutex[my_rank]);
8    printf("Thread %ld > %s\n", my_rank, messages[my_rank]);
9    . . .
```

Now suppose we have two threads, and thread 0 gets so far ahead of thread 1 that it reaches the second call to pthread_mutex_lock in Line 7 before thread 1 reaches the first in Line 2. Then, of course, it will acquire the lock and continue to the printf statement. This will result in thread 0 dereferencing a null pointer and it will crash.

There *are* other approaches to solving this problem with mutexes. See, for example, Exercise 4.8. However, POSIX also provides a somewhat different means of controlling access to critical sections: **semaphores**. Let's take a look at them.

A semaphore can be thought of as a special type of **unsigned int**, so they take on the values 0, 1, 2, ... In many cases, we'll only be interested in using them when they take on the values 0 and 1. A semaphore that only takes on these values is called a *binary* semaphore. Very roughly speaking, 0 corresponds to a locked mutex, and 1 corresponds to an unlocked mutex. To use a binary semaphore as a mutex, you *initialize* it to 1: "unlocked." Before the critical section you want to protect, you place a call to the function sem_wait. A thread that executes sem_wait will block if the semaphore is 0. If the semaphore is nonzero, it will *decrement* the semaphore and proceed. After executing the code in the critical section, a thread calls sem_post, which *increments* the semaphore, and a thread waiting in sem_wait can proceed.

Semaphores were first defined by the computer scientist Edsger Dijkstra in [15]. The name is taken from the mechanical device that railroads use to control which train can use a track. The device consists of an arm attached by a pivot to a post. When the arm points down, approaching trains can proceed, and when the arm is perpendicular to the post, approaching trains must stop and wait. The track corresponds to the critical section: when the arm is down corresponds to a semaphore of 1, and when the arm is up corresponds to a semaphore of 0. The sem_wait and sem_post calls correspond to signals sent by the train to the semaphore controller.

For our current purposes, the crucial difference between semaphores and mutexes is that there is no ownership associated with a semaphore. The main thread can initialize all of the semaphores to 0—that is, "locked"—and then any thread can execute a sem_post on any of the semaphores. Similarly, any thread can execute sem_wait on any of the semaphores. Thus, if we use semaphores, our Send_msg function can be written as shown in Program 4.8.

The syntax of the various semaphore functions is

```
int sem_init(
      sem_t*      semaphore_p    /* out */,
      int         shared         /* in  */,
      unsigned    initial_val    /* in  */);

int sem_destroy(sem_t*   semaphore_p   /* in/out */);
int sem_post(sem_t*      semaphore_p   /* in/out */);
int sem_wait(sem_t*      semaphore_p   /* in/out */);
```

The second argument to sem_init controls whether the semaphore is shared among threads or processes. In our examples, we'll be sharing the semaphore among threads, so the constant 0 can be passed in.

Note that semaphores are part of the POSIX standard, but *not* part of Pthreads. Hence it is necessary to ensure your operating system does indeed support semaphores, and then add the following preprocessor directive to any program that uses them[6]:

---

[6] Some systems, including macOS, don't support this version of semaphores. However, they may support something called "named" semaphores. The functions sem_wait and sem_post can be used in the same

```
1   /* 'messages' is allocated and initialized to NULL in main */
2   /* 'semaphores' is allocated and initialized to            */
3   /* 0 (locked) in main                                      */
4   void *Send_msg(void* rank) {
5      long my_rank = (long) rank;
6      long dest = (my_rank + 1) % thread_count;
7      char* my_msg = malloc(MSG_MAX*sizeof(char));
8
9      sprintf(my_msg, "Hello to %ld from %ld", dest, my_rank);
10     messages[dest] = my_msg;
11     /* ''Unlock'' the semaphore of dest: */
12     sem_post(&semaphores[dest]);
13
14     /* Wait for our semaphore to be unlocked */
15     sem_wait(&semaphores[my_rank]);
16     printf("Thread %ld > %s\n", my_rank, messages[my_rank]);
17
18     return NULL;
19  }  /* Send_msg */
```

Program 4.8: Using semaphores so that threads can send messages.

```
#include <semaphore.h>
```

Finally, note that the message-sending problem didn't involve a critical section. The problem wasn't that there was a block of code that could only be executed by one thread at a time. Rather, thread my_rank couldn't proceed until thread source had finished creating the message. This type of synchronization, when a thread can't proceed until another thread has taken some action, is sometimes called **producer–consumer synchronization**. For example, imagine a *producer* thread that generates tasks and places them in a fixed-size queue (or *bounded buffer*) for a *consumer* thread to execute. In this case, the consumer blocks until at least one task is ready, at which point it will be signaled by the producer. Once signaled, the work is carried out by the thread in isolation; no critical section is involved. This paradigm is seen in stream processing, web servers, and so on; in the case of a web server, the producer thread could listen for incoming request URIs and place them in the queue, while the consumer would be responsible for reading the corresponding file from disk (e.g., http://server/file.txt might be located at /www/file.txt on the web server's file system) and sending data back to the client that requested the URI.

As mentioned earlier, binary semaphores (those that only take on the values 0 and 1) are fairly typical. However, *counting* semaphores can also be useful in scenar-

---

way. However, sem_init should be replaced by sem_open, and sem_destroy should be replaced by sem_close and sem_unlink. See the book's website for an example.

ios where we wish to restrict access to a finite resource. One common example is an application design pattern that involves limiting the number of threads used by a program to be no more than the number of cores available on a given machine. Consider a program with a workload of N tasks, where N is much greater than the available cores. In this case, the main thread is responsible for distributing the workload and would initialize its semaphore with the number of cores available, and then call `sem_wait` before starting each worker thread with `pthread_create`. Once the counter reaches 0, the main thread will block; the machine has a task running for each core and the program must wait for a thread to finish before starting more. When a thread does finish its task, it will call `sem_post` to signal that the main thread can create another worker thread. For this approach to be efficient, the amount of time spent on each task much be longer than the thread creation overhead because N total threads will be started during the program's execution. For an approach that reuses existing threads in a *thread pool*, see Programming Assignment 4.5.

4.8 Barriers and condition variables

Let's take a look at another problem in shared-memory programming: synchronizing the threads by making sure that they all are at the same point in a program. Such a point of synchronization is called a **barrier**, because no thread can proceed beyond the barrier until all the threads have reached it.

Barriers have numerous applications. As we discussed in Chapter 2, if we're timing some part of a multithreaded program, we'd like for all the threads to start the timed code at the same instant, and then report the time taken by the last thread to finish, i.e., the "slowest" thread. So we'd like to do something like this:

```
/* Shared */
double elapsed_time;

. . .
/* Private */
double my_start, my_finish, my_elapsed;

. . .
Synchronize threads;
Store current time in my_start;
/* Execute timed code */
. . .
Store current time in my_finish;
my_elapsed = my_finish - my_start;

elapsed = Maximum of my_elapsed values;
```

Using this approach, we're sure that all of the threads will record `my_start` at approximately the same time.

Another very important use of barriers is in debugging. As you've probably already seen, it can be very difficult to determine *where* an error is occurring in a parallel program. We can, of course, have each thread print a message indicating which point it's reached in the program, but it doesn't take long for the volume of the output to become overwhelming. Barriers provide an alternative:

```
point in program we want to reach;
barrier;
if (my_rank == 0) {
   printf("All threads reached this point\n");
   fflush(stdout);
}
```

Many implementations of Pthreads don't provide barriers, so if our code is to be portable, we need to develop our own implementation. There are a number of options; we'll look at three. The first two only use constructs that we've already studied. The third uses a new type of Pthreads object: a *condition variable*.

4.8.1 Busy-waiting and a mutex

Implementing a barrier using busy-waiting and a mutex is straightforward: we use a shared counter protected by the mutex. When the counter indicates that every thread has entered the critical section, threads can leave the busy-wait loop.

```
/* Shared and initialized by the main thread */
int counter;    /* Initialize to 0 */
int thread_count;
pthread_mutex_t barrier_mutex;
. . .

void* Thread_work(. . .) {
   . . .
   /* Barrier */
   pthread_mutex_lock(&barrier_mutex);
   counter++;
   pthread_mutex_unlock(&barrier_mutex);
   while (counter < thread_count);
   . . .
}
```

Of course, this implementation will have the same problems that our other busy-wait codes had: we'll waste CPU cycles when threads are in the busy-wait loop, and, if we run the program with more threads than cores, we may find that the performance of the program seriously degrades.

Another issue is the shared variable counter. What happens if we want to implement a second barrier and we try to reuse the counter? When the first barrier is completed, counter will have the value thread_count. Unless we can somehow reset counter, the **while** condition we used for our first barrier counter < thread_count will be false, and the barrier won't cause the threads to block. Furthermore, any attempt to reset counter to zero is almost certainly doomed to failure. If the last thread to enter

the loop tries to reset it, some thread in the busy-wait may never see the fact that
counter == thread_count, and that thread may hang in the busy-wait. If some thread
tries to reset the counter after the barrier, some other thread may enter the second barrier before the counter is reset and its increment to the counter will be lost. This will
have the unfortunate effect of causing all the threads to hang in the second busy-wait
loop. So if we want to use this barrier, we need one counter variable for each instance
of the barrier.

4.8.2 Semaphores

A natural question is whether we can implement a barrier with semaphores, and, if so,
whether we can reduce the number of problems we encountered with busy-waiting.
The answer to the first question is yes:

```
/* Shared variables */
int counter;            /* Initialize to 0 */
sem_t count_sem;        /* Initialize to 1 */
sem_t barrier_sem;      /* Initialize to 0 */
. . .
void* Thread_work (...) {
    . . .
    /* Barrier */
    sem_wait(&count_sem);
    if (counter == thread_count −1) {
        counter = 0;
        sem_post(&count_sem);
        for (j = 0; j < thread_count −1; j++)
            sem_post(&barrier_sem);
    } else {
        counter ++;
        sem_post(&count_sem);
        sem_wait(&barrier_sem);
    }
    . . .
}
```

As with the busy-wait barrier, we have a counter that we use to determine how
many threads have entered the barrier. We use two semaphores: count_sem protects
the counter, and barrier_sem is used to block threads that have entered the barrier.
The count_sem semaphore is initialized to 1 (that is, "unlocked"), so the first thread to
reach the barrier will be able to proceed past the call to sem_wait. Subsequent threads,
however, will block until they can have exclusive access to the counter. When a thread
has exclusive access to the counter, it checks to see if counter < thread_count-1. If it
is, the thread increments counter, relinquishes the lock (sem_post(&count_sem)), and
blocks in sem_wait(&barrier_sem). On the other hand, if counter == thread_count-1,

the thread is the last to enter the barrier, so it can reset `counter` to zero and "unlock" `count_sem` by calling `sem_post(&count_sem)`. Now, it wants to notify all the other threads that they can proceed, so it executes `sem_post(&barrier_sem)` for each of the `thread_count`-1 threads that are blocked in `sem_wait(&barrier_sem)`.

Note that it doesn't matter if the thread executing the loop of calls to `sem_post(&barrier_sem)` races ahead and executes multiple calls to `sem_post` before a thread can be unblocked from `sem_wait(&barrier_sem)`. Recall that a semaphore is an **unsigned int**, and the calls to `sem_post` increment it, while the calls to `sem_wait` decrement it—unless it's already 0. If it's 0, the calling threads will block until it's positive again. Therefore it doesn't matter if the thread executing the loop of calls to `sem_post(&barrier_sem)` gets ahead of the threads blocked in the calls to `sem_wait(&barrier_sem)`, because eventually the blocked threads will see that `barrier_sem` is positive, and they'll decrement it and proceed.

It should be clear that this implementation of a barrier is superior to the busy-wait barrier, since the threads don't need to consume CPU cycles when they're blocked in `sem_wait`. Can we reuse the data structures from the first barrier if we want to execute a second barrier?

The `counter` can be reused, since we were careful to reset it before releasing any of the threads from the barrier. Also, `count_sem` can be reused, since it is reset to 1 before any threads can leave the barrier. This leaves `barrier_sem`. Since there's exactly one `sem_post` for each `sem_wait`, it might appear that the value of `barrier_sem` will be 0 when the threads start executing a second barrier. However, suppose we have two threads, and thread 0 is blocked in `sem_wait(&barrier_sem)` in the first barrier, while thread 1 is executing the loop of calls to `sem_post`. Also suppose that the operating system has seen that thread 0 is idle, and descheduled it out. Then thread 1 can go on to the second barrier. Since `counter == 0`, it will execute the **else** clause. After incrementing `counter`, it executes `sem_post(&count_sem)`, and then executes `sem_wait(&barrier_sem)`.

However, if thread 0 is still descheduled, it will not have decremented `barrier_sem`. Thus when thread 1 reaches `sem_wait(&barrier_sem)`, `barrier_sem` will still be 1, so it will simply decrement `barrier_sem` and proceed. This will have the unfortunate consequence that when thread 0 starts executing again, it will still be blocked in the *first* `sem_wait(&barrier_sem)`, and thread 1 will proceed through the second barrier before thread 0 has entered it. Reusing `barrier_sem` therefore results in a race condition.

4.8.3 Condition variables

A somewhat better approach to creating a barrier in Pthreads is provided by *condition variables*. A **condition variable** is a data object that allows a thread to suspend execution until a certain event or *condition* occurs. When the event or condition occurs another thread can *signal* the thread to "wake up." A condition variable is *always* associated with a mutex.

Typically, condition variables are used in constructs similar to this pseudocode:

```
lock mutex;
if condition has occurred
   signal thread(s);
else {
   unlock the mutex and block;
   /* when thread is unblocked, mutex is relocked */
}
unlock mutex;
```

Condition variables in Pthreads have type `pthread_cond_t`. The function

```
int pthread_cond_signal(
      pthread_cond_t*  cond_var_p  /* in/out */);
```

will unblock *one* of the blocked threads, and

```
int pthread_cond_broadcast(
      pthread_cond_t* cond_var_p  /* in/out */);
```

will unblock *all* of the blocked threads. This is one advantage of condition variables; recall that we needed a **for** loop calling `sem_post` to achieve similar functionality with semaphores. The function

```
int pthread_cond_wait(
         pthread_cond_t*    cond_var_p  /* in/out */,
         pthread_mutex_t*   mutex_p     /* in/out */);
```

will unlock the mutex referred to by `mutex_p` and cause the executing thread to block until it is unblocked by another thread's call to `pthread_cond_signal` or `pthread_cond_broadcast`. When the thread is unblocked, it reacquires the mutex. So in effect, `pthread_cond_wait` implements the following sequence of functions:

```
pthread_mutex_unlock(&mutex_p);
wait_on_signal(&cond_var_p);
pthread_mutex_lock(&mutex_p);
```

The following code implements a barrier with a condition variable:

```
/* Shared */
int counter = 0;
pthread_mutex_t mutex;
pthread_cond_t cond_var;

. . .
void* Thread_work(. . .) {
   . . .
   /* Barrier */
   pthread_mutex_lock(&mutex);
   counter++;
   if (counter == thread_count) {
      counter = 0;
      pthread_cond_broadcast(&cond_var);
   } else {
```

```
        while (pthread_cond_wait(&cond_var, &mutex) != 0);
    }
    pthread_mutex_unlock(&mutex);
    . . .
}
```

Note that it is possible that events other than the call to `pthread_cond_broadcast` can cause a suspended thread to unblock (see, for example, Butenhof [7], page 80). This is called a *spurious wake-up*. Hence, the call to `pthread_cond_wait` should usually be placed in a **while** loop. If the thread is unblocked by some event other than a call to `pthread_cond_signal` or `pthread_cond_broadcast`, then the return value of `pthread_cond_wait` will be nonzero, and the unblocked thread will call `pthread_cond_wait` again.

If a single thread is being awakened, it's also a good idea to check that the condition has, in fact, been satisfied before proceeding. In our example, if a single thread were being released from the barrier with a call to `pthread_cond_signal`, then that thread should verify that `counter == 0` before proceeding. This can be dangerous with the broadcast, though. After being awakened, some thread may race ahead and change the condition, and if each thread is checking the condition, a thread that awakened later may find the condition is no longer satisfied and go back to sleep.

Note that in order for our barrier to function correctly, it's essential that the call to `pthread_cond_wait` unlock the mutex. If it didn't unlock the mutex, then only one thread could enter the barrier; all of the other threads would block in the call to `pthread_mutex_lock`, the first thread to enter the barrier would block in the call to `pthread_cond_wait`, and our program would hang.

Also note that the semantics of mutexes require that the mutex be relocked before we return from the call to `pthread_cond_wait`. We obtained the lock when we returned from the call to `pthread_mutex_lock`. Hence, we should at some point relinquish the lock through a call to `pthread_mutex_unlock`.

Like mutexes and semaphores, condition variables should be initialized and destroyed. In this case, the functions are

```
int pthread_cond_init(
        pthread_cond_t*            cond_p        /* out */,
        const pthread_condattr_t*  cond_attr_p   /* in  */);

int pthread_cond_destroy(
        pthread_cond_t*  cond_p  /* in/out */);
```

We won't be using the second argument to `pthread_cond_init`—as with mutexes, the default the attributes are fine for our purposes—so we'll call it with second argument set to NULL. As usual, there is also a *static* version of the initializer if we are planning to use the default attributes:

```
pthread_cond_t cond = PTHREAD_COND_INITIALIZER;
```

Condition variables are often quite useful whenever a thread needs to wait for something. When protected application state cannot be represented by an unsigned integer counter, condition variables may be preferable to semaphores.

4.8.4 Pthreads barriers

Before proceeding we should note that the Open Group, the standards group that is continuing to develop the POSIX standard, does define a barrier interface for Pthreads. However, as we noted earlier, it is not universally available, so we haven't discussed it in the text. See Exercise 4.10 for some of the details of the API.

4.9 Read-write locks

Let's take a look at the problem of controlling access to a large, shared data structure, which can be either simply searched or updated by the threads. For the sake of explicitness, let's suppose the shared data structure is a sorted, singly-linked list of **int**s, and the operations of interest are Member, Insert, and Delete.

4.9.1 Sorted linked list functions

The list itself is composed of a collection of list *nodes*, each of which is a struct with two members: an **int** and a pointer to the next node. We can define such a struct with the definition

```
struct list_node_s {
    int data;
    struct list_node_s* next;
}
```

A typical list is shown in Fig. 4.4. A pointer, head_p, with type **struct** list_node_s* refers to the first node in the list. The next member of the last node is NULL (which is indicated by a slash (/) in the next member).

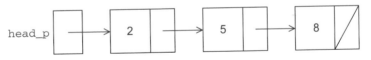

FIGURE 4.4

A linked list.

The Member function (Program 4.9) uses a pointer to traverse the list until it either finds the desired value or determines that the desired value cannot be in the list. Since the list is sorted, the latter condition occurs when the curr_p pointer is NULL or when the data member of the current node is larger than the desired value.

The Insert function (Program 4.10) begins by searching for the correct position in which to insert the new node. Since the list is sorted, it must search until it finds a node whose data member is greater than the value to be inserted. When it finds this node, it needs to insert the new node in the position *preceding* the node that's been found. Since the list is singly-linked, we can't "back up" to this position without traversing the list a second time. There are several approaches to dealing with this;

```
1   int   Member(int value,  struct list_node_s* head_p) {
2      struct list_node_s* curr_p = head_p;
3
4      while (curr_p != NULL && curr_p->data < value)
5         curr_p = curr_p->next;
6
7      if (curr_p == NULL || curr_p->data > value) {
8         return 0;
9      } else {
10        return 1;
11     }
12  }   /* Member */
```

Program 4.9: The Member function.

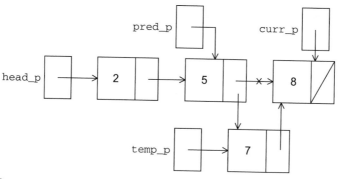

FIGURE 4.5

Inserting a new node into a list.

the approach we use is to define a second pointer pred_p, which, in general, refers to the predecessor of the current node. When we exit the loop that searches for the position to insert, the next member of the node referred to by pred_p can be updated so that it refers to the new node. (See Fig. 4.5.)

The Delete function (Program 4.11) is similar to the Insert function in that it also needs to keep track of the predecessor of the current node while it's searching for the node to be deleted. The predecessor node's next member can then be updated after the search is completed. (See Fig. 4.6.)

4.9.2 A multithreaded linked list

Now let's try to use these functions in a Pthreads program. To share access to the list, we can define head_p to be a global variable. This will simplify the function headers for Member, Insert, and Delete, since we won't need to pass in either head_p or a

```
1   int Insert(int value, struct list_node_s** head_pp) {
2       struct list_node_s* curr_p = *head_pp;
3       struct list_node_s* pred_p = NULL;
4       struct list_node_s* temp_p;
5
6       while (curr_p != NULL && curr_p->data < value) {
7           pred_p = curr_p;
8           curr_p = curr_p->next;
9       }
10
11      if (curr_p == NULL || curr_p->data > value) {
12          temp_p = malloc(sizeof(struct list_node_s));
13          temp_p->data = value;
14          temp_p->next = curr_p;
15          if (pred_p == NULL)  /* New first node */
16              *head_pp = temp_p;
17          else
18              pred_p->next = temp_p;
19          return 1;
20      } else { /* Value already in list */
21          return 0;
22      }
23  } /* Insert */
```

Program 4.10: The Insert function.

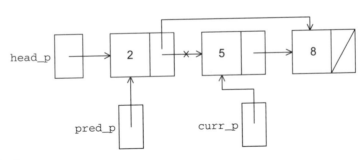

FIGURE 4.6

Deleting a node from the list.

pointer to head_p, we'll only need to pass in the value of interest. What now are the consequences of having multiple threads simultaneously execute the three functions?

Since multiple threads can simultaneously *read* a memory location without conflict, it should be clear that multiple threads can simultaneously execute Member. On the other hand, Delete and Insert also *write* to memory locations, so there may be problems if we try to execute either of these operations at the same time as another

```
1   int Delete(int value, struct list_node_s** head_pp) {
2      struct list_node_s* curr_p = *head_pp;
3      struct list_node_s* pred_p = NULL;
4
5      while (curr_p != NULL && curr_p->data < value) {
6         pred_p = curr_p;
7         curr_p = curr_p->next;
8      }
9
10     if (curr_p != NULL && curr_p->data == value) {
11        if (pred_p == NULL) { /* Deleting first node in list */
12           *head_pp = curr_p->next;
13           free(curr_p);
14        } else {
15           pred_p->next = curr_p->next;
16           free(curr_p);
17        }
18        return 1;
19     } else {   /* Value isn't in list */
20        return 0;
21     }
22  }  /* Delete */
```

Program 4.11: The Delete function.

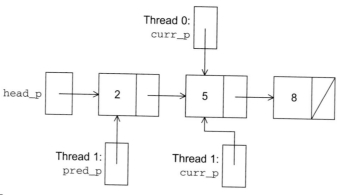

FIGURE 4.7

Simultaneous access by two threads.

operation. As an example, suppose that thread 0 is executing Member(5) at the same time that thread 1 is executing Delete(5), and the current state of the list is shown in Fig. 4.7. An obvious problem is that if thread 0 is executing Member(5), it is going to report that 5 is in the list, when, in fact, it may be deleted even before thread 0 returns.

A second obvious problem is if thread 0 is executing Member(8), thread 1 may free the memory used for the node storing 5 before thread 0 can advance to the node storing 8. Although typical implementations of free don't overwrite the freed memory, if the memory is reallocated before thread 0 advances, there can be serious problems. For example, if the memory is reallocated for use in something other than a list node, what thread 0 "thinks" is the next member may be set to utter garbage, and after it executes

```
curr_p = curr_p->next;
```

dereferencing curr_p may result in a segmentation violation.

More generally, we can run into problems if we try to simultaneously execute another operation while we're executing an Insert or a Delete. It's OK for multiple threads to simultaneously execute Member—that is, *read* the list nodes—but it's unsafe for multiple threads to access the list if at least one of the threads is executing an Insert or a Delete—that is, is *writing* to the list nodes (see Exercise 4.12).

How can we deal with this problem? An obvious solution is to simply lock the list any time that a thread attempts to access it. For example, a call to each of the three functions can be protected by a mutex, so we might execute

```
Pthread_mutex_lock(&list_mutex);
Member(value);
Pthread_mutex_unlock(&list_mutex);
```

instead of simply calling Member(value).

An equally obvious problem with this solution is that we are serializing access to the list, and if the vast majority of our operations are calls to Member, we'll fail to exploit this opportunity for parallelism. On the other hand, if most of our operations are calls to Insert and Delete, then this may be the best solution, since we'll need to serialize access to the list for most of the operations, and this solution will certainly be easy to implement.

An alternative to this approach involves "finer-grained" locking. Instead of locking the entire list, we could try to lock individual nodes. We would add, for example, a mutex to the list node struct:

```
struct list_node_s {
    int data;
    struct list_node_s* next;
    pthread_mutex_t mutex;
}
```

Now each time we try to access a node we must first lock the mutex associated with the node. Note that this will also require that we have a mutex associated with the head_p pointer. So, for example, we might implement Member as shown in Program 4.12. Admittedly this implementation is *much* more complex than the original Member function. It is also much slower, since, in general, each time a node is accessed, a mutex must be locked and unlocked. At a minimum it will add two function calls to the node access, but it can also add a substantial delay if a thread has to wait

```
int   Member(int value) {
   struct list_node_s* temp_p;

   pthread_mutex_lock(&head_p_mutex);
   temp_p = head_p;
   while (temp_p != NULL && temp_p->data < value) {
      if (temp_p->next != NULL)
         pthread_mutex_lock(&(temp_p->next->mutex));
      if (temp_p == head_p)
         pthread_mutex_unlock(&head_p_mutex);
      pthread_mutex_unlock(&(temp_p->mutex));
      temp_p = temp_p->next;
   }

   if (temp_p == NULL || temp_p->data > value) {
      if (temp_p == head_p)
         pthread_mutex_unlock(&head_p_mutex);
      if (temp_p != NULL)
         pthread_mutex_unlock(&(temp_p->mutex));
      return 0;
   } else {
      if (temp_p == head_p)
         pthread_mutex_unlock(&head_p_mutex);
      pthread_mutex_unlock(&(temp_p->mutex));
      return 1;
   }
}  /* Member */
```

Program 4.12: Implementation of `Member` with one mutex per list node.

for a lock. A further problem is that the addition of a mutex field to each node will substantially increase the amount of storage needed for the list. On the other hand, the finer-grained locking might be a closer approximation to what we want. Since we're only locking the nodes of current interest, multiple threads can simultaneously access different parts of the list, regardless of which operations they're executing.

4.9.3 Pthreads read-write locks

Neither of our multithreaded linked lists exploits the potential for simultaneous access to *any* node by threads that are executing `Member`. The first solution only allows one thread to access the entire list at any instant, and the second only allows one thread to access any given node at any instant. An alternative is provided by Pthreads' **read-write locks**. A read-write lock is somewhat like a mutex except that it provides *two* lock functions. The first lock function locks the read-write lock for *reading*, while the second locks it for *writing*. Multiple threads can thereby simultaneously obtain the lock by calling the read-lock function, while only one thread can obtain the lock by calling the write-lock function. Thus if any threads own the lock for reading, any threads that want to obtain the lock for writing will block in the call to the write-lock function. Furthermore, if any thread owns the lock for writing, any threads that want to obtain the lock for reading or writing will block in their respective locking functions.

Using Pthreads read-write locks, we can protect our linked list functions with the following code (we're ignoring function return values):

```
pthread_rwlock_rdlock(&rwlock);
Member(value);
pthread_rwlock_unlock(&rwlock);
. . .
pthread_rwlock_wrlock(&rwlock);
Insert(value);
pthread_rwlock_unlock(&rwlock);
. . .
pthread_rwlock_wrlock(&rwlock);
Delete(value);
pthread_rwlock_unlock(&rwlock);
```

The syntax for the new Pthreads functions is

```
int pthread_rwlock_rdlock(
    pthread_rwlock_t*  rwlock_p  /* in/out */);


int pthread_rwlock_wrlock(
    pthread_rwlock_t*  rwlock_p  /* in/out */);


int pthread_rwlock_unlock(
    pthread_rwlock_t*  rwlock_p  /* in/out */);
```

As their names suggest, the first function locks the read-write lock for reading, the second locks it for writing, and the last unlocks it.

As with mutexes, read-write locks should be initialized before use and destroyed after use. The following function can be used for initialization:

```
int pthread_rwlock_init(
    pthread_rwlock_t*             rwlock_p   /* out */,
    const pthread_rwlockattr_t*   attr_p     /* in  */);


/* And, the static version: */
pthread_rwlock_t rwlock = PTHREAD_RWLOCK_INITIALIZER;
```

Also as with mutexes, we'll not use the second argument, so we'll just pass NULL. The following function can be used for destruction of a read-write lock:

```
int pthread_rwlock_destroy(
    pthread_rwlock_t*  rwlock_p  /* in/out */);
```

4.9.4 Performance of the various implementations

Of course, we really want to know which of the three implementations is "best," so we included our implementations in a small program in which the main thread first inserts a user-specified number of randomly generated keys into an empty list. After being started by the main thread, each thread carries out a user-specified number of operations on the list. The user also specifies the percentages of each type of operation (Member, Insert, Delete). However, which operation occurs when and on which key is determined by a random number generator. Thus, for example, the user might specify that 1000 keys should be inserted into an initially empty list and a total of 100,000 operations are to be carried out by the threads. Further, she might specify that 80% of the operations should be Member, 15% should be Insert, and the remaining 5% should be Delete. However, since the operations are randomly generated, it might happen that the threads execute a total of, say, 79,000 calls to Member, 15,500 calls to Insert, and 5500 calls to Delete.

Tables 4.3 and 4.4 show the times (in seconds) that it took for 100,000 operations on a list that was initialized to contain 1000 keys. Both sets of data were taken on a system containing four dual-core processors.

Table 4.3 Linked list times: 100,000 ops/thread, 99.9% Member, 0.05% Insert, 0.05% Delete.

Implementation	Number of Threads			
	1	2	4	8
Read-Write Locks	0.213	0.123	0.098	0.115
One Mutex for Entire List	0.211	0.450	0.385	0.457
One Mutex per Node	1.680	5.700	3.450	2.700

Table 4.4 Linked list times: 100,000 ops/thread, 80% Member, 10% Insert, 10% Delete.

Implementation	Number of Threads			
	1	2	4	8
Read-Write Locks	2.48	4.97	4.69	4.71
One Mutex for Entire List	2.50	5.13	5.04	5.11
One Mutex per Node	12.00	29.60	17.00	12.00

Table 4.3 shows the times when 99.9% of the operations are Member and the remaining 0.1% are divided equally between Insert and Delete. Table 4.4 shows the times when 80% of the operations are Member, 10% are Insert, and 10% are Delete. Note that in both tables when one thread is used, the run-times for the read-write locks and the single-mutex implementations are about the same. This makes sense: the operations are serialized, and since there is no contention for the read-write lock or the mutex, the overhead associated with both implementations should consist of a function call before the list operation and a function call after the operation. On the

other hand, the implementation that uses one mutex per node is *much* slower. This also makes sense, since each time a single node is accessed, there will be two function calls—one to lock the node mutex and one to unlock it. Thus there's considerably more overhead for this implementation.

The inferiority of the implementation that uses one mutex per node persists when we use multiple threads. There is far too much overhead associated with all the locking and unlocking to make this implementation competitive with the other two implementations.

Perhaps the most striking difference between the two tables is the relative performance of the read-write lock implementation and the single-mutex implementation when multiple threads are used. When there are very few Inserts and Deletes, the read-write lock implementation is far better than the single-mutex implementation. Since the single-mutex implementation will serialize all the operations, this suggests that if there are very few Inserts and Deletes, the read-write locks do a very good job of allowing concurrent access to the list. On the other hand, if there are a relatively large number of Inserts and Deletes (for example, 10% each), there's very little difference between the performance of the read-write lock implementation and the single-mutex implementation. Thus, for linked list operations, read-write locks *can* provide a considerable increase in performance, but only if the number of Inserts and Deletes is quite small.

Also notice that if we use one mutex or one mutex per node, the program is *always* as fast or faster when it's run with one thread. Furthermore, when the number of Inserts and Deletes is relatively large, the read-write lock program is also faster with one thread. This isn't surprising for the one mutex implementation, since effectively accesses to the list are serialized. For the read-write lock implementation, it appears that when there are a substantial number of write locks, there is too much contention for the locks and overall performance deteriorates significantly.

In summary, the read-write lock implementation is superior to the single mutex and one mutex per node implementations. However, unless the number of Inserts and Deletes is small, a serial implementation will be superior.

4.9.5 Implementing read-write locks

The original Pthreads specification didn't include read-write locks, so some of the early texts describing Pthreads include implementations of read-write locks (see, for example, [7]). A typical implementation[7] defines a data structure that uses two condition variables—one for "readers" and one for "writers"—and a mutex. The structure also contains members that indicate

1. how many readers own the lock, that is, are currently reading,
2. how many readers are waiting to obtain the lock,

---

[7] This discussion follows the basic outline of Butenhof's implementation [7].

3. whether a writer owns the lock, and

4. how many writers are waiting to obtain the lock.

The mutex protects the read-write lock data structure: whenever a thread calls one of the functions (read-lock, write-lock, unlock), it first locks the mutex, and whenever a thread completes one of these calls, it unlocks the mutex. After acquiring the mutex, the thread checks the appropriate data members to determine how to proceed. As an example, if it wants read-access, it can check to see if there's a writer that currently owns the lock. If not, it increments the number of active readers and proceeds. If a writer is active, it increments the number of readers waiting and starts a condition wait on the reader condition variable. When it's awakened, it decrements the number of readers waiting, increments the number of active readers, and proceeds. The write-lock function has an implementation that's similar to the read-lock function.

The action taken in the unlock function depends on whether the thread was a reader or a writer. If the thread was a reader, there are no currently active readers, *and* there's a writer waiting, then it can signal a writer to proceed before returning. If, on the other hand, the thread was a writer, there can be both readers and writers waiting, so the thread needs to decide whether it will give preference to readers or writers. Since writers must have exclusive access, it is likely that it is much more difficult for a writer to obtain the lock. Many implementations therefore give writers preference. Programming Assignment 4.6 explores this further.

4.10 Caches, cache-coherence, and false sharing[8]

Recall that for a number of years now, computers have been able to execute operations involving only the processor much faster than they can access data in main memory. If a processor must read data from main memory for each operation, it will spend most of its time simply waiting for the data from memory to arrive. Also recall that to address this problem, chip designers have added blocks of relatively fast memory to processors. This faster memory is called **cache memory**.

The design of cache memory takes into consideration the principles of **temporal and spatial locality**: if a processor accesses main memory location x at time t, then it is likely that at times close to t it will access main memory locations close to x. Thus if a processor needs to access main memory location x, rather than transferring only the contents of x to/from main memory, a block of memory containing x is transferred from/to the processor's cache. Such a block of memory is called a **cache line** or **cache block**.

In Section 2.3.5, we saw that the use of cache memory can have a huge impact on shared memory. Let's recall why. First, consider the following situation: Suppose

---

[8] This material is also covered in Chapter 5. So if you've already read that chapter, you may want to skim this section.

x is a shared variable with the value five, and both thread 0 and thread 1 read x from memory into their (separate) caches, because both want to execute the statement

```
my_y = x;
```

Here, my_y is a private variable defined by both threads. Now suppose thread 0 executes the statement

```
x++;
```

Finally, suppose that thread 1 now executes

```
my_z = x;
```

where my_z is another private variable.

What's the value in my_z? Is it five? Or is it six? The problem is that there are (at least) three copies of x: the one in main memory, the one in thread 0's cache, and the one in thread 1's cache. When thread 0 executed x++, what happened to the values in main memory and thread 1's cache? This is the **cache coherence** problem, which we discussed in Chapter 2. We saw there that most systems insist that the caches be made aware that changes have been made to data they are caching. The line in the cache of thread 1 would have been marked *invalid* when thread 0 executed x++, and before assigning my_z = x, the core running thread 1 would see that its value of x was out of date. Thus the core running thread 0 would have to update the copy of x in main memory (either now or earlier), and the core running thread 1 would get the line with the updated value of x from main memory. For further details, see Chapter 2.

The use of cache coherence can have a dramatic effect on the performance of shared-memory systems. To illustrate this, recall our Pthreads matrix-vector multiplication example: the main thread initialized an $m \times n$ matrix A and an n-dimensional vector $\mathbf{x}$. Each thread was responsible for computing m/t components of the product vector $\mathbf{y} = A\mathbf{x}$. (As usual, t is the number of threads.) The data structures representing A, $\mathbf{x}$, $\mathbf{y}$, m, and n were all shared. For ease of reference, we reproduce the code in Program 4.13.

If T_{serial} is the run-time of the serial program, and T_{parallel} is the run-time of the parallel program, recall that the *efficiency* E of the parallel program is the speedup S divided by the number of threads:

$$E = \frac{S}{t} = \frac{\left(\frac{T_{\text{serial}}}{T_{\text{parallel}}}\right)}{t} = \frac{T_{\text{serial}}}{t \times T_{\text{parallel}}}.$$

Since $S \leq t$, $E \leq 1$. Table 4.5 shows the run-times and efficiencies of our matrix-vector multiplication with different sets of data and differing numbers of threads.

In each case, the total number of floating point additions and multiplications is 64,000,000; so an analysis that only considers arithmetic operations would predict that a single thread running the code would take the same amount of time for all three inputs. However, it's clear that this is *not* the case. With one thread, the 8,000,000 × 8 system requires about 14% more time than the 8000 × 8000 system,

```
1   void *Pth_mat_vect(void* rank) {
2      long my_rank = (long) rank;
3      int i, j;
4      int local_m = m/thread_count;
5      int my_first_row = my_rank*local_m;
6      int my_last_row = (my_rank+1)*local_m - 1;
7
8      for (i = my_first_row; i <= my_last_row; i++) {
9         y[i] = 0.0;
10        for (j = 0; j < n; j++)
11           y[i] += A[i][j]*x[j];
12     }
13
14     return NULL;
15  } /* Pth_mat_vect */
```

Program 4.13: Pthreads matrix-vector multiplication.

Table 4.5 Run-times and efficiencies of matrix-vector multiplication (times are in seconds).

| | Matrix Dimension | | | | | |
| | 8,000,000 × 8 | | 8000 × 8000 | | 8 × 8,000,000 | |
Threads	Time	Eff.	Time	Eff.	Time	Eff.
1	0.393	1.000	0.345	1.000	0.441	1.000
2	0.217	0.906	0.188	0.918	0.300	0.735
4	0.139	0.707	0.115	0.750	0.388	0.290

and the $8 \times 8,000,000$ system requires about 28% more time than the 8000×8000 system. Both of these differences are at least partially attributable to cache performance

Recall that a *write-miss* occurs when a core tries to update a variable that's not in the cache, and it has to access main memory. A cache profiler (such as Valgrind [51]) shows that when the program is run with the $8,000,000 \times 8$ input, it has far more cache write-misses than either of the other inputs. The bulk of these occur in Line 9. Since the number of elements in the vector y is far greater in this case (8,000,000 vs. 8000 or 8), and each element must be initialized, it's not surprising that this line slows down the execution of the program with the $8,000,000 \times 8$ input.

Also recall that a *read-miss* occurs when a core tries to read a variable that's not in the cache, and it has to access main memory. A cache profiler shows that when the program is run with the $8 \times 8,000,000$ input, it has far more cache read-misses than either of the other inputs. These occur in Line 11, and a careful study of this program (see Exercise 4.16) shows that the main source of the differences is due to the reads of x. Once again, this isn't surprising, since for this input, x has 8,000,000 elements, versus only 8000 or 8 for the other inputs.

It should be noted that there may be other factors that are affecting the relative performance of the single-threaded program with the differing inputs. For example, we haven't taken into consideration whether virtual memory (see Subsection 2.2.4) has affected the performance of the program with the different inputs. How frequently does the CPU need to access the page table in main memory?

Of more interest to us, though, is the tremendous difference in efficiency as the number of threads is increased. The two-thread efficiency of the program with the $8 \times 8,000,000$ input is nearly 20% less than the efficiency of the program with the $8,000,000 \times 8$ and the 8000×8000 inputs. The four-thread efficiency of the program with the $8 \times 8,000,000$ input is nearly 60% less than the program's efficiency with the $8,000,000 \times 8$ input and *more* than 60% less than the program's efficiency with the 8000×8000 input. These dramatic decreases in efficiency are even more remarkable when we note that with one thread the program is much slower with $8 \times 8,000,000$ input. Therefore the numerator in the formula for the efficiency:

$$\text{Parallel Efficiency} = \frac{\text{Serial Run-Time}}{(\text{Number of Threads}) \times (\text{Parallel Run-Time})}$$

will be much larger. Why, then, is the multithreaded performance of the program so much worse with the $8 \times 8,000,000$ input?

In this case, once again, the answer has to do with cache. Let's take a look at the program when we run it with four threads. With the $8,000,000 \times 8$ input, y has 8,000,000 components, so each thread is assigned 2,000,000 components. With the 8000×8000 input, each thread is assigned 2000 components of y, and with the $8 \times 8,000,000$ input, each thread is assigned 2 components. On the system we used, a cache line is 64 bytes. Since the type of y is **double**, and a **double** is 8 bytes, a single cache line can store 8 **doubles**.

Cache coherence is enforced at the "cache-line level." That is, each time any value in a cache line is written, if the line is also stored in another processor's cache, the entire *line* will be invalidated—not just the value that was written. The system we're using has two dual-core processors, and each processor has its own cache. Suppose for the moment that threads 0 and 1 are assigned to one of the processors and threads 2 and 3 are assigned to the other. Also suppose that for the $8 \times 8,000,000$ problem all of y is stored in a single cache line. Then every write to some element of y will invalidate the line in the other processor's cache. For example, each time thread 0 updates y[0] in the statement

```
y[i] += A[i][j]*x[j];
```

If thread 2 or 3 is executing this code, it will have to reload y. Each thread will update each of its components 8,000,000 times. We see that with this assignment of threads to processors and components of y to cache lines, all the threads will have to reload y *many* times. This is going to happen in spite of the fact that only one thread accesses any one component of y—for example, only thread 0 accesses y[0].

Each thread will update its assigned components of y a total of 16,000,000 times. It appears that many if not most of these updates are forcing the threads to access main memory. This is called **false sharing**. Suppose two threads with separate caches access different variables that belong to the same cache line. Further suppose at least one of the threads updates its variable. Then even though neither thread has written to a variable that the other thread is using, the cache controller invalidates the entire cache line and forces the threads to get the values of the variables from main memory. The threads aren't sharing anything (except a cache line), but the behavior of the threads with respect to memory access is the same as if they were sharing a variable, hence the name *false sharing*.

Why is false sharing not a problem with the other inputs? Let's look at what happens with the 8000 × 8000 input. Suppose thread 2 is assigned to one of the processors and thread 3 is assigned to another. (We don't actually know which threads are assigned to which processors, but it turns out—see Exercise 4.17—that it doesn't matter.) Thread 2 is responsible for computing

$$y[4000], \ y[4001], \ . \ . \ . \ , \ y[5999],$$

and thread 3 is responsible for computing

$$y[6000], \ y[6001], \ . \ . \ . \ , \ y[7999].$$

If a cache line contains 8 consecutive **doubles**, the only possibility for false sharing is on the interface between their assigned elements. If, for example, a single cache line contains

$$y[5996], \ y[5997], \ y[5998], \ y[5999],$$
$$y[6000], \ y[6001], \ y[6002], \ y[6003],$$

then it's conceivable that there might be false sharing of this cache line. However, thread 2 will access

$$y[5996], \ y[5997], \ y[5998], \ y[5999]$$

at the *end* of its **for** i loop, while thread 3 will access

$$y[6000], \ y[6001], \ y[6002], \ y[6003]$$

at the *beginning* of its **for** i loop. So it's very likely that when thread 2 accesses (say) $y[5996]$, thread 3 will be long done with all four of

$$y[6000], \ y[6001], \ y[6002], \ y[6003].$$

Similarly, when thread 3 accesses, say, $y[6003]$, it's very likely that thread 2 won't be anywhere near starting to access

$$y[5996], \ y[5997], \ y[5998], \ y[5999].$$

It's therefore unlikely that false sharing of the elements of y will be a significant problem with the 8000 × 8000 input. Similar reasoning suggests that false sharing of y is unlikely to be a problem with the 8,000,000 × 8 input. Also note that we don't need to worry about false sharing of A or x, since their values are never updated by the matrix-vector multiplication code.

This brings up the question of how we might avoid false sharing in our matrix-vector multiplication program. One possible solution is to "pad" the y vector with dummy elements to ensure that any update by one thread won't affect another thread's cache line. Another alternative is to have each thread use its own private storage during the multiplication loop, and then update the shared storage when they're done. See Exercise 4.19.

4.11 Thread-safety[9]

Let's look at another potential problem that occurs in shared-memory programming: *thread-safety*. A block of code is **thread-safe** if it can be simultaneously executed by multiple threads without causing problems.

As an example, suppose we want to use multiple threads to "tokenize" a file. Let's suppose that the file consists of ordinary English text, and that the tokens are just contiguous sequences of characters separated from the rest of the text by white space—a space, a tab, or a newline. A simple approach to this problem is to divide the input file into lines of text and assign the lines to the threads in a round-robin fashion: the first line goes to thread 0, the second goes to thread 1, ..., the tth goes to thread t, the $t + 1$st goes to thread 0, and so on.

We can serialize access to the lines of input using semaphores. Then, after a thread has read a single line of input, it can tokenize the line. One way to do this is to use the strtok function in string.h, which has the following prototype:

```
char* strtok(
        char*           string        /* in/out */,
        const char*     separators     /* in      */);
```

Its usage is a little unusual: the first time it's called the string argument should be the text to be tokenized, so in our example it should be the line of input. For subsequent calls, the first argument should be NULL. The idea is that in the first call, strtok caches a pointer to string, and for subsequent calls it returns successive tokens taken from the cached copy. The characters that delimit tokens should be passed in separators. We should pass in the string " \t\n" as the separators argument.

Given these assumptions, we can write the thread function shown in Program 4.14. The main thread has initialized an array of t semaphores—one for each thread. Thread 0's semaphore is initialized to 1. All the other semaphores are initialized to 0. So the code in Lines 9 to 11 will force the threads to sequentially access the lines of input. Thread 0 will immediately read the first line, but all the other threads will block in sem_wait. When thread 0 executes the sem_post, thread 1 can read a line

---

[9] This material is also covered in Chapter 5. So if you've already read that chapter, you may want to skim this section.

```
1   void *Tokenize(void* rank) {
2      long my_rank = (long) rank;
3      int count;
4      int next = (my_rank + 1) % thread_count;
5      char *fg_rv;
6      char my_line[MAX];
7      char *my_string;
8
9      sem_wait(&sems[my_rank]);
10     fg_rv = fgets(my_line, MAX, stdin);
11     sem_post(&sems[next]);
12     while (fg_rv != NULL) {
13        printf("Thread %ld > my line = %s", my_rank, my_line);
14
15        count = 0;
16        my_string = strtok(my_line, " \t\n");
17        while ( my_string != NULL ) {
18           count++;
19           printf("Thread %ld > string %d = %s\n",
20                  my_rank, count, my_string);
21           my_string = strtok(NULL, " \t\n");
22        }
23
24        sem_wait(&sems[my_rank]);
25        fg_rv = fgets(my_line, MAX, stdin);
26        sem_post(&sems[next]);
27     }
28
29     return NULL;
30  }  /* Tokenize */
```

Program 4.14: A first attempt at a multithreaded tokenizer.

of input. After each thread has read its first line of input (or end-of-file), any additional input is read in lines 24 to 26. The fgets function reads a single line of input and lines 15 to 22 identify the tokens in the line. When we run the program with a single thread, it correctly tokenizes the input stream. The first time we run it with two threads and the input

> Pease porridge hot.
> Pease porridge cold.
> Pease porridge in the pot
> Nine days old.

the output is also correct. However, the second time we run it with this input, we get the following output:

```
Thread 0 > my line = Pease porridge hot.
Thread 0 > string 1 = Pease
Thread 0 > string 2 = porridge
Thread 0 > string 3 = hot.
Thread 1 > my line = Pease porridge cold.
Thread 0 > my line = Pease porridge in the pot
Thread 0 > string 1 = Pease
Thread 0 > string 2 = porridge
Thread 0 > string 3 = in
Thread 0 > string 4 = the
Thread 0 > string 5 = pot
Thread 1 > string 1 = Pease
Thread 1 > my line = Nine days old.
Thread 1 > string 1 = Nine
Thread 1 > string 2 = days
Thread 1 > string 3 = old.
```

What happened? Recall that strtok caches the input line. It does this by declaring a variable to have **static** storage class. This causes the value stored in this variable to persist from one call to the next. Unfortunately for us, this cached string is shared, not private. Thus thread 0's call to strtok with the third line of the input has apparently overwritten the contents of thread 1's call with the second line.

The strtok function is *not* thread-safe: if multiple threads call it simultaneously, the output it produces may not be correct. Regrettably, it's not uncommon for C library functions to fail to be thread-safe. For example, neither the random number generator rand in stdlib.h nor the time conversion function localtime in time.h is guaranteed to be thread-safe. In some cases, the C standard specifies an alternate, thread-safe, version of a function. In fact, there is a thread-safe version of strtok:

```
char* strtok_r (
        char*           string          /* in/out */,
        const char*     separators      /* in     */,
        char**          saveptr_p       /* in/out */);
```

The "_r" indicates the function is *reentrant*, which is sometimes used as a synonym for thread-safe.[10] The first two arguments have the same purpose as the arguments to strtok. The saveptr_p argument is used by strtok_r for keeping track of where the

---

[10] However, the distinction is a bit more nuanced; being reentrant means a function can be interrupted and called again (reentered) in different parts of a program's control flow and still execute correctly. This can happen due to nested calls to the function or a trap/interrupt sent from the operating system. Since strtok uses a single static pointer to track its state while parsing, multiple calls to the function from different parts of a program's control flow will corrupt the string—therefore it is *not* reentrant. It's worth noting that although reentrant functions, such as strtok_r, can also be thread safe, there is no guarantee a reentrant function will *always* be thread safe—and vice versa. It's best to consult the documentation if there's any doubt.

function is in the input string; it serves the purpose of the cached pointer in `strtok`. We can correct our original `Tokenize` function by replacing the calls to `strtok` with calls to `strtok_r`. We simply need to declare a **char**∗ variable to pass in for the third argument, and replace the calls in line 16 and line 21 with the calls

```
my_string = strtok_r(my_line, " \t\n", &saveptr);
   . . .
my_string = strtok_r(NULL, " \t\n", &saveptr);
```

respectively.

4.11.1 Incorrect programs can produce correct output

Notice that our original version of the tokenizer program shows an especially insidious form of program error: the first time we ran it with two threads, the program produced correct output. It wasn't until a later run that we saw an error. This, unfortunately, is not a rare occurrence in parallel programs. It's especially common in shared-memory programs. Since, for the most part, the threads are running independently of each other, as we noted earlier, the exact sequence of statements executed is nondeterministic. For example, we can't say when thread 1 will first call `strtok`. If its first call takes place after thread 0 has tokenized its first line, then the tokens identified for the first line should be correct. However, if thread 1 calls `strtok` before thread 0 has finished tokenizing its first line, it's entirely possible that thread 0 may not identify all the tokens in the first line. Therefore it's especially important in developing shared-memory programs to resist the temptation to assume that since a program produces correct output, it must be correct. We always need to be wary of race conditions.

4.12 Summary

Like MPI, Pthreads is a library of functions that programmers can use to implement parallel programs. Unlike MPI, Pthreads is used to implement shared-memory parallelism.

A **thread** in shared-memory programming is analogous to a process in distributed-memory programming. However, a thread is often lighter-weight than a full-fledged process.

We saw that in Pthreads programs, all the threads have access to global variables, while local variables usually are private to the thread running the function. To use Pthreads, we should include the `pthread.h` header file, and, when we compile our program, it may be necessary to link our program with the Pthread library by adding `−lpthread` to the command line. We saw that we can use the functions `pthread_create` and `pthread_join`, respectively, to start and stop a thread function.

When multiple threads are executing, the order in which the statements are executed by the different threads is usually nondeterministic. When nondeterminism results from multiple threads attempting to access a shared resource, such as a shared variable or a shared file, at least one of the accesses is an update, and the accesses can result in an error, we have a **race condition**. One of our most important tasks in writing shared-memory programs is identifying and correcting race conditions. A **critical section** is a block of code that updates a shared resource that can only be updated by one thread at a time, so the execution of code in a critical section should, effectively, be executed as serial code. Thus we should try to design our programs so that they use them as infrequently as possible, and the critical sections we do use should be as short as possible.

We looked at three basic approaches to avoiding conflicting access to critical sections: busy-waiting, mutexes, and semaphores. **Busy-waiting** can be done with a flag variable and a **while** loop with an empty body. It can be very wasteful of CPU cycles. It can also be unreliable if compiler optimization is turned on, so mutexes and semaphores are generally preferable.

A **mutex** can be thought of as a lock on a critical section, since mutexes arrange for *mutually exclusive* access to a critical section. In Pthreads, a thread attempts to obtain a mutex with a call to `pthread_mutex_lock`, and it relinquishes the mutex with a call to `pthread_mutex_unlock`. When a thread attempts to obtain a mutex that is already in use, it *blocks* in the call to `pthread_mutex_lock`. This means that it remains idle in the call to `pthread_mutex_lock` until the system gives it the lock.

A **semaphore** is an **unsigned int** together with two operations: `sem_wait` and `sem_post`. If the semaphore is positive, a call to `sem_wait` simply decrements the semaphore, but if the semaphore is zero, the calling thread blocks until the semaphore is positive, at which point the semaphore is decremented and the thread returns from the call. The `sem_post` operation increments the semaphore; a semaphore can be used as a mutex with `sem_wait` corresponding to `pthread_mutex_lock` and `sem_post` corresponding to `pthread_mutex_unlock`. However, semaphores are more powerful than mutexes, since they can be initialized to any nonnegative value. Furthermore, since there is no "ownership" of a semaphore, any thread can "unlock" a locked semaphore. We saw that semaphores can be easily used to implement **producer–consumer synchronization**. In producer–consumer synchronization, a "consumer" thread waits for some condition or data created by a "producer" thread before proceeding. Semaphores are not part of Pthreads. To use them, we need to include the `semaphore.h` header file.

A **barrier** is a point in a program at which the threads block until all of the threads have reached it. We saw several different means for constructing barriers. One of them used a **condition variable**. A condition variable is a special Pthreads object that can be used to suspend execution of a thread until a condition has occurred. When the condition has occurred, another thread can awaken the suspended thread with a condition signal or a condition broadcast.

The last Pthreads construct we looked at was a **read-write lock**. A read-write lock is used when it's safe for multiple threads to simultaneously *read* a data structure, but if a thread needs to modify or *write* to the data structure, then only that thread can access the data structure during the modification.

We recalled that modern microprocessor architectures use caches to reduce memory access times, so typical architectures have special hardware to ensure that the caches on the different chips are **coherent**. Since the unit of cache coherence, a **cache line** or **cache block**, is usually larger than a single word of memory, this can have the unfortunate side effect that two threads may be accessing different memory locations, but when the two locations belong to the same cache line, the cache-coherence hardware acts as if the threads were accessing the same memory location. Thus, if one of the threads updates its memory location, and then the other thread tries to read its memory location, it will have to retrieve the value from main memory. That is, the hardware is forcing the thread to act as if it were actually sharing the memory location. Hence, this is called **false sharing**, and it can seriously degrade the performance of a shared-memory program.

Some C functions cache data between calls by declaring variables to be **static**. This can cause errors when multiple threads call the function; since static storage is shared among the threads, one thread can overwrite another thread's data. Such a function is not **thread-safe**, and, unfortunately, there are several such functions in the C library. Sometimes, however, there is a thread-safe variant.

When we looked at the program that used the function that wasn't thread-safe, we saw a particularly insidious problem: when we ran the program with multiple threads and a fixed set of input, it sometimes produced correct output, even though the program was erroneous. This means that even if a program produces correct output during testing, there's no guarantee that it is in fact correct—it's up to us to identify possible race conditions.

4.13 Exercises

4.1 When we discussed matrix-vector multiplication, we assumed that both m and n, the number of rows and the number of columns, respectively, were evenly divisible by t, the number of threads. How do the formulas for the assignments change if this is *not* the case?

4.2 If we decide to physically divide a data structure among the threads, that is, if we decide to make various members local to individual threads, we need to consider at least three issues:

a. How are the members of the data structure used by the individual threads?

b. Where and how is the data structure initialized?

c. Where and how is the data structure used after its members are computed?

We briefly looked at the first issue in the matrix-vector multiplication function. We saw that the entire vector x was used by all of the threads, so it seemed pretty clear that it should be shared. However, for both the matrix A and the product vector y, just looking at (a) seemed to suggest that A and y should have their components distributed among the threads. Let's take a closer look at this.

What would we have to do to divide A and y among the threads? Dividing y wouldn't be difficult—each thread could allocate a block of memory that could be used for storing its assigned components. Presumably, we could do the same for A—each thread could allocate a block of memory for storing its assigned rows. Modify the matrix-vector multiplication program so that it distributes both of these data structures. Can you "schedule" the input and output so that the threads can read in A and print out y? How does distributing A and y affect the run-time of the matrix-vector multiplication? (Don't include input or output in your run-time.)

4.3 Recall that the compiler is unaware that an ordinary C program is multi-threaded, and as a consequence, it may make optimizations that can interfere with busy-waiting. (Note that compiler optimizations should *not* affect mu-texes, condition variables, or semaphores.) An alternative to completely turn-ing off compiler optimizations is to identify some shared variables with the C keyword **volatile**. This tells the compiler that these variables may be updated by multiple threads and, as a consequence, it shouldn't apply optimizations to statements involving them. As an example, recall our busy-wait solution to the race condition when multiple threads attempt to add a private variable into a shared variable:

```
/* x and flag are shared, y is private          */
/* x and flag are initialized to 0 by main thread */

y = Compute(my_rank);
while (flag != my_rank);
x = x + y;
flag++;
```

It's impossible to tell by looking at this code that the order of the **while** statement and the x = x + y statement is important; if this code were single-threaded, the order of these two statements wouldn't affect the outcome of the code. But if the compiler determined that it could improve register usage by interchanging the order of these two statements, the resulting code would be erroneous.

If, instead of defining

```
int flag;
int x;
```

we define

```
int volatile flag;
int volatile x;
```

then the compiler will know that both x and flag can be updated by other threads, so it shouldn't try reordering the statements.

With the `gcc` compiler, the default behavior is no optimization. You can make certain of this by adding the option −00 to the command line. Try running the π calculation program that uses busy-waiting (`pth_pi_busy.c`) without optimization. How does the result of the multithreaded calculation compare to the single-threaded calculation? Now try running it with optimization; if you're using `gcc`, replace the −00 option with −02. If you found an error, how many threads did you use?

Which variables should be made volatile in the π calculation? Change these variables so that they're volatile and rerun the program with and without optimization. How do the results compare to the single-threaded program?

4.4 The performance of the π calculation program that uses mutexes remains roughly constant once we increase the number of threads beyond the number of available CPUs. What does this suggest about how the threads are scheduled on the available processors?

4.5 Modify the mutex version of the π calculation program so that the critical section is in the **for** loop. How does the performance of this version compare to the performance of the original busy-wait version? How might we explain this?

4.6 Modify the mutex version of the π calculation program so that it uses a semaphore instead of a mutex. How does the performance of this version compare with the mutex version?

4.7 Modify the Pthreads hello, world program to launch an unlimited number of threads—you can effectively ignore the `thread_count` and instead call `pthread_create` in an infinite loop (e.g., for (thread = 0; ; thread++)). Note that in most cases the program will *not* create an unlimited number of threads; you'll observe that the "Hello from thread" messages stop after some time, depending on the configuration of your system. How many threads were created before the messages stopped?

Observe that while nothing is being printed, the program is still running. To determine why no new threads are being created, check the return value of the call to `pthread_create` (hint: use the `perror` function to get a human-readable description of the problem, or look up the error codes). What is the cause of this bug?

Finally, modify the **for** loop containing `pthread_create` to detach each new thread using `pthread_detach`. How many threads are created now?

4.8 Although producer–consumer synchronization is easy to implement with semaphores, it's also possible to implement it with mutexes. The basic idea is to have the producer and the consumer share a mutex. A flag variable that's initialized to `false` by the main thread indicates whether there's anything to consume. With two threads we'd execute something like this:

```
while (1) {
    pthread_mutex_lock(&mutex);
    if (my_rank == consumer) {
        if (message_available) {
```

```
            print message;
            pthread_mutex_unlock(&mutex);
            break;

        }
    } else { /* my_rank == producer */
        create message;
        message_available = 1;
        pthread_mutex_unlock(&mutex);
        break;

    }
    pthread_mutex_unlock(&mutex);

}
```

So if the consumer gets into the loop first, it will see there's no message available and return to the call to `pthread_mutex_lock`. It will continue this process until the producer creates the message. Write a Pthreads program that implements this version of producer–consumer synchronization with two threads. Can you generalize this so that it works with 2k threads—odd-ranked threads are consumers and even-ranked threads are producers? Can you generalize this so that each thread is both a producer and a consumer? For example, suppose that thread q "sends" a message to thread $(q + 1)$ mod t and "receives" a message from thread $(q - 1 + t)$ mod t? Does this use busy-waiting?

4.9 If a program uses more than one mutex, and the mutexes can be acquired in different orders, the program can **deadlock**. That is, threads may block forever waiting to acquire one of the mutexes. As an example, suppose that a program has two shared data structures—for example, two arrays or two linked lists—each of which has an associated mutex. Further suppose that each data structure can be accessed (read or modified) after acquiring the data structure's associated mutex.

  a. Suppose the program is run with two threads. Further suppose that the following sequence of events occurs:

Time	Thread 0	Thread 1
0	`pthread_mutex_lock(&mut0)`	`pthread_mutex_lock(&mut1)`
1	`pthread_mutex_lock(&mut1)`	`pthread_mutex_lock(&mut0)`

  What happens?
  b. Would this be a problem if the program used busy-waiting (with two flag variables) instead of mutexes?
  c. Would this be a problem if the program used semaphores instead of mutexes?

4.10 Some implementations of Pthreads define barriers. The function

```
int pthread_barrier_init(
    pthread_barrier_t*              barrier_p    /* out */,
    const pthread_barrierattr_t*    attr_p       /* in  */,
    unsigned                        count        /* in  */);
```

initializes a barrier object, `barrier_p`. As usual, we'll ignore the second argument and just pass in `NULL`. The last argument indicates the number of threads that must reach the barrier before they can continue. The barrier itself is a call to the function

```
int pthread_barrier_wait(
    pthread_barrier_t* barrier_p  /* in/out */);
```

As with most other Pthreads objects, there is a destroy function

```
int pthread_barrier_destroy(
    pthread_barrier_t* barrier_p  /* in/out */);
```

Modify one of the barrier programs from the book's website so that it uses a Pthreads barrier. Find a system with a Pthreads implementation that includes barrier and run your program with various numbers of threads. How does its performance compare to the other implementations?

4.11 Modify one of the programs you wrote in the Programming Assignments that follow so that it uses the scheme outlined in Section 4.8 to time itself. To get the time that has elapsed since some point in the past, you can use the macro `GET_TIME` defined in the header file `timer.h` on the book's website. Note that this will give *wall clock* time, not CPU time. Also note that since it's a macro, it can operate directly on its argument. For example, to implement

```
Store current time in my_start;
```

you would use

```
GET_TIME(my_start);
```

not

```
GET_TIME(&my_start);
```

How will you implement the barrier? How will you implement the following pseudocode?

```
elapsed = Maximum of my_elapsed values;
```

4.12 Give an example of a linked list and a sequence of memory accesses to the linked list in which the following pairs of operations can potentially result in problems:
   a. Two deletes executed simultaneously
   b. An insert and a delete executed simultaneously
   c. A member and a delete executed simultaneously
   d. Two inserts executed simultaneously
   e. An insert and a member executed simultaneously.

4.13 The linked list operations `Insert` and `Delete` consist of two distinct "phases." In the first phase, both operations search the list for either the position of the new node or the position of the node to be deleted. If the outcome of the first phase so indicates, in the second phase a new node is inserted or an existing node is deleted. In fact, it's quite common for linked list programs to split each of these operations into two function calls. For both operations, the first phase involves only read-access to the list; only the second phase modifies the list. Would it be safe to lock the list using a read-lock for the first phase? And then to lock the list using a write-lock for the second phase? Explain your answer.

4.14 Download the various threaded linked list programs from the website. In our examples, we ran a fixed percentage of searches and split the remaining percentage among inserts and deletes.

   a. Rerun the experiments with all searches and all inserts.

   b. Rerun the experiments with all searches and all deletes.

   Is there a difference in the overall run-times? Is insert or delete more expensive?

4.15 Recall that in C a function that takes a two-dimensional array argument must specify the number of columns in the argument list. Thus it is quite common for C programmers to only use one-dimensional arrays, and to write explicit code for converting pairs of subscripts into a single dimension. Modify the Pthreads matrix-vector multiplication so that it uses a one-dimensional array for the matrix and calls a matrix-vector multiplication function. How does this change affect the run-time?

4.16 Download the source file `pth_mat_vect_rand_split.c` from the book's website. Find a program that does cache profiling (for example, Valgrind [51]) and compile the program according to the instructions in the cache profiler documentation. (with Valgrind you will want a symbol table and full optimization; e.g., `gcc -g -O2 . . .`). Now run the program according to the instructions in the cache profiler documentation, using input $k \times (k \cdot 10^6)$, $(k \cdot 10^3) \times (k \cdot 10^3)$, and $(k \cdot 10^6) \times k$. Choose k so large that the number of level 2 cache misses is of the order 10^6 for at least one of the input sets of data.

   a. How many level 1 cache write-misses occur with each of the three inputs?

   b. How many level 2 cache write-misses occur with each of the three inputs?

   c. Where do most of the write-misses occur? For which input data does the program have the most write-misses? Can you explain why?

   d. How many level 1 cache read-misses occur with each of the three inputs?

   e. How many level 2 cache read-misses occur with each of the three inputs?

   f. Where do most of the read-misses occur? For which input data does the program have the most read-misses? Can you explain why?

   g. Run the program with each of the three inputs, but without using the cache profiler. With which input is the program the fastest? With which input is the program the slowest? Can your observations about cache misses help explain the differences? How?

4.17 Recall the matrix-vector multiplication example with the 8000×8000 input. Suppose that the program is run with four threads, and thread 0 and thread 2 are assigned to different processors. If a cache line contains 64 bytes or eight **doubles**, is it possible for false sharing between threads 0 and 2 to occur for any part of the vector y? Why? What about if thread 0 and thread 3 are assigned to different processors—is it possible for false sharing to occur between them for any part of y?

4.18 Recall the matrix-vector multiplication example with an $8 \times 8{,}000{,}000$ matrix. Suppose that **doubles** use 8 bytes of memory and that a cache line is 64 bytes. Also suppose that our system consists of two dual-core processors.

    a. What is the minimum number of cache lines that are needed to store the vector y?

    b. What is the maximum number of cache lines that are needed to store the vector y?

    c. If the boundaries of cache lines always coincide with the boundaries of 8-byte **doubles**, in how many different ways can the components of y be assigned to cache lines?

    d. If we only consider which pairs of threads share a processor, in how many different ways can four threads be assigned to the processors in our computer? Here we're assuming that cores on the same processor share cache.

    e. Is there an assignment of components to cache lines and threads to processors that will result in no false sharing in our example? In other words, is it possible that the threads assigned to one processor will have their components of y in one cache line, and the threads assigned to the other processor will have their components in a different cache line?

    f. How many assignments of components to cache lines and threads to processors are there?

    g. Of these assignments, how many will result in no false sharing?

4.19 a. Modify the matrix-vector multiplication program so that it pads the vector y when there's a possibility of false sharing. The padding should be done so that if the threads execute in lock-step, there's no possibility that a single cache line containing an element of y will be shared by two or more threads. Suppose, for example, that a cache line stores eight `doubles` and we run the program with four threads. If we allocate storage for at least 48 **doubles** in y, then, on each pass through the **for** i loop, there's no possibility that two threads will simultaneously access the same cache line.

    b. Modify the matrix-vector multiplication so that each thread uses private storage for its part of y during the **for** i loop. When a thread is done computing its part of y, it should copy its private storage into the shared variable.

    c. How does the performance of these two alternatives compare to the original program? How do they compare to each other?

4.20 Although `strtok_r` is thread-safe, it has the rather unfortunate property that it gratuitously modifies the input string. Write a tokenizer that is thread-safe and doesn't modify the input string.

4.14 Programming assignments

4.1 Write a Pthreads program that implements the histogram program in Chapter 2.

4.2 Suppose we toss darts randomly at a square dartboard, whose bullseye is at the origin, and whose sides are two feet in length. Suppose also that there's a circle inscribed in the square dartboard. The radius of the circle is 1 foot, and its area is π square feet. If the points that are hit by the darts are uniformly distributed (and we always hit the square), then the number of darts that hit inside the circle should approximately satisfy the equation

$$\frac{\text{number in circle}}{\text{total number of tosses}} = \frac{\pi}{4},$$

since the ratio of the area of the circle to the area of the square is $\pi/4$.
We can use this formula to estimate the value of π with a random number generator:

```
number_in_circle = 0;
for (toss = 0; toss < number_of_tosses; toss++) {
    x = random double between −1 and 1;
    y = random double between −1 and 1;
    distance_squared = x*x + y*y;
    if (distance_squared <= 1)
        number_in_circle++;
}
pi_estimate = 4*number_in_circle
        / ((double) number_of_tosses);
```

This is called a "Monte Carlo" method, since it uses randomness (the dart tosses).
Write a Pthreads program that uses a Monte Carlo method to estimate π. The main thread should read in the total number of tosses and print the estimate. You may want to use **long long ints** for the number of hits in the circle and the number of tosses, since both may have to be very large to get a reasonable estimate of π.

4.3 Write a Pthreads program that implements the trapezoidal rule. Use a shared variable for the sum of all the threads' computations. Use busy-waiting, mutexes, and semaphores to enforce mutual exclusion in the critical section. What advantages and disadvantages do you see with each approach?

4.4 Write a Pthreads program that finds the average time required by your system to create and terminate a thread. Does the number of threads affect the average time? If so, how?

4.5 Write a Pthreads program that implements a "task queue." The main thread begins by starting a user-specified number of threads that immediately go to sleep in a condition wait. The main thread generates blocks of tasks to be carried out by the other threads; each time it generates a new block of tasks, it awakens a thread with a condition signal. When a thread finishes executing its block of tasks, it should return to a condition wait. When the main thread completes generating tasks, it sets a global variable indicating that there will be no more tasks, and awakens all the threads with a condition broadcast. For the sake of explicitness, make your tasks linked list operations.

4.6 Write a Pthreads program that uses two condition variables and a mutex to implement a read-write lock. Download the online linked list program that uses Pthreads read-write locks, and modify it to use your read-write locks. Now compare the performance of the program when readers are given preference with the program when writers are given preference. Can you make any generalizations?

Shared-memory programming with OpenMP

5

Like Pthreads, OpenMP is an API for shared-memory MIMD programming. The "MP" in OpenMP stands for "multiprocessing," a term that is synonymous with shared-memory MIMD computing. Thus OpenMP is designed for systems in which each thread or process can potentially have access to all available memory, and when we're programming with OpenMP, we view our system as a collection of autonomous cores or CPUs, all of which have access to main memory, as in Fig. 5.1.

Although OpenMP and Pthreads are both APIs for shared-memory programming, they have many fundamental differences. Pthreads requires that the programmer explicitly specify the behavior of each thread. OpenMP, on the other hand, sometimes allows the programmer to simply state that a block of code should be executed in parallel, and the precise determination of the tasks and which thread should execute them is left to the compiler and the run-time system. This suggests a further difference between OpenMP and Pthreads; Pthreads (like MPI) is a library of functions that can be linked to a C program, so any Pthreads program can be used with any C compiler, provided the system has a Pthreads library. OpenMP, on the other hand, requires compiler support for some operations, and hence it's entirely possible that

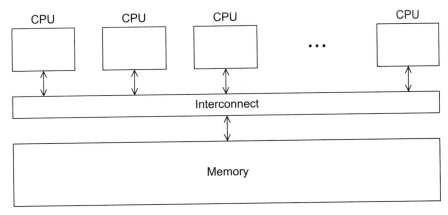

FIGURE 5.1

A shared-memory system.

An Introduction to Parallel Programming. https://doi.org/10.1016/B978-0-12-804605-0.00012-9

you may run across a C compiler that can't compile OpenMP programs into parallel programs.

These differences also suggest why there are two standard APIs for shared-memory programming: Pthreads is lower level and provides us with the power to program virtually any conceivable thread behavior. This power, however, comes with some associated cost—it's up to us to specify every detail of the behavior of each thread. OpenMP, on the other hand, allows the compiler and run-time system to determine some of the details of thread behavior, so it can be simpler to code some parallel behaviors using OpenMP. The cost is that some low-level thread interactions can be more difficult to program.

OpenMP was developed by a group of programmers and computer scientists who believed that writing large-scale high-performance programs using APIs, such as Pthreads, was too difficult, and they defined the OpenMP specification so that shared-memory programs could be developed at a higher level. In fact, OpenMP was explicitly designed to allow programmers to *incrementally* parallelize existing serial programs; this is virtually impossible with MPI and fairly difficult with Pthreads.

In this chapter, we'll learn the basics of OpenMP. We'll learn how to write a program that can use OpenMP, and we'll learn how to compile and run OpenMP programs. Next, we'll learn how to exploit one of the most powerful features of OpenMP: its ability to parallelize serial **for** loops with only small changes to the source code. We'll then look at some other features of OpenMP: task-parallelism and explicit thread synchronization. We'll also look at some standard problems in shared-memory programming: the effect of cache memories on shared-memory programming and problems that can be encountered when serial code—especially a serial library—is used in a shared-memory program.

5.1 Getting started

OpenMP provides what's known as a "directives-based" shared-memory API. In C and C++, this means that there are special preprocessor instructions known as `pragmas`. Pragmas are typically added to a system to allow behaviors that aren't part of the basic C specification. Compilers that don't support the `pragmas` are free to ignore them. This allows a program that uses the `pragmas` to run on platforms that don't support them. So, in principle, if you have a carefully written OpenMP program, it can be compiled and run on any system with a C compiler, regardless of whether the compiler supports OpenMP. If OpenMP is not supported, then the directives are simply ignored and the code will execute sequentially.

Pragmas in C and C++ start with

```
#pragma
```

As usual, we put the pound sign, #, in column 1, and like other preprocessor directives, we shift the remainder of the directive so that it is aligned with the rest of the code. `Pragmas` (like all preprocessor directives) are, by default, one line in length, so if a pragma won't fit on a single line, the newline needs to be "escaped"—that is, preceded by a backslash \. The details of what follows the **#pragma** depend entirely on which extensions are being used.

Let's take a look at a *very* simple example, a "hello, world" program that uses OpenMP. (See Program 5.1.)

```
1   #include <stdio.h>
2   #include <stdlib.h>
3   #include <omp.h>
4
5   void Hello(void);  /* Thread function */
6
7   int main(int argc, char* argv[]) {
8      /* Get number of threads from command line */
9      int thread_count = strtol(argv[1], NULL, 10);
10
11  #   pragma omp parallel num_threads(thread_count)
12     Hello();
13
14     return 0;
15  }  /* main */
16
17  void Hello(void) {
18     int my_rank = omp_get_thread_num();
19     int thread_count = omp_get_num_threads();
20
21     printf("Hello from thread %d of %d\n",
22           my_rank, thread_count);
23
24  }  /* Hello */
```

Program 5.1: A "hello, world" program that uses OpenMP.

5.1.1 Compiling and running OpenMP programs

To compile this with gcc we need to include the −fopenmp option[1]:

```
$ gcc −g −Wall −fopenmp −o omp_hello omp_hello.c
```

To run the program, we specify the number of threads on the command line. For example, we might run the program with four threads and type

---

[1] Some older versions of gcc may not include OpenMP support. Other compilers will, in general, use different command-line options to specify that the source is an OpenMP program. For details on our assumptions about compiler use, see Section 2.9.

```
$ ./omp_hello 4
```

If we do this, the output might be

```
Hello from thread 0 of 4
Hello from thread 1 of 4
Hello from thread 2 of 4
Hello from thread 3 of 4
```

However, it should be noted that the threads are competing for access to stdout, so there's no guarantee that the output will appear in thread-rank order. For example, the output might also be

```
Hello from thread 1 of 4
Hello from thread 2 of 4
Hello from thread 0 of 4
Hello from thread 3 of 4
```

or

```
Hello from thread 3 of 4
Hello from thread 1 of 4
Hello from thread 2 of 4
Hello from thread 0 of 4
```

or any other permutation of the thread ranks.

If we want to run the program with just one thread, we can type

```
$ ./omp_hello 1
```

and we would get the output

```
Hello from thread 0 of 1
```

5.1.2 The program

Let's take a look at the source code. In addition to a collection of directives, OpenMP consists of a library of functions and macros, so we usually need to include a header file with prototypes and macro definitions. The OpenMP header file is omp.h, and we include it in Line 3.

In our Pthreads programs, we specified the number of threads on the command line. We'll also usually do this with our OpenMP programs. In Line 9 we therefore use the strtol function from stdlib.h to get the number of threads. Recall that the syntax of this function is

```
long strtol(
        const char*    number_p        /* in  */,
        char**         end_p           /* out */,
        int            base            /* in  */);
```

The first argument is a string—in our example, it's the command-line argument, a string—and the last argument is the numeric base in which the string is represented—in our example, it's base 10. We won't make use of the second argument, so we'll just pass in a NULL pointer. The return value is the command-line argument converted to a C **long int**.

If you've done a little C programming, there's nothing really new up to this point. When we start the program from the command line, the operating system starts a single-threaded process, and the process executes the code in the main function. However, things get interesting in Line 11. This is our first OpenMP directive, and we're using it to specify that the program should start some threads. Each thread should execute the Hello function, and when the threads return from the call to Hello, they should be terminated, and the process should then terminate when it executes the **return** statement.

That's a lot of bang for the buck (or code). If you studied the Pthreads chapter, you'll recall that we had to write a lot of code to achieve something similar: we needed to allocate storage for a special struct for each thread, we used a **for** loop to start all the threads, and we used another **for** loop to terminate the threads. Thus it's immediately evident that OpenMP provides a higher-level abstraction than Pthreads provides.

We've already seen that pragmas in C and C++ start with

```
#   pragma
```

OpenMP pragmas always begin with

```
#   pragma omp
```

Our first directive is a parallel directive, and, as you might have guessed, it specifies that the **structured block** of code that follows should be executed by multiple threads. A structured block is a C statement or a compound C statement with one point of entry and one point of exit, although calls to the function exit are allowed. This definition simply prohibits code that branches into or out of the middle of the structured block.

Recall that **thread** is short for *thread of execution*. The name is meant to suggest a sequence of statements executed by a program. Threads are typically started or **forked** by a process, and they share most of the resources of the process that starts them—for example, access to stdin and stdout—but each thread has its own stack and program counter. When a thread completes execution, it **joins** the process that started it. This terminology comes from diagrams that show threads as directed lines. (See Fig. 5.2.) For more details see Chapters 2 and 4.

At its most basic the parallel directive is simply

```
#   pragma omp parallel
```

and the number of threads that run the following structured block of code will be determined by the run-time system. The algorithm used is fairly complicated; see the OpenMP Standard [47] for details. However, if there are no other threads started, the system will typically run one thread on each available core.

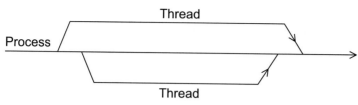

FIGURE 5.2

A process forking and joining two threads.

As we noted earlier, we'll usually specify the number of threads on the command line, so we'll modify our `parallel` directives with the `num_threads` *clause*. A **clause** in OpenMP is just some text that modifies a directive. The `num_threads` clause can be added to a `parallel` directive. It allows the programmer to specify the number of threads that should execute the following block:

```
#   pragma omp parallel num_threads(thread_count)
```

It should be noted that there may be system-defined limitations on the number of threads that a program can start. The OpenMP Standard doesn't guarantee that this will actually start `thread_count` threads. However, most current systems can start hundreds or even thousands of threads, so unless we're trying to start *a lot* of threads, we will almost always get the desired number of threads.

What actually happens when the program gets to the `parallel` directive? Prior to the `parallel` directive, the program is using a single thread, the process started when the program started execution. When the program reaches the `parallel` directive, the original thread continues executing and `thread_count` − 1 additional threads are started. In OpenMP parlance, the collection of threads executing the `parallel` block—the original thread and the new threads—is called a **team**. OpenMP thread terminology includes the following:

- **master**: the first thread of execution, or *thread 0*.
- **parent**: thread that encountered a `parallel` directive and started a team of threads. In many cases, the parent is also the master thread.
- **child**: each thread started by the parent is considered a *child* thread.

Each thread in the team executes the block following the directive, so in our example, each thread calls the `Hello` function.

When the block of code is completed—in our example, when the threads return from the call to `Hello`—there's an **implicit barrier**. This means that a thread that has completed the block of code will wait for all the other threads in the team to complete the block—in our example, a thread that has completed the call to `Hello` will wait for all the other threads in the team to return. When all the threads have completed the block, the child threads will terminate and the parent thread will continue executing the code that follows the block. In our example, the parent thread will execute the **return** statement in Line 14, and the program will terminate.

Since each thread has its own stack, a thread executing the `Hello` function will create its own private, local variables in the function. In our example, when the function is called, each thread will get its rank or ID and the number of threads in the team by calling the OpenMP functions `omp_get_thread_num` and `omp_get_num_threads`, respectively. The rank or ID of a thread is an **int** that is in the range $0, 1, \ldots,$ `thread_count` $- 1$. The syntax for these functions is

```
int omp_get_thread_num(void);
int omp_get_num_threads(void);
```

Since `stdout` is shared among the threads, each thread can execute the `printf` statement, printing its rank and the number of threads.

As we noted earlier, there is no scheduling of access to `stdout`, so the actual order in which the threads print their results is nondeterministic.

5.1.3 Error checking

To make the code more compact and more readable, our program doesn't do any error checking. Of course, this is dangerous, and, in practice, it's a *very* good idea—one might even say mandatory—to try to anticipate errors and check for them. In this example, we should definitely check for the presence of a command-line argument, and, if there is one, after the call to `strtol`, we should check that the value is positive. We might also check that the number of threads actually created by the `parallel` directive is the same as `thread_count`, but in this simple example, this isn't crucial.

A second source of potential problems is the compiler. If the compiler doesn't support OpenMP, it will just ignore the `parallel` directive. However, the attempt to include `omp.h` and the calls to `omp_get_thread_num` and `omp_get_num_threads` *will* cause errors. To handle these problems, we can check whether the preprocessor macro `_OPENMP` is defined. If this is defined, we can include `omp.h` and make the calls to the OpenMP functions. We might make the modifications that follow to our program.

Instead of simply including `omp.h`:

```
#include <omp.h>
```

we can check for the definition of `_OPENMP` before trying to include it:

```
#ifdef _OPENMP
#   include <omp.h>
#endif
```

Also, instead of just calling the OpenMP functions, we can first check whether `_OPENMP` is defined:

```
#   ifdef _OPENMP
        int my_rank = omp_get_thread_num();
        int thread_count = omp_get_num_threads();
#   else
        int my_rank = 0;
        int thread_count = 1;
#   endif
```

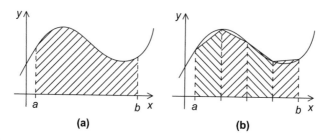

FIGURE 5.3

The trapezoidal rule.

Here, if OpenMP isn't available, we assume that the `Hello` function will be single-threaded. Thus the single thread's rank will be 0, and the number of threads will be 1.

The book's website contains the source for a version of this program that makes these checks. To make our code as clear as possible, we'll usually show little, if any, error checking in the code displayed in the text. We'll also assume that OpenMP is available and supported by the compiler.

5.2 The trapezoidal rule

Let's take a look at a somewhat more useful (and more complicated) example: the trapezoidal rule for estimating the area under a curve. Recall from Section 3.2 that if $y = f(x)$ is a reasonably nice function, and $a < b$ are real numbers, then we can estimate the area between the graph of $f(x)$, the vertical lines $x = a$ and $x = b$, and the x-axis by dividing the interval $[a, b]$ into n subintervals and approximating the area over each subinterval by the area of a trapezoid. See Fig. 5.3.

Also recall that if each subinterval has the same length and if we define $h = (b - a)/n$, $x_i = a + ih$, $i = 0, 1, \ldots, n$, then our approximation will be

$$h[\, f(x_0)/2 + f(x_1) + f(x_2) + \cdots + f(x_{n-1}) + f(x_n)/2\,].$$

Thus we can implement a serial algorithm using the following code:

```
/* Input:   a, b, n */
h = (b-a)/n;
approx = (f(a) + f(b))/2.0;
for (i = 1; i <= n-1; i++) {
    x_i = a + i*h;
    approx += f(x_i);
}
approx = h*approx;
```

See Section 3.2.1 for details.

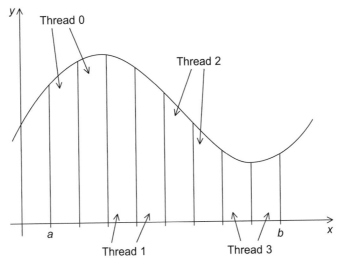

FIGURE 5.4

Assignment of trapezoids to threads.

5.2.1 A first OpenMP version

Recall that we applied Foster's parallel program design methodology to the trapezoidal rule as described in the following list (see Section 3.2.2):

1. We identified two types of jobs:
   a. Computation of the areas of individual trapezoids, and
   b. Adding the areas of trapezoids.
2. There is no communication among the jobs in the first collection, but each job in the first collection communicates with job 1b.
3. We assumed that there would be many more trapezoids than cores, so we aggregated jobs by assigning a contiguous block of trapezoids to each thread (and a single thread to each core).[2] Effectively, this partitioned the interval $[a, b]$ into larger subintervals, and each thread simply applied the serial trapezoidal rule to its subinterval. See Fig. 5.4.

We aren't quite done, however, since we still need to add up the threads' results. An obvious solution is to use a shared variable for the sum of all the threads' results, and each thread can add its (private) result into the shared variable. We would like to have each thread execute a statement that looks something like

```
global_result += my_result;
```

---

[2] Since we were discussing MPI, we actually used *processes* instead of threads.

However, as we've already seen, this can result in an erroneous value for `global_result`—if two (or more) threads attempt to simultaneously execute this statement, the result will be unpredictable. For example, suppose that `global_result` has been initialized to 0, thread 0 has computed `my_result = 1`, and thread 1 has computed `my_result = 2`. Furthermore, suppose that the threads execute the statement `global_result += my_result` according to the following timetable:

Time	Thread 0	Thread 1
0	`global_result = 0` to register	finish `my_result`
1	`my_result = 1` to register	`global_result = 0` to register
2	add `my_result` to `global_result`	`my_result = 2` to register
3	store `global_result = 1`	add `my_result` to `global_result`
4		store `global_result = 2`

We see that the value computed by thread 0 (`my_result = 1`) is overwritten by thread 1.

Of course, the actual sequence of events might be different, but unless one thread finishes the computation `global_result += my_result` before the other starts, the result will be incorrect. Recall that this is an example of a **race condition**: multiple threads are attempting to access a shared resource, at least one of the accesses is an update, and the accesses can result in an error. Also recall that the code that causes the race condition, `global_result += my_result`, is called a **critical section**. A critical section is code executed by multiple threads that updates a shared resource, and the shared resource can only be updated by one thread at a time.

We therefore need some mechanism to make sure that once one thread has started executing `global_result += my_result`, no other thread can start executing this code until the first thread has finished. In Pthreads we used mutexes or semaphores. In OpenMP we can use the `critical` directive

```
#   pragma omp critical
      global_result += my_result;
```

This directive tells the compiler that the system needs to arrange for the threads to have **mutually exclusive** access to the following structured block of code.[3] That is, only one thread can execute the following structured block at a time. The code for this version is shown in Program 5.2. We've omitted any error checking. We've also omitted code for the function $f(x)$.

In the `main` function, prior to Line 17, the code is single-threaded, and it simply gets the number of threads and the input (a, b, and n). In Line 17 the `parallel` directive specifies that the `Trap` function should be executed by `thread_count` threads. After

---

[3] You are likely used to seeing blocks preceded by a control flow statement (for example, **if**, **for**, **while**, and so on). As you'll soon see, this needn't always be the case; if we wanted to define a critical section that spanned the next two lines of code, we would simply enclose it in curly braces.

```
1   #include <stdio.h>
2   #include <stdlib.h>
3   #include <omp.h>
4
5   void Trap(double a, double b, int n, double* global_result_p);
6
7   int main(int argc, char* argv[]) {
8      /* We'll store our result in global_result: */
9      double   global_result = 0.0;
10     double   a, b;  /* Left and right endpoints   */
11     int      n;     /* Total number of trapezoids */
12     int      thread_count;
13
14     thread_count = strtol(argv[1], NULL, 10);
15     printf("Enter a, b, and n\n");
16     scanf("%lf %lf %d", &a, &b, &n);
17  #  pragma omp parallel num_threads(thread_count)
18     Trap(a, b, n, &global_result);
19
20     printf("With n = %d trapezoids, our estimate\n", n);
21     printf("of the integral from %f to %f = %.14e\n",
22         a, b, global_result);
23     return 0;
24  }  /* main */
25
26  void Trap(double a, double b, int n, double* global_result_p) {
27     double h, x, my_result;
28     double local_a, local_b;
29     int  i, local_n;
30     int my_rank = omp_get_thread_num();
31     int thread_count = omp_get_num_threads();
32
33     h = (b-a)/n;
34     local_n = n/thread_count;
35     local_a = a + my_rank*local_n*h;
36     local_b = local_a + local_n*h;
37     my_result = (f(local_a) + f(local_b))/2.0;
38     for (i = 1; i <= local_n-1; i++) {
39       x = local_a + i*h;
40       my_result += f(x);
41     }
42     my_result = my_result*h;
43
44  #  pragma omp critical
45     *global_result_p += my_result;
46  }  /* Trap */
```

Program 5.2: First OpenMP trapezoidal rule program.

returning from the call to `Trap`, any new threads that were started by the `parallel` directive are terminated, and the program resumes execution with only one thread. The one thread prints the result and terminates.

In the `Trap` function, each thread gets its rank and the total number of threads in the team started by the `parallel` directive. Then each thread determines the following:

1. The length of the bases of the trapezoids (Line 33),
2. The number of trapezoids assigned to each thread (Line 34),
3. The left and right endpoints of its interval (Lines 35 and 36, respectively)
4. Its contribution to `global_result` (Lines 37–42).

The threads finish by adding in their individual results to `global_result` in Lines 44–45.

We use the prefix `local_` for some variables to emphasize that their values may differ from the values of corresponding variables in the `main` function—for example, `local_a` may differ from `a`, although it is the *thread's* left endpoint.

Notice that unless n is evenly divisible by `thread_count`, we'll use fewer than n trapezoids for `global_result`. For example, if $n = 14$ and `thread_count = 4`, each thread will compute

```
local_n = n/thread_count = 14/4 = 3.
```

Thus each thread will only use 3 trapezoids, and `global_result` will be computed with $4 \times 3 = 12$ trapezoids instead of the requested 14. So in the error checking (which isn't shown), we check that n is evenly divisible by `thread_count` by doing something like this:

```
if (n % thread_count != 0) {
    fprintf(stderr,
        "n must be evenly divisible by thread_count\n");
    exit(0);
}
```

Since each thread is assigned a block of `local_n` trapezoids, the length of each thread's interval will be `local_n*h`, so the left endpoints will be

```
thread 0:   a + 0*local_n*h
thread 1:   a + 1*local_n*h
thread 2:   a + 2*local_h*h
      . . .
```

So in Line 35, we assign

```
local_a = a + my_rank*local_n*h;
```

Furthermore, since the length of each thread's interval will be `local_n*h`, its right endpoint will just be

```
local_b = local_a + local_n*h;
```

5.3 Scope of variables

In serial programming, the *scope* of a variable consists of those parts of a program in which the variable can be used. For example, a variable declared at the beginning of a C function has "function-wide" scope, that is, it can only be accessed in the body of the function. On the other hand, a variable declared at the beginning of a `.c` file but outside any function has "file-wide" scope, that is, any function in the file in which the variable is declared can access the variable. In OpenMP, the **scope** of a variable refers to the set of threads that can access the variable in a `parallel` block. A variable that can be accessed by all the threads in the team has **shared** scope, while a variable that can only be accessed by a single thread has **private** scope.

In the "hello, world" program, the variables used by each thread (`my_rank` and `thread_count`) were declared in the `Hello` function, which is called inside the `parallel` block. Consequently, the variables used by each thread are allocated from the thread's (private) stack, and hence all of the variables have private scope. This is *almost* the case in the trapezoidal rule program; since the `parallel` block is just a function call, all of the variables used by each thread in the `Trap` function are allocated from the thread's stack.

However, the variables that are declared in the `main` function (`a`, `b`, `n`, `global_result`, and `thread_count`) are all accessible to all the threads in the team started by the `parallel` directive. Hence, the *default* scope for variables declared before a `parallel` block is shared. In fact, we've made implicit use of this: each thread in the team gets the values of `a`, `b`, and `n` from the call to `Trap`. Since this call takes place in the `parallel` block, it's essential that each thread has access to `a`, `b`, and `n` when their values are copied into the corresponding formal arguments.

Furthermore, in the `Trap` function, although `global_result_p` is a private variable, it refers to the variable `global_result` which was declared in `main` before the `parallel` directive, and the value of `global_result` is used to store the result that's printed out after the `parallel` block. Thus in the code

```
*global_result_p += my_result;
```

it's essential that `*global_result_p` have shared scope. If it were private to each thread, there would be no need for the `critical` directive. Furthermore, if it were private, we would have a hard time determining the value of `global_result` in `main` after completion of the `parallel` block.

To summarize, then, variables that have been declared before a `parallel` directive have shared scope among the threads in the team, while variables declared in the block (e.g., local variables in functions) have private scope. Furthermore, the value of a shared variable at the beginning of the `parallel` block is the same as the value before the block, and, after completion of the `parallel` block, the value of the variable is the value at the end of the block.

We'll shortly see that the *default* scope of a variable can change with other directives, and that OpenMP provides clauses to modify the default scope.

5.4 The reduction clause

If we developed a serial implementation of the trapezoidal rule, we'd probably use a slightly different function prototype. Rather than

```
void Trap(
    double a,
    double b,
    int n,
    double* global_result_p);
```

we would probably define

```
double Trap(double a, double b, int n);
```

and our function call would be

```
global_result = Trap(a, b, n);
```

This is somewhat easier to understand and probably more attractive to all but the most fanatical believers in pointers.

We resorted to the pointer version, because we needed to add each thread's local calculation to get global_result. However, we might prefer the following function prototype:

```
double Local_trap(double a, double b, int n);
```

With this prototype, the body of Local_trap would be the same as the Trap function in Program 5.2, except that there would be no critical section. Rather, each thread would return its part of the calculation, the final value of its my_result variable. If we made this change, we might try modifying our parallel block so that it looks like this:

```
    global_result = 0.0;
#   pragma omp parallel num_threads(thread_count)
    {
#       pragma omp critical
        global_result += Local_trap(double a, double b, int n);
    }
```

Can you see a problem with this code? It should give the correct result. However, since we've specified that the critical section is

```
global_result += Local_trap(double a, double b, int n);
```

the call to Local_trap can only be executed by one thread at a time, and, effectively, we're forcing the threads to execute the trapezoidal rule sequentially. If we check the run-time of this version, it may actually be *slower* with multiple threads than one thread (see Exercise 5.3).

We can avoid this problem by declaring a private variable inside the parallel block and moving the critical section after the function call:

```
      global_result = 0.0;
#   pragma omp parallel num_threads(thread_count)
    {
        double my_result = 0.0;   /* private */
        my_result += Local_trap(double a, double b, int n);
#       pragma omp critical
        global_result += my_result;
    }
```

Now the call to Local_trap is outside the critical section, and the threads can execute their calls simultaneously. Furthermore, since my_result is declared in the parallel block, it's private, and before the critical section each thread will store its part of the calculation in its my_result variable.

OpenMP provides a cleaner alternative that also avoids serializing execution of Local_trap: we can specify that global_result is a *reduction* variable. A **reduction operator** is an associative binary operation (such as addition or multiplication), and a **reduction** is a computation that repeatedly applies the same reduction operator to a sequence of operands to get a single result. Furthermore, all of the intermediate results of the operation should be stored in the same variable: the **reduction variable**. For example, if A is an array of n **int**s, the computation

```
    int sum = 0;
    for (i = 0; i < n; i++)
        sum += A[i];
```

is a reduction in which the reduction operator is addition.

In OpenMP it may be possible to specify that the result of a reduction is a reduction variable. To do this, a reduction clause can be added to a parallel directive. In our example, we can modify the code as follows:

```
      global_result = 0.0;
#   pragma omp parallel num_threads(thread_count) \
        reduction(+: global_result)
      global_result += Local_trap(double a, double b, int n);
```

First note that the parallel directive is two lines long. Recall that C preprocessor directives are, by default, only one line long, so we need to "escape" the newline character by putting a backslash (\) immediately before it.

The code specifies that global_result is a reduction variable, and the plus sign ("+") indicates that the reduction operator is addition. Effectively, OpenMP creates a private variable for each thread, and the run-time system stores each thread's result in this private variable. OpenMP also creates a critical section, and the values stored in the private variables are added in this critical section. Thus the calls to Local_trap can take place in parallel.

The syntax of the reduction clause is

```
    reduction(<operator>: <variable list>)
```

In C, operator can be any one of the operators +, *, −, &, |, ^, &&, ||. You may wonder whether the use of subtraction is problematic, though, since subtraction isn't associative or commutative. For example, the serial code

```
result = 0;
for (i = 1; i <= 4; i++)
    result -= i;
```

stores the value -10 in result. However, if we split the iterations among two threads, with thread 0 subtracting 1 and 2 and thread 1 subtracting 3 and 4, then thread 0 will compute -3 and thread 1 will compute -7. This results in an incorrect calculation, $-3 - (-7) = 4$. Luckily, the OpenMP standard states that partial results of a subtraction reduction are *added* to form the final value, so the reduction will work as intended.

It should also be noted that if a reduction variable is a **float** or a **double**, the results may differ slightly when different numbers of threads are used. This is due to the fact that floating point arithmetic isn't associative. For example, if a, b, and c are **float**s, then $(a + b) + c$ may not be exactly equal to $a + (b + c)$. See Exercise 5.5.

When a variable is included in a reduction clause, the variable itself is shared. However, a private variable is created for each thread in the team. In the parallel block each time a thread executes a statement involving the variable, it uses the private variable. When the parallel block ends, the values in the private variables are combined into the shared variable. Thus our latest version of the code

```
    global_result = 0.0;
#   pragma omp parallel num_threads(thread_count) \
        reduction(+: global_result)
    global_result += Local_trap(double a, double b, int n);
```

effectively executes code that is identical to our previous version:

```
    global_result = 0.0;
#   pragma omp parallel num_threads(thread_count)
    {
        double my_result = 0.0;  /* private */
        my_result += Local_trap(double a, double b, int n);
#       pragma omp critical
        global_result += my_result;
    }
```

One final point to note is that the threads' private variables are initialized to 0. This is analogous to our initializing my_result to zero. In general, the private variables created for a reduction clause are initialized to the *identity value* for the operator. For example, if the operator is multiplication, the private variables would be initialized to 1. See Table 5.1 for the entire list.

Table 5.1 Identity values for
the various reduction operators in OpenMP.

Operator	Identity Value
+	0
*	1
-	0
&	~0
|	0
^	0
&&	1
||	0

5.5 The `parallel for` **directive**

As an alternative to our explicit parallelization of the trapezoidal rule, OpenMP provides the `parallel for` directive. Using it, we can parallelize the serial trapezoidal
rule

```
h = (b−a)/n;
approx = (f(a) + f(b))/2.0;
for (i = 1; i <= n−1; i++)
    approx += f(a + i*h);
approx = h*approx;
```

by simply placing a directive immediately before the **for** loop:

```
h = (b−a)/n;
approx = (f(a) + f(b))/2.0;
#   pragma omp parallel for num_threads(thread_count) \
        reduction(+: approx)
    for (i = 1; i <= n−1; i++)
        approx += f(a + i*h);
    approx = h*approx;
```

Like the `parallel` directive, the `parallel for` directive forks a team of threads to execute the following structured block. However, the structured block following the
`parallel for` directive must be a **for** loop. Furthermore, with the `parallel for` directive the system parallelizes the **for** loop by dividing the iterations of the loop among
the threads. So the `parallel for` directive is therefore very different from the `parallel`
directive, because in a block that is preceded by a `parallel` directive, in general, the
work must be divided among the threads by the threads themselves.

In a **for** loop that has been parallelized with a `parallel for` directive, the default
partitioning of the iterations among the threads is up to the system. However, most
systems use roughly a block partitioning, that is, if there are m iterations, then roughly
the first $m/$`thread_count` are assigned to thread 0, the next $m/$`thread_count` are
assigned to thread 1, and so on.

Note that it was essential that we made `approx` a reduction variable. If we hadn't, it would have been an ordinary shared variable, and the body of the loop

```
approx += f(a + i*h);
```

would be an unprotected critical section, leading to inconsistent values of `approx`.

However, speaking of scope, the default scope for all variables in a `parallel` directive is shared, but in our `parallel` **for** if the loop variable `i` were shared, the variable update, `i++`, would also be an unprotected critical section. Hence, in a loop that is parallelized with a `parallel` **for** directive the default scope of the loop variable is *private*; in our code, each thread in the team has its own copy of `i`.

5.5.1 Caveats

This is truly wonderful: It may be possible to parallelize a serial program that consists of one large **for** loop by just adding a single `parallel` **for** directive. It may be possible to incrementally parallelize a serial program that has many **for** loops by successively placing `parallel` **for** directives before each loop.

However, things may not be quite as rosy as they seem. There are several caveats associated with the use of the `parallel` **for** directive. First, OpenMP will only parallelize **for** loops—it won't parallelize **while** loops or **do**−**while** loops directly. This may not seem to be too much of a limitation, since any code that uses a **while** loop or a **do**−**while** loop can be converted to equivalent code that uses a **for** loop instead. However, OpenMP will only parallelize **for** loops for which the number of iterations can be determined:

- from the **for** statement itself (that is, the code **for** (. . . ; . . . ; . . .)), and
- prior to execution of the loop.

For example, the "infinite loop"

```
for ( ; ; ) {
    . . .
}
```

cannot be parallelized. Similarly, the loop

```
for (i = 0; i < n; i++) {
    if ( . . . ) break;
    . . .
}
```

cannot be parallelized, since the number of iterations can't be determined from the **for** statement alone. This **for** loop is also not a structured block, since the **break** adds another point of exit from the loop.

In fact, OpenMP will only parallelize **for** loops that are in **canonical form**. Loops in canonical form take one of the forms shown in Program 5.3. The variables and expressions in this template are subject to some fairly obvious restrictions:

$$\textbf{for} \left(\text{index} = \text{start} \quad ; \quad \begin{array}{l} \text{index} < \text{end} \\ \text{index} <= \text{end} \\ \text{index} >= \text{end} \\ \text{index} > \text{end} \end{array} \quad ; \quad \begin{array}{l} \text{index++} \\ \text{++index} \\ \text{index-} \\ \text{-index} \\ \text{index += incr} \\ \text{index -= incr} \\ \text{index = index + incr} \\ \text{index = incr + index} \\ \text{index = index - incr} \end{array} \right)$$

Program 5.3: Legal forms for parallelizable **for** statements.

- The variable index must have integer or pointer type (e.g., it can't be a **float**).
- The expressions start, end, and incr must have a compatible type. For example, if index is a pointer, then incr must have integer type.
- The expressions start, end, and incr must not change during execution of the loop.
- During execution of the loop, the variable index can only be modified by the "increment expression" in the **for** statement.

These restrictions allow the run-time system to determine the number of iterations prior to execution of the loop.

The sole exception to the rule that the run-time system must be able to determine the number of iterations prior to execution is that there *can* be a call to exit in the body of the loop.

5.5.2 Data dependences

If a **for** loop fails to satisfy one of the rules outlined in the preceding section, the compiler will simply reject it. For example, suppose we try to compile a program with the following linear search function:

```
1    int Linear_search(int key, int A[], int n) {
2        int i;
3        /* thread_count is global */
4    #   pragma omp parallel for num_threads(thread_count)
5        for (i = 0; i < n; i++)
6            if (A[i] == key) return i;
7        return -1;  /* key not in list */
8    }
```

The gcc compiler reports:

```
Line 6: error: invalid exit from OpenMP structured block
```

A more insidious problem occurs in loops in which the computation in one iteration depends on the results of one or more previous iterations. As an example, consider the following code, which computes the first n Fibonacci numbers:

```
fibo[0] = fibo[1] = 1;
for (i = 2; i < n; i++)
    fibo[i] = fibo[i-1] + fibo[i-2];
```

Although we may be suspicious that something isn't quite right, let's try parallelizing the **for** loop with a parallel **for** directive:

```
    fibo[0] = fibo[1] = 1;
#   pragma omp parallel for num_threads(thread_count)
    for (i = 2; i < n; i++)
        fibo[i] = fibo[i-1] + fibo[i-2];
```

The compiler will create an executable without complaint. However, if we try running it with more than one thread, we may find that the results are, at best, unpredictable. For example, on one of our systems (if we try using two threads to compute the first 10 Fibonacci numbers), we sometimes get

$$1\ 1\ 2\ 3\ 5\ 8\ 13\ 21\ 34\ 55,$$

which is correct. However, we also occasionally get

$$1\ 1\ 2\ 3\ 5\ 8\ 0\ 0\ 0\ 0.$$

What happened? It appears that the run-time system assigned the computation of fibo[2], fibo[3], fibo[4], and fibo[5] to one thread, while fibo[6], fibo[7], fibo[8], and fibo[9] were assigned to the other. (Remember, the loop starts with $i = 2$.) In some runs of the program, everything is fine, because the thread that was assigned fibo[2], fibo[3], fibo[4], and fibo[5] finishes its computations before the other thread starts. However, in other runs, the first thread has evidently not computed fibo[4] and fibo[5] when the second computes fibo[6]. It appears that the system has initialized the entries in fibo to 0, and the second thread is using the values fibo[4] = 0 and fibo[5] = 0 to compute fibo[6]. It then goes on to use fibo[5] = 0 and fibo[6] = 0 to compute fibo[7], and so on.

We see two important points here:

1. OpenMP compilers don't check for dependences among iterations in a loop that's being parallelized with a parallel **for** directive. It's up to us, the programmers, to identify these dependences.
2. A loop in which the results of one or more iterations depend on other iterations *cannot*, in general, be correctly parallelized by OpenMP without using features such as the Tasking API. (See Section 5.10).

The dependence of the computation of fibo[6] on the computation of fibo[5] is called a **data dependence**. Since the value of fibo[5] is calculated in one iteration, and the result is used in a subsequent iteration, the dependence is sometimes called a **loop-carried dependence**.

5.5.3 Finding loop-carried dependences

Perhaps the first thing to observe is that when we're attempting to use a `parallel` **for** directive, we only need to worry about loop-carried dependences. We don't need to worry about more general data dependences. For example, in the loop

```
1      for (i = 0; i < n; i++) {
2          x[i] = a + i*h;
3          y[i] = exp(x[i]);
4      }
```

there is a data dependence between Lines 2 and 3. However, there is no problem with the parallelization

```
1   #   pragma omp parallel for num_threads(thread_count)
2      for (i = 0; i < n; i++) {
3          x[i] = a + i*h;
4          y[i] = exp(x[i]);
5      }
```

since the computation of `x[i]` and its subsequent use will always be assigned to the same thread.

Also observe that at least one of the statements must write or update the variable in order for the statements to represent a dependence, so to detect a loop-carried dependence, we should only concern ourselves with variables that are updated by the loop body. That is, we should look for variables that are read or written in one iteration, and written in another. Let's look at a couple of examples.

5.5.4 Estimating π

One way to get a numerical approximation to π is to use many terms in the formula[4]

$$\pi = 4\left[1 - \frac{1}{3} + \frac{1}{5} - \frac{1}{7} + \cdots\right] = 4\sum_{k=0}^{\infty} \frac{(-1)^k}{2k+1}.$$

We can implement this formula in serial code with

```
1          double factor = 1.0;
2          double sum = 0.0;
3          for (k = 0; k < n; k++) {
4              sum += factor/(2*k+1);
5              factor = -factor;
6          }
7          pi_approx = 4.0*sum;
```

---

[4] This is by no means the best method for approximating π, since it requires a *lot* of terms to get a reasonably accurate result. However, in this case, lots of terms will be better to demonstrate the effects of parallelism, and we're more interested in the formula itself than the actual estimate.

(Why is it important that `factor` is a **double** instead of an **int** or a **long**?)

How can we parallelize this with OpenMP? We might at first be inclined to do something like this:

```
1       double factor = 1.0;
2       double sum = 0.0;
3  #    pragma omp parallel for num_threads(thread_count) \
4          reduction(+:sum)
5       for (k = 0; k < n; k++) {
6          sum += factor/(2*k+1);
7          factor = -factor;
8       }
9       pi_approx = 4.0*sum;
```

However, it's pretty clear that the update to `factor` in Line 7 in iteration `k` and the subsequent increment of `sum` in Line 6 in iteration `k+1` is an instance of a loop-carried dependence. If iteration `k` is assigned to one thread and iteration `k+1` is assigned to another thread, there's no guarantee that the value of `factor` in Line 6 will be correct. In this case, we can fix the problem by examining the series

$$\sum_{k=0}^{\infty} \frac{(-1)^k}{2k+1}.$$

We see that in iteration k, the value of `factor` should be $(-1)^k$, which is $+1$ if k is even and -1 if k is odd, so if we replace the code

```
1          sum += factor/(2*k+1);
2          factor = -factor;
```

by

```
1          if (k % 2 == 0)
2             factor = 1.0;
3          else
4             factor = -1.0;
5          sum += factor/(2*k+1);
```

or, if you prefer the ?: operator,

```
1          factor = (k % 2 == 0) ? 1.0 : -1.0;
2          sum += factor/(2*k+1);
```

we will eliminate the loop dependence.

However, things still aren't quite right. If we run the program on one of our systems with just two threads and $n = 1000$, the result is consistently wrong. For example,

```
1       With n = 1000 terms and 2 threads,
2          Our estimate of pi = 2.97063289263385
3       With n = 1000 terms and 2 threads,
4          Our estimate of pi = 3.22392164798593
```

On the other hand, if we run the program with only one thread, we always get

```
1    With n = 1000 terms and 1 threads,
2       Our estimate of pi = 3.14059265383979
```

What's wrong here?

Recall that in a block that has been parallelized by a `parallel for` directive, by default any variable declared before the loop—with the sole exception of the loop variable—is shared among the threads. So `factor` is shared and, for example, thread 0 might assign it the value 1, but before it can use this value in the update to `sum`, thread 1 could assign it the value −1. Therefore, in addition to eliminating the loop-carried dependence in the calculation of `factor`, we need to ensure that each thread has its own copy of `factor`. That is, to make our code correct, we need to also ensure that `factor` has private scope. We can do this by adding a `private` clause to the `parallel for` directive.

```
1        double sum = 0.0;
2    #   pragma omp parallel for num_threads(thread_count) \
3              reduction(+:sum) private(factor)
4        for (k = 0; k < n; k++) {
5           if (k % 2 == 0)
6              factor = 1.0;
7           else
8              factor = -1.0;
9           sum += factor/(2*k+1);
10       }
```

The `private` clause specifies that for each variable listed inside the parentheses, a private copy is to be created for each thread. Thus, in our example, each of the `thread_count` threads will have its own copy of the variable `factor`, and hence the updates of one thread to `factor` won't affect the value of `factor` in another thread.

It's important to remember that the value of a variable with private scope is unspecified at the beginning of a `parallel` block or a `parallel for` block. Its value is also unspecified after completion of a `parallel` or `parallel for` block. So, for example, the output of the first `printf` statement in the following code is unspecified, since it prints the private variable `x` before it's explicitly initialized. Similarly, the output of the final `printf` is unspecified, since it prints `x` after the completion of the `parallel` block.

```
1    int x = 5;
2    #  pragma omp parallel num_threads(thread_count) \
3          private(x)
4       {
5          int my_rank = omp_get_thread_num();
6          printf("Thread %d > before initialization, x = %d\n",
7                 my_rank, x);
8          x = 2*my_rank + 2;
```

```
 9          printf("Thread %d > after initialization, x = %d\n",
10                  my_rank, x);
11      }
12      printf("After parallel block, x = %d\n", x);
```

5.5.5 More on scope

Our problem with the variable `factor` is a common one. We usually need to think about the scope of each variable in a `parallel` block or a `parallel` **for** block. Therefore, rather than letting OpenMP decide on the scope of each variable, it's a very good practice for us as programmers to specify the scope of each variable in a block. In fact, OpenMP provides a clause that will explicitly require us to do this: the **default** clause. If we add the clause

> **default**(none)

to our `parallel` or `parallel` **for** directive, then the compiler will require that we specify the scope of each variable we use in the block and that has been declared outside the block. (Variables that are declared within the block are always private, since they are allocated on the thread's stack.)

For example, using a **default**(none) clause, our calculation of π could be written as follows:

```
           double sum = 0.0;
  #        pragma omp parallel for num_threads(thread_count) \
              default(none) reduction(+:sum) private(k, factor) \
              shared(n)
           for (k = 0; k < n; k++) {
              if (k % 2 == 0)
                 factor = 1.0;
              else
                 factor = -1.0;
              sum += factor/(2*k+1);
           }
```

In this example, we use four variables in the **for** loop. With the default clause, we need to specify the scope of each. As we've already noted, `sum` is a reduction variable (which has properties of both private and shared scope). We've also already noted that `factor` and the loop variable `k` should have private scope. Variables that are never updated in the `parallel` or `parallel` **for** block, such as `n` in this example, can be safely shared. Recall that unlike `private` variables, `shared` variables have the same value in the `parallel` or `parallel` **for** block that they had before the block, and their value after the block is the same as their last value in the block. Thus if `n` were initialized before the block to 1000, it would retain this value in the `parallel` **for** statement, and since the value isn't changed in the **for** loop, it would retain this value after the loop has completed.

5.6 More about loops in OpenMP: sorting
5.6.1 Bubble sort

Recall that the serial *bubble sort* algorithm for sorting a list of integers can be implemented as follows:

```
for (list_length = n; list_length >= 2; list_length--)
    for (i = 0; i < list_length - 1; i++)
        if (a[i] > a[i+1]) {
            tmp = a[i];
            a[i] = a[i+1];
            a[i+1] = tmp;
        }
```

Here, a stores n **int**s and the algorithm sorts them in increasing order. The outer loop first finds the largest element in the list and stores it in a[n−1]; it then finds the next-to-the-largest element and stores it in a[n−2], and so on. So, effectively, the first pass is working with the full n-element list. The second is working with all of the elements, except the largest; it's working with an $n - 1$-element list, and so on.

The inner loop compares consecutive pairs of elements in the current list. When a pair is out of order (a[i] > a[i+1]) it swaps them. This process of swapping will move the largest element to the last slot in the "current" list, that is, the list consisting of the elements

$$a[0], \ a[1], \ . \ . \ . \ , \ a[list\_length - 1]$$

It's pretty clear that there's a loop-carried dependence in the outer loop; in any iteration of the outer loop the contents of the current list depend on the previous iterations of the outer loop. For example, if at the start of the algorithm a = {3, 4, 1, 2}, then the second iteration of the outer loop should work with the list {3, 1, 2}, since the 4 should be moved to the last position by the first iteration. But if the first two iterations are executing simultaneously, it's possible that the effective list for the second iteration will contain 4.

The loop-carried dependence in the inner loop is also fairly easy to see. In iteration i, the elements that are compared depend on the outcome of iteration $i - 1$. If in iteration $i - 1$, a[i−1] and a[i] are not swapped, then iteration i should compare a[i] and a[i+1]. If, on the other hand, iteration $i - 1$ swaps a[i−1] and a[i], then iteration i should be comparing the original a[i−1] (which is now a[i]) and a[i+1]. For example, suppose the current list is {3,1,2}. Then when $i = 1$, we should compare 3 and 2, but if the $i = 0$ and the $i = 1$ iterations are happening simultaneously, it's entirely possible that the $i = 1$ iteration will compare 1 and 2.

It's also not at all clear how we might remove either loop-carried dependence without completely rewriting the algorithm. It's important to keep in mind that even though we can always find loop-carried dependences, it may be difficult or impossible to remove them. The parallel **for** directive is not a universal solution to the problem of parallelizing **for** loops.

Table 5.2 Serial odd-even transposition sort.

Phase	Subscript in Array			
	0	**1**	**2**	**3**
0	9 ⇔	7	8 ⇔	6
	7	9	6	8
1	7	9 ⇔	6	8
	7	6	9	8
2	7 ⇔	6	9 ⇔	8
	6	7	8	9
3	6	7 ⇔	8	9
	6	7	8	9

5.6.2 Odd-even transposition sort

Odd-even transposition sort is a sorting algorithm that's similar to bubble sort, but it has considerably more opportunities for parallelism. Recall from Section 3.7.1 that serial odd-even transposition sort can be implemented as follows:

```
for (phase = 0; phase < n; phase++)
    if (phase % 2 == 0)
        for (i = 1; i < n; i += 2)
            if (a[i-1] > a[i]) Swap(&a[i-1],&a[i]);
    else
        for (i = 1; i < n-1; i += 2)
            if (a[i] > a[i+1]) Swap(&a[i], &a[i+1]);
```

The list a stores n **int**s, and the algorithm sorts them into increasing order. During an "even phase" (phase % 2 == 0), each odd-subscripted element, a[i], is compared to the element to its "left," a[i−1], and if they're out of order, they're swapped. During an "odd" phase, each odd-subscripted element is compared to the element to its right, and if they're out of order, they're swapped. A theorem guarantees that after n phases, the list will be sorted.

As a brief example, suppose a = {9, 7, 8, 6}. Then the phases are shown in Table 5.2. In this case, the final phase wasn't necessary, but the algorithm doesn't bother checking whether the list is already sorted before carrying out each phase.

It's not hard to see that the outer loop has a loop-carried dependence. As an example, suppose as before that a = {9, 7, 8, 6}. Then in phase 0 the inner loop will compare elements in the pairs (9, 7) and (8, 6), and both pairs are swapped. So for phase 1, the list should be {7, 9, 6, 8}, and during phase 1 the elements in the pair (9, 6) should be compared and swapped. However, if phase 0 and phase 1 are executed simultaneously, the pair that's checked in phase 1 might be (7, 8), which is in order. Furthermore, it's not clear how one might eliminate this loop-carried dependence, so it would appear that parallelizing the outer **for** loop isn't an option.

The *inner* **for** loops, however, don't appear to have any loop-carried dependences. For example, in an even phase loop variable i will be odd, so for two distinct values of i, say $i = j$ and $i = k$, the pairs $\{j - 1, j\}$ and $\{k - 1, k\}$ will be disjoint. The comparison and possible swaps of the pairs (a[j−1], a[j]) and (a[k−1], a[k]) can therefore proceed simultaneously.

```
1      for (phase = 0; phase < n; phase++) {
2          if (phase % 2 == 0)
3  #           pragma omp parallel for num_threads(thread_count) \
4                   default(none) shared(a, n) private(i, tmp)
5              for (i = 1; i < n; i += 2) {
6                  if (a[i−1] > a[i]) {
7                      tmp = a[i−1];
8                      a[i−1] = a[i];
9                      a[i] = tmp;
10                 }
11             }
12         else
13  #           pragma omp parallel for num_threads(thread_count) \
14                   default(none) shared(a, n) private(i, tmp)
15             for (i = 1; i < n−1; i += 2) {
16                 if (a[i] > a[i+1]) {
17                     tmp = a[i+1];
18                     a[i+1] = a[i];
19                     a[i] = tmp;
20                 }
21             }
22     }
```

Program 5.4: First OpenMP implementation of odd-even sort.

Thus we could try to parallelize odd-even transposition sort using the code shown in Program 5.4, but there are a couple of potential problems. First, although any iteration of, say, one even phase doesn't depend on any other iteration of that phase, we've already noted that this is not the case for iterations in phase p and phase $p + 1$. We need to be sure that all the threads have finished phase p before any thread starts phase $p + 1$. However, like the parallel directive, the parallel **for** directive has an implicit barrier at the end of the loop, so none of the threads will proceed to the next phase, phase $p + 1$, until all of the threads have completed the current phase, phase p.

A second potential problem is the overhead associated with forking and joining the threads. The OpenMP implementation *may* fork and join thread_count threads on *each* pass through the body of the outer loop. The first row of Table 5.3 shows runtimes for 1, 2, 3, and 4 threads on one of our systems when the input list contained 20,000 elements.

These aren't terrible times, but let's see if we can do better. Each time we execute one of the inner loops, we use the same number of threads, so it would seem to be superior to fork the threads once and reuse the same team of threads for each execution of the inner loops. Not surprisingly, OpenMP provides directives that allow us to do just this. We can fork our team of `thread_count` threads *before* the outer loop with a `parallel` directive. Then, rather than forking a new team of threads with each execution of one of the inner loops, we use a **for** directive, which tells OpenMP to parallelize the **for** loop with the existing team of threads. This modification to the original OpenMP implementation is shown in Program 5.5.

```
1  #   pragma omp parallel num_threads(thread_count) \
2          default(none) shared(a, n) private(i, tmp, phase)
3       for (phase = 0; phase < n; phase++) {
4          if (phase % 2 == 0)
5  #          pragma omp for
6              for (i = 1; i < n; i += 2) {
7                  if (a[i-1] > a[i]) {
8                      tmp = a[i-1];
9                      a[i-1] = a[i];
10                     a[i] = tmp;
11                 }
12             }
13         else
14 #          pragma omp for
15             for (i = 1; i < n-1; i += 2) {
16                 if (a[i] > a[i+1]) {
17                     tmp = a[i+1];
18                     a[i+1] = a[i];
19                     a[i] = tmp;
20                 }
21             }
22      }
```

Program 5.5: Second OpenMP implementation of odd-even sort.

The **for** directive, unlike the `parallel` **for** directive, doesn't fork any threads. It uses whatever threads have already been forked in the enclosing `parallel` block. There *is* an implicit barrier at the end of the loop. The results of the code—the final list—will therefore be the same as the results obtained from the original parallelized code.

Run-times for this second version of odd-even sort are in the second row of Table 5.3. When we're using two or more threads, the version that uses two **for** directives is at least 17% faster than the version that uses two `parallel` **for** directives, so for this system the slight effort involved in making the change is well worth it.

Table 5.3 Odd-even sort with two `parallel` **for** directives and two **for** directives. Times are in seconds.

thread_count	1	2	3	4
Two `parallel` **for** directives	0.770	0.453	0.358	0.305
Two **for** directives	0.732	0.376	0.294	0.239

5.7 Scheduling loops

When we first encountered the `parallel` **for** directive, we saw that the exact assignment of loop iterations to threads is system dependent. However most OpenMP implementations use roughly a block partitioning: if there are n iterations in the serial loop, then in the parallel loop the first n/`thread_count` are assigned to thread 0, the next n/`thread_count` are assigned to thread 1, and so on. It's not difficult to think of situations in which this assignment of iterations to threads would be less than optimal. For example, suppose we want to parallelize the loop

```
sum = 0.0;
for (i = 0; i <= n; i++)
   sum += f(i);
```

Also suppose that the time required by the call to f is proportional to the size of the argument `i`. Then a block partitioning of the iterations will assign much more work to thread `thread_count` $-$ 1 than it will assign to thread 0. A better assignment of work to threads might be obtained with a **cyclic** partitioning of the iterations among the threads. In a cyclic partitioning, the iterations are assigned, one at a time, in a "round-robin" fashion to the threads. Suppose $t =$ `thread_count`. Then a cyclic partitioning will assign the iterations as follows:

Thread	Iterations
0	$0, n/t, 2n/t, \ldots$
1	$1, n/t + 1, 2n/t + 1, \ldots$
$\vdots$	$\vdots$
$t - 1$	$t - 1, n/t + t - 1, 2n/t + t - 1, \ldots$

To get a feel for how drastically this can affect performance, we wrote a program in which we defined

```
double f(int i) {
   int j, start = i*(i+1)/2, finish = start + i;
   double return_val = 0.0;

   for (j = start; j <= finish; j++) {
      return_val += sin(j);
   }
   return return_val;
} /* f */
```

The call $f(i)$ calls the sin function i times, and, for example, the time to execute $f(2i)$ requires approximately twice as much time as the time to execute $f(i)$.

When we ran the program with $n = 10{,}000$ and one thread, the run-time was 3.67 seconds. When we ran the program with two threads and the default assignment— iterations 0–5000 on thread 0 and iterations 5001–10,000 on thread 1—the run-time was 2.76 seconds. This is a speedup of only 1.33. However, when we ran the program with two threads and a cyclic assignment, the run-time was decreased to 1.84 seconds. This is a speedup of 1.99 over the one-thread run and a speedup of 1.5 over the two-thread block partition!

We can see that a good assignment of iterations to threads can have a very significant effect on performance. In OpenMP, assigning iterations to threads is called **scheduling**, and the schedule clause can be used to assign iterations in either a parallel **for** or a **for** directive.

5.7.1 The schedule **clause**

In our example, we already know how to obtain the default schedule: we just add a parallel **for** directive with a reduction clause:

```
      sum = 0.0;
#     pragma omp parallel for num_threads(thread_count) \
         reduction(+:sum)
      for (i = 0; i <= n; i++)
         sum += f(i);
```

To get a cyclic schedule, we can add a schedule clause to the parallel **for** directive:

```
      sum = 0.0;
#     pragma omp parallel for num_threads(thread_count) \
         reduction(+:sum) schedule(static,1)
      for (i = 0; i <= n; i++)
         sum += f(i);
```

In general, the schedule clause has the form

```
   schedule(<type> [, <chunksize>])
```

The type can be any one of the following:

- static. The iterations can be assigned to the threads before the loop is executed.
- dynamic or guided. The iterations are assigned to the threads while the loop is executing, so after a thread completes its current set of iterations, it can request more from the run-time system.
- auto. The compiler and/or the run-time system determine the schedule.
- runtime. The schedule is determined at run-time based on an environment variable (more on this later).

schedule(static)

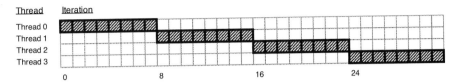

schedule(static, 2)

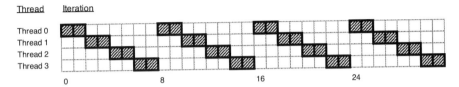

schedule(dynamic, 2)

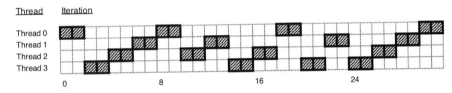

schedule(guided)

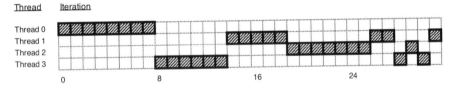

FIGURE 5.5

Scheduling visualization for the `static`, `dynamic`, and `guided` schedule types with 4 threads and 32 iterations. The first static schedule uses the default `chunksize`, whereas the second uses a `chunksize` of **2**. The exact distribution of work across threads will vary between different executions of the program for the `dynamic` and `guided` schedule types, so this visualization shows one of many possible scheduling outcomes.

The `chunksize` is a positive integer. In OpenMP parlance, a **chunk** of iterations is a block of iterations that would be executed consecutively in the serial loop. The number of iterations in the block is the `chunksize`. Only `static`, `dynamic`, and `guided` schedules can have a `chunksize`. This determines the details of the schedule, but its exact interpretation depends on the `type`. Fig. 5.5 provides a visualization of how work is scheduled using the static, dynamic, and guided types.

5.7.2 The `static` schedule type

For a `static` schedule, the system assigns chunks of `chunksize` iterations to each thread in a round-robin fashion. As an example, suppose we have 12 iterations, $0, 1, \ldots, 11$, and three threads. Then if `schedule(static, 1)` is used, in the `parallel for` or **for** directive, we've already seen that the iterations will be assigned as

$$
\begin{aligned}
&\text{Thread } 0: \quad 0, 3, 6, 9 \\
&\text{Thread } 1: \quad 1, 4, 7, 10 \\
&\text{Thread } 2: \quad 2, 5, 8, 11
\end{aligned}
$$

If `schedule(static, 2)` is used, then the iterations will be assigned as

$$
\begin{aligned}
&\text{Thread } 0: \quad 0, 1, 6, 7 \\
&\text{Thread } 1: \quad 2, 3, 8, 9 \\
&\text{Thread } 2: \quad 4, 5, 10, 11
\end{aligned}
$$

If `schedule(static, 4)` is used, the iterations will be assigned as

$$
\begin{aligned}
&\text{Thread } 0: \quad 0, 1, 2, 3 \\
&\text{Thread } 1: \quad 4, 5, 6, 7 \\
&\text{Thread } 2: \quad 8, 9, 10, 11
\end{aligned}
$$

The default schedule is defined by your particular implementation of OpenMP, but in most cases it is equivalent to the clause

```
schedule(static, total_iterations / thread_count)
```

It is also worth noting that the `chunksize` can be omitted. If omitted, the `chunksize` is approximately `total_iterations / thread_count`.

The `static` schedule is a good choice when each loop iteration takes roughly the same amount of time to compute. It also has the advantage that threads in subsequent loops with the same number of iterations will be assigned to the same ranges; this can improve the speed of memory accesses, particularly on NUMA systems (see Chapter 2).

5.7.3 The `dynamic` and `guided` schedule types

In a `dynamic` schedule, the iterations are also broken up into chunks of `chunksize` consecutive iterations. Each thread executes a chunk, and when a thread finishes a chunk, it requests another one from the run-time system. This continues until all the iterations are completed. The `chunksize` can be omitted. When it is omitted, a `chunksize` of 1 is used.

The primary difference between `static` and `dynamic` schedules is that the `dynamic` schedule assigns ranges to threads on a first-come, first-served basis. This can be advantageous if loop iterations do not take a uniform amount of time to compute (some

Table 5.4 Assignment of trapezoidal rule iterations 1–9999 using a guided schedule with two threads.

Thread	Chunk	Size of Chunk	Remaining Iterations
0	1 – 5000	5000	4999
1	5001 – 7500	2500	2499
1	7501 – 8750	1250	1249
1	8751 – 9375	625	624
0	9376 – 9687	312	312
1	9688 – 9843	156	156
0	9844 – 9921	78	78
1	9922 – 9960	39	39
1	9961 – 9980	20	19
1	9981 – 9990	10	9
1	9991 – 9995	5	4
0	9996 – 9997	2	2
1	9998 – 9998	1	1
0	9999 – 9999	1	0

algorithms are more compute-intensive in later iterations, for instance). However, since the ranges are not allocated ahead of time, there is some overhead associated with assigning them dynamically at run-time. Increasing the chunk size strikes a balance between the performance characteristics of static and dynamic scheduling; with larger chunk sizes, fewer dynamic assignments will be made.

The guided schedule is similar to dynamic in that each thread also executes a chunk and requests another one when it's finished. However, in a guided schedule, as chunks are completed, the size of the new chunks decreases. For example, on one of our systems, if we run the trapezoidal rule program with the parallel **for** directive and a schedule(guided) clause, then when $n = 10,000$ and thread_count $= 2$, the iterations are assigned as shown in Table 5.4. We see that the size of the chunk is approximately the number of iterations remaining divided by the number of threads. The first chunk has size $9999/2 \approx 5000$, since there are 9999 unassigned iterations. The second chunk has size $4999/2 \approx 2500$, and so on.

In a guided schedule, if no chunksize is specified, the size of the chunks decreases down to 1. If chunksize is specified, it decreases down to chunksize, with the exception that the very last chunk can be smaller than chunksize. The guided schedule can improve the balance of load across threads when later iterations are more compute-intensive.

5.7.4 The runtime schedule type

To understand schedule(runtime), we need to digress for a moment and talk about **environment variables**. As the name suggests, environment variables are named values that can be accessed by a running program. That is, they're available in the program's *environment.* Some commonly used environment variables are PATH, HOME, and SHELL. The PATH variable specifies which directories the shell should search when it's looking for an executable and is usually defined in both Unix and Windows. The HOME variable specifies the location of the user's home directory, and the SHELL variable specifies the location of the executable for the user's shell. These are usually defined in Unix systems. In both Unix-like systems (e.g., Linux and macOS) and Windows, environment variables can be examined and specified on the command line. In Unix-like systems, you can use the shell's command line. In Windows systems, you can use the command line in an integrated development environment.

As an example, if we're using the bash shell (one of the most common Unix shells), we can examine the value of an environment variable by typing:

```
$ echo $PATH
```

and we can use the export command to set the value of an environment variable:

```
$ export TEST_VAR="hello"
```

These commands also work on ksh, sh, and zsh. For details about how to examine and set environment variables for your particular system, check the man pages for your shell, or consult with your system administrator or local expert.

When schedule(runtime) is specified, the system uses the environment variable OMP_SCHEDULE to determine at run-time how to schedule the loop. The OMP_SCHEDULE environment variable can take on any of the values that can be used for a static, dynamic, or guided schedule. For example, suppose we have a parallel **for** directive in a program and it has been modified by schedule(runtime). Then if we use the bash shell, we can get a cyclic assignment of iterations to threads by executing the command

```
$ export OMP_SCHEDULE="static,1"
```

Now, when we start executing our program, the system will schedule the iterations of the **for** loop as if we had the clause schedule(static,1) modifying the parallel **for** directive. This can be very useful for testing a variety of scheduling configurations.

The following bash shell script demonstrates how one might take advantage of this environment variable to test a range of schedules and chunk sizes. It runs a matrix-vector multiplication program that has a parallel **for** directive with the schedule(runtime) clause.

```bash
#!/usr/bin/env bash

declare -a schedules=("static" "dynamic" "guided")
declare -a chunk_sizes=("" 1000 100 10 1)

for schedule in "${schedules[@]}"; do
    echo "Schedule: ${schedule}"
    for chunk_size in "${chunk_sizes[@]}"; do
        echo "  Chunk Size: ${chunk_size}"
        sched_param="${schedule}"
```

```
        if [[ "${chunk_size}" != "" ]]; then
            # A blank string indicates we want
            # the default chunk size
            sched_param="${schedule},${chunk_size}"
        fi

        # Run the program with OMP_SCHEDULE set:
        OMP_SCHEDULE="${sched_param}" ./omp_mat_vect 4 2500 2500
    done
    echo
done
```

5.7.5 Which schedule?

If we have a **for** loop that we're able to parallelize, how do we decide which type of schedule we should use and what the chunksize should be? As you may have guessed, there *is* some overhead associated with the use of a schedule clause. Furthermore, the overhead is greater for dynamic schedules than static schedules, and the overhead associated with guided schedules is the greatest of the three. Thus if we're getting satisfactory performance without a schedule clause, we should go no further. However, if we suspect that the performance of the default schedule can be substantially improved, we should probably experiment with some different schedules.

In the example at the beginning of this section, when we switched from the default schedule to schedule(**static**,1), the speedup of the two-threaded execution of the program increased from 1.33 to 1.99. Since it's *extremely* unlikely that we'll get speedups that are significantly better than 1.99, we can just stop here, at least if we're only going to use two threads with 10,000 iterations. If we're going to be using varying numbers of threads and varying numbers of iterations, we need to do more experimentation, and it's entirely possible that we'll find that the optimal schedule depends on both the number of threads and the number of iterations.

It can also happen that we'll decide that the performance of the default schedule isn't very good, and we'll proceed to search through a large array of schedules and iteration counts only to conclude that our loop doesn't parallelize very well and *no* schedule is going to give us much improved performance. For an example of this, see Programming Assignment 5.4.

There are some situations in which it's a good idea to explore some schedules before others:

- If each iteration of the loop requires roughly the same amount of computation, then it's likely that the default distribution will give the best performance.
- If the cost of the iterations decreases (or increases) linearly as the loop executes, then a static schedule with small chunksizes will probably give the best performance.
- If the cost of each iteration can't be determined in advance, then it may make sense to explore a variety of scheduling options. The schedule(runtime) clause can

be used here, and the different options can be explored by running the program with different assignments to the environment variable OMP_SCHEDULE.

5.8 Producers and consumers

Let's take a look at a parallel problem that isn't amenable to parallelization using a parallel **for** or **for** directive.

5.8.1 Queues

Recall that a **queue** is a list abstract datatype in which new elements are inserted at the "rear" of the queue and elements are removed from the "front" of the queue. A queue can thus be viewed as an abstraction of a line of customers waiting to pay for their groceries in a supermarket. The elements of the list are the customers. New customers go to the end or "rear" of the line, and the next customer to check out is the customer standing at the "front" of the line.

When a new entry is added to the rear of a queue, we sometimes say that the entry has been "enqueued," and when an entry is removed from the front of a queue, we sometimes say that the entry has been "dequeued."

Queues occur frequently in computer science. For example, if we have a number of processes, each of which wants to store some data on a hard drive, then a natural way to ensure that only one process writes to the disk at a time is to have the processes form a queue, that is, the first process that wants to write gets access to the drive first, the second process gets access to the drive next, and so on.

A queue is also a natural data structure to use in many multithreaded applications. For example, suppose we have several "producer" threads and several "consumer" threads. The producer threads might "produce" requests for data from a server— for example, current stock prices—while the consumer threads might "consume" the request by finding or generating the requested data—the current stock prices. The producer threads could enqueue the requested prices, and the consumer threads could dequeue them. In this example, the process wouldn't be completed until the consumer threads had given the requested data to the producer threads.

5.8.2 Message-passing

Another natural application would be implementing message-passing on a shared-memory system. Each thread could have a shared-message queue, and when one thread wanted to "send a message" to another thread, it could enqueue the message in the destination thread's queue. A thread could receive a message by dequeuing the message at the head of its message queue.

Let's implement a relatively simple message-passing program, in which each thread generates random integer "messages" and random destinations for the messages. After creating the message, the thread enqueues the message in the appropriate message queue. After sending a message, a thread checks its queue to see if it has re-ceived a message. If it has, it dequeues the first message in its queue and prints it out. Each thread alternates between sending and trying to receive messages. We'll let the user specify the number of messages each thread should send. When a thread is done sending messages, it receives messages until all the threads are done, at which point all the threads quit. Pseudocode for each thread might look something like this:

```
for (sent_msgs = 0; sent_msgs < send_max; sent_msgs++) {
    Send_msg();
    Try_receive();
}

while (!Done())
    Try_receive();
```

5.8.3 Sending messages

Note that accessing a message queue to enqueue a message is probably a critical section. Although we haven't looked into the details of the implementation of the message queue, it seems likely that we'll want to have a variable that keeps track of the rear of the queue. For example, if we use a singly linked list with the tail of the list corresponding to the rear of the queue, then, to efficiently enqueue, we would want to store a pointer to the rear. When we enqueue a new message, we'll need to check and update the rear pointer. If two threads try to do this simultaneously, we may lose a message that has been enqueued by one of the threads. (It might help to draw a picture!) The results of the two operations will conflict, and hence enqueuing a message will form a critical section.

Pseudocode for the Send_msg() function might look something like this:

```
    mesg = random();
    dest = random() % thread_count;
#   pragma omp critical
    Enqueue(queue, dest, my_rank, mesg);
```

Note that this allows a thread to send a message to itself.

5.8.4 Receiving messages

The synchronization issues for receiving a message are a little different. Only the owner of the queue (that is, the destination thread) will dequeue from a given message queue. As long as we dequeue one message at a time, if there are at least two messages in the queue, a call to Dequeue can't possibly conflict with any calls to Enqueue. So if we keep track of the size of the queue, we can avoid any synchronization (for example, critical directives), as long as there are at least two messages.

Now you may be thinking, "What about the variable storing the size of the queue?" This would be a problem if we simply store the size of the queue. However, if we store two variables, enqueued and dequeued, then the number of messages in the queue is

```
queue_size = enqueued - dequeued
```

and the only thread that will update dequeued is the owner of the queue. Observe that one thread can update enqueued at the same time that another thread is using it to compute queue_size. To see this, let's suppose thread q is computing queue_size. It will either get the old value of enqueued or the new value. It *may* therefore compute a queue_size of 0 or 1 when queue_size should actually be 1 or 2, respectively, but in our program this will only cause a modest delay. Thread q will try again later if queue_size is 0 when it should be 1, and it will execute the critical section directive unnecessarily if queue_size is 1 when it should be 2.

Thus we can implement Try_receive as follows:

```
    queue_size = enqueued - dequeued;
    if (queue_size == 0) return;
    else if (queue_size == 1)
#       pragma omp critical
        Dequeue(queue, &src, &mesg);
    else
        Dequeue(queue, &src, &mesg);
    Print_message(src, mesg);
```

5.8.5 Termination detection

We also need to think about implementation of the Done function. First note that the following "obvious" implementation will have problems:

```
queue_size = enqueued - dequeued;
if (queue_size == 0)
    return TRUE;
else
    return FALSE;
```

If thread u executes this code, it's entirely possible that some thread—call it thread v—will send a message to thread u *after* u has computed queue_size = 0. Of course, after thread u computes queue_size = 0, it will terminate and the message sent by thread v will never be received.

However, in our program, after each thread has completed the **for** loop, it won't send any new messages. Thus if we add a counter done_sending, and each thread increments this after completing its **for** loop, then we *can* implement Done as follows:

```
queue_size = enqueued - dequeued;
if (queue_size == 0 && done_sending == thread_count)
    return TRUE;
else
    return FALSE;
```

5.8.6 **Startup**

When the program begins execution, a single thread, the master thread, will get command-line arguments and allocate an array of message queues, one for each thread. This array needs to be shared among the threads, since any thread can send to any other thread, and hence any thread can enqueue a message in any of the queues. Given that a message queue will (at a minimum) store

- a list of messages,
- a pointer or index to the rear of the queue,
- a pointer or index to the front of the queue,
- a count of messages enqueued, and
- a count of messages dequeued,

it makes sense to store the queue in a struct, and to reduce the amount of copying when passing arguments, it also makes sense to make the message queue an array of pointers to structs. Thus once the array of queues is allocated by the master thread, we can start the threads using a `parallel` directive, and each thread can allocate storage for its individual queue.

An important point here is that one or more threads may finish allocating their queues before some other threads. If this happens, the threads that finish first could start trying to enqueue messages in a queue that hasn't been allocated and cause the program to crash. We therefore need to make sure that none of the threads starts sending messages until all the queues are allocated. Recall that we've seen that several OpenMP directives provide implicit barriers when they're completed, that is, no thread will proceed past the end of the block until all the threads in the team have completed the block. In this case, though, we'll be in the middle of a `parallel` block, so we can't rely on an implicit barrier from some other OpenMP construct—we need an *explicit* barrier. Fortunately, OpenMP provides one:

```
# pragma omp barrier
```

When a thread encounters the barrier, it blocks until all the threads in the team have reached the barrier. After all the threads have reached the barrier, all the threads in the team can proceed.

5.8.7 **The** `atomic` **directive**

After completing its sends, each thread increments `done_sending` before proceeding to its final loop of receives. Clearly, incrementing `done_sending` is a critical section, and we could protect it with a `critical` directive. However, OpenMP provides a potentially higher performance directive: the `atomic` directive[5]:

```
# pragma omp atomic
```

---

[5] OpenMP provides several clauses that modify the behavior of the `atomic` directive. We're describing the default `atomic` directive, which is the same as an `atomic` directive with an `update` clause. See [47].

Unlike the `critical` directive, it can only protect critical sections that consist of a single C assignment statement. Further, the statement must have one of the following forms:

```
x <op>= <expression>;
x++;
++x;
x--;
--x;
```

Here `<op>` can be one of the binary operators

$$+, *, -, /, \&, \wedge, |, <<, \text{ or } >>.$$

It's also important to remember that `<expression>` must not reference `x`.

It should be noted that only the load and store of `x` are guaranteed to be protected. For example, in the code

```
#       pragma omp atomic
        x += y++;
```

a thread's update to `x` will be completed before any other thread can begin updating `x`. However, the update to `y` may be unprotected and the results may be unpredictable.

The idea behind the `atomic` directive is that many processors provide a special load-modify-store instruction, and a critical section that only does a load-modify-store can be protected much more efficiently by using this special instruction rather than the constructs that are used to protect more general critical sections.

5.8.8 Critical sections and locks

To finish our discussion of the message-passing program, we need to take a more careful look at OpenMP's specification of the `critical` directive. In our earlier examples, our programs had at most one critical section, and the `critical` directive forced mutually exclusive access to the section by all the threads. In this program, however, the use of critical sections is more complex. If we simply look at the source code, we'll see three blocks of code preceded by a `critical` or an `atomic` directive:

- `done_sending++;`
- `Enqueue(q_p, my_rank, mesg);`
- `Dequeue(q_p, &src, &mesg);`

However, we don't need to enforce exclusive access across all three of these blocks of code. We don't even need to enforce completely exclusive access within `Enqueue` and `Dequeue`. For example, it would be fine for, say, thread 0 to enqueue a message in thread 1's queue at the same time that thread 1 is enqueuing a message in thread 2's queue. But for the second and third blocks—the blocks protected by `critical` directives—this is exactly what OpenMP does. From OpenMP's point of view our program has two distinct critical sections: the critical section protected by the `atomic` directive, (done_sending++), and the "composite" critical section in which we enqueue and dequeue messages.

Since enforcing mutual exclusion among threads serializes execution, this default behavior of OpenMP—treating all critical blocks as part of one composite critical section—can be highly detrimental to our program's performance. OpenMP *does* provide the option of adding a name to a critical directive:

```
# pragma omp critical(name)
```

When we do this, two blocks protected with critical directives with different names *can* be executed simultaneously. However, the names are set during compilation, and we want a different critical section for each thread's queue. Therefore we need to set the names at run-time, and in our setting, when we want to allow simultaneous access to the same block of code by threads accessing different queues, the named critical directive isn't sufficient.

The alternative is to use **locks**.[6] A lock consists of a data structure and functions that allow the programmer to explicitly enforce mutual exclusion in a critical section. The use of a lock can be roughly described by the following pseudocode:

```
/* Executed by one thread */
Initialize the lock data structure;
. . .
/* Executed by multiple threads */
Attempt to lock or set the lock data structure;
Critical section;
Unlock or unset the lock data structure;
. . .
/* Executed by one thread */
Destroy the lock data structure;
```

The lock data structure is shared among the threads that will execute the critical section. One of the threads (e.g., the master thread) will initialize the lock, and when all the threads are done using the lock, one of the threads should destroy it.

Before a thread enters the critical section, it attempts to *set* the lock by calling the lock function. If no other thread is executing code in the critical section, it *acquires* the lock and proceeds into the critical section past the call to the lock function. When the thread finishes the code in the critical section, it calls an unlock function, which *releases* or *unsets* the lock and allows another thread to acquire the lock.

While a thread owns the lock, no other thread can enter the critical section. If another thread attempts to enter the critical section, it will *block* when it calls the lock function. If multiple threads are blocked in a call to the lock function, then when the thread in the critical section releases the lock, one of the blocked threads returns from the call to the lock, and the others remain blocked.

---

[6] If you've studied the Pthreads chapter, you've already learned about locks, and you can skip ahead to the syntax for OpenMP locks.

OpenMP has two types of locks: **simple** locks and **nested** locks. A simple lock can only be set once before it is unset, while a nested lock can be set multiple times by the same thread before it is unset. The type of an OpenMP simple lock is `omp_lock_t`, and the simple lock functions that we'll be using are

```
void omp_init_lock(omp_lock_t*    lock_p    /* out */);
void omp_set_lock(omp_lock_t*     lock_p    /* in/out */);
void omp_unset_lock(omp_lock_t*   lock_p    /* in/out */);
void omp_destroy_lock(omp_lock_t* lock_p    /* in/out */);
```

The type and the functions are specified in `omp.h`. The first function initializes the lock so that it's unlocked, that is, no thread owns the lock. The second function attempts to set the lock. If it succeeds, the calling thread proceeds; if it fails, the calling thread blocks until the lock becomes available. The third function unsets the lock so another thread can acquire it. The fourth function makes the lock uninitialized. We'll only use simple locks. For information about nested locks, see [9], [10], or [47].

5.8.9 Using locks in the message-passing program

In our earlier discussion of the limitations of the `critical` directive, we saw that in the message-passing program, we wanted to ensure mutual exclusion in each individual message queue, not in a particular block of source code. Locks allow us to do this. If we include a data member with type `omp_lock_t` in our queue struct, we can simply call `omp_set_lock` each time we want to ensure exclusive access to a message queue. So the code

```
#   pragma omp critical
    /* q_p = msg_queues[dest] */
    Enqueue(q_p, my_rank, mesg);
```

can be replaced with

```
/* q_p = msg_queues[dest] */
omp_set_lock(&q_p->lock);
Enqueue(q_p, my_rank, mesg);
omp_unset_lock(&q_p->lock);
```

Similarly, the code

```
#   pragma omp critical
    /* q_p = msg_queues[my_rank] */
    Dequeue(q_p, &src, &mesg);
```

can be replaced with

```
/* q_p = msg_queues[my_rank] */
omp_set_lock(&q_p->lock);
Dequeue(q_p, &src, &mesg);
omp_unset_lock(&q_p->lock);
```

Now when a thread tries to send or receive a message, it can only be blocked by a thread attempting to access the same message queue, since different message queues have different locks. In our original implementation, only one thread could send at a time, regardless of the destination.

Note that it would also be possible to put the calls to the lock functions in the queue functions `Enqueue` and `Dequeue`. However, to preserve the performance of `Dequeue`, we would also need to move the code that determines the size of the queue (enqueued − dequeued) to `Dequeue`. Without it, the `Dequeue` function will lock the queue every time it is called by `Try_receive`. In the interest of preserving the structure of the code we've already written, we'll leave the calls to `omp_set_lock` and `omp_unset_lock` in the `Send` and `Try_receive` functions.

Since we're now including the lock associated with a queue in the queue struct, we can add initialization of the lock to the function that initializes an empty queue. Destruction of the lock can be done by the thread that owns the queue before it frees the queue.

5.8.10 Critical **directives,** atomic **directives, or locks?**

Now that we have three mechanisms for enforcing mutual exclusion in a critical section, it's natural to wonder when one method is preferable to another. In general, the `atomic` directive has the potential to be the fastest method of obtaining mutual exclusion. Thus if your critical section consists of an assignment statement having the required form, it will probably perform at least as well with the `atomic` directive as the other methods. However, the OpenMP specification [47] allows the `atomic` directive to enforce mutual exclusion across *all* `atomic` directives in the program—this is the way the unnamed `critical` directive behaves. If this might be a problem—for example, you have multiple different critical sections protected by `atomic` directives—you should use named `critical` directives or locks. For example, suppose we have a program in which it's possible that one thread will execute the code on the left while another executes the code on the right.

```
#  pragma omp atomic          #  pragma omp atomic
   x++;                           y++;
```

Even if x and y are unrelated memory locations, it's possible that if one thread is executing x++, then no thread can simultaneously execute y++. It's important to note that the standard doesn't require this behavior. If two statements are protected by `atomic` directives and the two statements modify different variables, then there are implementations that treat the two statements as different critical sections. (See Exercise 5.10.) On the other hand, different statements that modify the same variable *will* be treated as if they belong to the same critical section, regardless of the implementation.

We've already seen some limitations to the use of `critical` directives. However, both named and unnamed `critical` directives are very easy to use. Furthermore, in the implementations of OpenMP that we've used there doesn't seem to be a very large difference between the performance of critical sections protected by a critical directive, and `critical` sections protected by locks, so if you can't use an `atomic` directive, but you can use a `critical` directive, you probably should. Thus the use of locks should probably be reserved for situations in which mutual exclusion is needed for a data structure rather than a block of code.

5.8.11 Some caveats

You should exercise caution when using the mutual exclusion techniques we've discussed. They can definitely cause serious programming problems. Here are a few things to be aware of:

1. You shouldn't mix the different types of mutual exclusion for a single critical section. For example, suppose a program contains the following two segments:

```
#   pragma omp atomic          #   pragma omp critical
        x += f(y);                     x = g(x);
```

The update to x on the right doesn't have the form required by the `atomic` directive, so the programmer used a `critical` directive. However, the `critical` directive won't exclude the action executed by the `atomic` block, and it's possible that the results will be incorrect. The programmer needs to either rewrite the function g so that its use can have the form required by the `atomic` directive or to protect both blocks with a `critical` directive.

2. There is no guarantee of **fairness** in mutual exclusion constructs. This means that it's possible that a thread can be blocked forever in waiting for access to a critical section. For example, in the code

```
    while (1) {
        . . .
#       pragma omp critical
        x = g(my_rank);
        . . .
    }
```

it's possible that, for example, thread 1 can block forever waiting to execute `x = g(my_rank)` while the other threads repeatedly execute the assignment. Of course, this wouldn't be an issue if the loop terminated.

3. It can be dangerous to "nest" mutual exclusion constructs. As an example, suppose a program contains the following two segments:

```
#   pragma omp critical
        y = f(x);
        . . .
    double f(double x) {
#       pragma omp critical
        z = g(x);    /* z is shared */
        . . .
    }
```

This is guaranteed to **deadlock**. When a thread attempts to enter the second critical section, it will block forever. If thread u is executing code in the first critical block, no thread can execute code in the second block. In particular, thread u can't execute this code. However, if thread u is blocked waiting to enter the second critical block, then it will never leave the first, and it will stay blocked forever.

In this example, we can solve the problem by using named critical sections. That is, we could rewrite the code as

```
#   pragma omp critical(one)
      y = f(x);
      . . .

      double f(double x) {
#       pragma omp critical(two)
          z = g(x);  /* z is global */
          . . .

      }
```

However, it's not difficult to come up with examples when naming won't help. For example, if a program has two named critical sections—say one and two—and threads can attempt to enter the critical sections in different orders, then deadlock can occur. For example, suppose thread u enters one at the same time that thread v enters two and u then attempts to enter two while v attempts to enter one:

Time	Thread u	Thread v
0	Enter crit. sect. one	Enter crit. sect. two
1	Attempt to enter two	Attempt to enter one
2	Block	Block

Then both u and v will block forever waiting to enter the critical sections. So it's not enough to just use different names for the critical sections—the programmer must ensure that different critical sections are always entered in the same order.

5.9 Caches, cache coherence, and false sharing[7]

Recall that for a number of years now, computers have been able to execute operations involving only the processor much faster than they can access data in main memory. If a processor must read data from main memory for each operation, it will spend most of its time simply waiting for the data from memory to arrive. Also recall that to address this problem, chip designers have added blocks of relatively fast memory to processors. This faster memory is called **cache memory**.

---

[7] This material is also covered in Chapter 4. So if you've already read that chapter, you may want to just skim this section.

Table 5.5 Memory and cache accesses.

Time	Memory	Th 0	Th 0 cache	Th 1	Th 1 cache
0	x = 5	Load x	—	Load x	—
1	x = 5	—	x = 5	—	x = 5
2	x = 5	x++	x = 5	—	x = 5
3	???	—	x = 6	my_z = x	???

The design of cache memory takes into consideration the principles of **temporal and spatial locality**: if a processor accesses main memory location x at time t, then it is likely that at times close to t it will access main memory locations close to x. Thus if a processor needs to access main memory location x, rather than transferring only the contents of x to/from main memory, a block of memory containing x is transferred from/to the processor's cache. Such a block of memory is called a **cache line** or **cache block**.

In Section 2.3.5, we saw that the use of cache memory can have a huge impact on shared memory. Let's recall why. First, consider the following situation: Suppose x is a shared variable with the value five, and both thread 0 and thread 1 read x from memory into their (separate) caches, because both want to execute the statement

```
my_y = x;
```

Here, `my_y` is a private variable defined by both threads. Now suppose thread 0 executes the statement

```
x++;
```

Finally, suppose that thread 1 now executes

```
my_z = x;
```

where `my_z` is another private variable. Table 5.5 illustrates the sequence of accesses.

What's the value in `my_z`? Is it five? Or is it six? The problem is that there are (at least) three copies of x: the one in main memory, the one in thread 0's cache, and the one in thread 1's cache. When thread 0 executed `x++`, what happened to the values in main memory and thread 1's cache? This is the **cache coherence** problem, which we discussed in Chapter 2. We saw there that most systems insist that the caches be made aware that changes have been made to data they are caching. The line in the cache of thread 1 would have been marked *invalid* when thread 0 executed `x++`, and before assigning `my_z = x`, the core running thread 1 would see that its value of x was out of date. Thus the core running thread 0 would have to update the copy of x in main memory (either now or earlier), and the core running thread 1 could get the line with the updated value of x from main memory. For further details, see Chapter 2.

The use of cache coherence can have a dramatic effect on the performance of shared-memory systems. To illustrate this, let's take a look at matrix-vector multiplication. Recall that if $A = (a_{ij})$ is an $m \times n$ matrix and $\mathbf{x}$ is a vector with n components, then their product $\mathbf{y} = A\mathbf{x}$ is a vector with m components, and its ith component y_i is

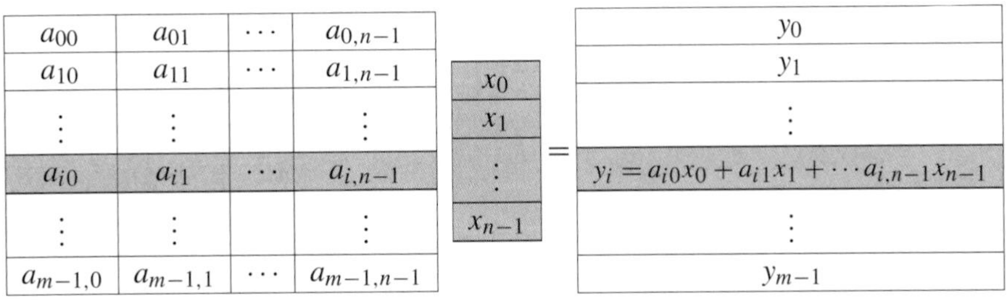

a_{00}	a_{01}	$\cdots$	$a_{0,n-1}$
a_{10}	a_{11}	$\cdots$	$a_{1,n-1}$
$\vdots$	$\vdots$		$\vdots$
a_{i0}	a_{i1}	$\cdots$	$a_{i,n-1}$
$\vdots$	$\vdots$		$\vdots$
$a_{m-1,0}$	$a_{m-1,1}$	$\cdots$	$a_{m-1,n-1}$

x_0
x_1
$\vdots$
x_{n-1}

=

y_0
y_1
$\vdots$
$y_i = a_{i0}x_0 + a_{i1}x_1 + \cdots a_{i,n-1}x_{n-1}$
$\vdots$
y_{m-1}

FIGURE 5.6

Matrix-vector multiplication.

found by forming the dot product of the ith row of A with **x**:

$$y_i = a_{i0}x_0 + a_{i1}x_1 + \cdots + a_{i,n-1}x_{n-1}.$$

See Fig. 5.6.

So if we store A as a two-dimensional array and **x** and **y** as one-dimensional arrays, we can implement serial matrix-vector multiplication with the following code:

```
for (i = 0; i < m; i++) {
    y[i] = 0.0;
    for (j = 0; j < n; j++)
        y[i] += A[i][j]*x[j];
}
```

There are no loop-carried dependences in the outer loop, since A and x are never updated and iteration i only updates y[i]. Thus we can parallelize this by dividing the iterations in the outer loop among the threads:

```
1  #   pragma omp parallel for num_threads(thread_count) \
2          default(none) private(i, j) shared(A, x, y, m, n)
3      for (i = 0; i < m; i++) {
4          y[i] = 0.0;
5          for (j = 0; j < n; j++)
6              y[i] += A[i][j]*x[j];
7      }
```

If T_{serial} is the run-time of the serial program, and T_{parallel} is the run-time of the parallel program, recall that the *efficiency* E of the parallel program is the speedup S divided by the number of threads:

$$E = \frac{S}{t} = \frac{\left(\dfrac{T_{\text{serial}}}{T_{\text{parallel}}}\right)}{t} = \frac{T_{\text{serial}}}{t \times T_{\text{parallel}}}.$$

Table 5.6 Run-times and efficiencies of matrix-vector multiplication (times are in seconds).

	Matrix Dimension					
	8,000,000 × 8		8000 × 8000		8 × 8,000,000	
Threads	Time	Eff.	Time	Eff.	Time	Eff.
1	0.322	1.000	0.264	1.000	0.333	1.000
2	0.219	0.735	0.189	0.698	0.300	0.555
4	0.141	0.571	0.119	0.555	0.303	0.275

Since $S \le t$, $E \le 1$. Table 5.6 shows the run-times and efficiencies of our matrix-vector multiplication with different sets of data and differing numbers of threads. In each case, the total number of floating point additions and multiplications is 64,000,000. An analysis that only considers arithmetic operations would predict that a single thread running the code would take the same amount of time for all three inputs. However, it's clear that this is *not* the case. The 8,000,000 × 8 system requires about 22% more time than the 8000 × 8000 system, and the 8 × 8,000,000 system requires about 26% more time than the 8000 × 8000 system. Both of these differences are at least partially attributable to cache performance.

Recall that a *write-miss* occurs when a core tries to update a variable that's not in cache, and it has to access main memory. A cache profiler (such as Valgrind [51]) shows that when the program is run with the 8,000,000 × 8 input, it has far more cache write-misses than either of the other inputs. The bulk of these occur in Line 4. Since the number of elements in the vector y is far greater in this case (8,000,000 vs. 8000 or 8), and each element must be initialized, it's not surprising that this line slows down the execution of the program with the 8,000,000 × 8 input.

Also recall that a *read-miss* occurs when a core tries to read a variable that's not in the cache, and it has to access main memory. A cache profiler shows that when the program is run with the 8 × 8,000,000 input, it has far more cache read-misses than either of the other inputs. These occur in Line 6, and a careful study of this program (see Exercise 5.12) shows that the main source of the differences is due to the reads of x. Once again, this isn't surprising, since for this input, x has 8,000,000 elements, versus only 8000 or 8 for the other inputs.

It should be noted that there may be other factors that are affecting the relative performance of the single-threaded program with the differing inputs. For example, we haven't taken into consideration whether virtual memory (see Subsection 2.2.4) has affected the performance of the program with the different inputs. How frequently does the CPU need to access the page table in main memory?

Of more interest to us, though, are the differences in efficiency as the number of threads is increased. The two-thread efficiency of the program with the 8 × 8,000,000 input is more than 20% less than the efficiency of the program with the 8,000,000 × 8 and the 8000 × 8000 inputs. The four-thread efficiency of the program with the 8 × 8,000,000 input is more than 50% less than the program's efficiency with the

8,000,000 × 8 and the 8000 × 8000 inputs. Why, then, is the multithreaded performance of the program so much worse with the 8 × 8,000,000 input?

In this case, once again, the answer has to do with cache. Let's take a look at the program when we run it with four threads. With the 8,000,000 × 8 input, y has 8,000,000 components, so each thread is assigned 2,000,000 components. With the 8000 × 8000 input, each thread is assigned 2000 components of y, and with the 8 × 8,000,000 input, each thread is assigned two components. On the system we used, a cache line is 64 bytes. Since the type of y is **double**, and a **double** is 8 bytes, a single cache line will store eight **double**s.

Cache coherence is enforced at the "cache-line level." That is, each time any value in a cache line is written, if the line is also stored in another core's cache, the entire *line* will be invalidated—not just the value that was written. The system we're using has two dual-core processors and each processor has its own cache. Suppose for the moment that threads 0 and 1 are assigned to one of the processors and threads 2 and 3 are assigned to the other. Also suppose that for the 8 × 8,000,000 problem all of y is stored in a single cache line. Then every write to some element of y will invalidate the line in the other processor's cache. For example, each time thread 0 updates y[0] in the statement

```
y[i] += A[i][j]*x[j];
```

if thread 2 or 3 is executing this code, it will have to reload y. Each thread will update each of its components 8,000,000 times. We see that with this assignment of threads to processors and components of y to cache lines, all the threads will have to reload y *many* times. This is going to happen in spite of the fact that only one thread accesses any one component of y—for example, only thread 0 accesses y[0].

Each thread will update its assigned components of y a total of 16,000,000 times. It appears that many if not most of these updates are forcing the threads to access main memory. This is called **false sharing**. Suppose two threads with separate caches access different variables that belong to the same cache line. Further suppose at least one of the threads updates its variable. Then even though neither thread has written to a shared variable, the cache controller invalidates the entire cache line and forces the other threads to get the values of the variables from main memory. The threads aren't sharing anything (except a cache line), but the behavior of the threads with respect to memory access is the same as if they were sharing a variable, hence the name *false sharing*.

Why is false sharing not a problem with the other inputs? Let's look at what happens with the 8000 × 8000 input. Suppose thread 2 is assigned to one of the processors and thread 3 is assigned to another. (We don't actually know which threads are assigned to which processors, but it turns out—see Exercise 5.13—that it doesn't matter.) Thread 2 is responsible for computing

y[4000], y[4001], . . . , y[5999],

and thread 3 is responsible for computing

$y[6000]$, $y[6001]$, . . . , $y[7999]$.

If a cache line contains eight consecutive **double**s, the only possibility for false sharing is on the interface between their assigned elements. If, for example, a single cache line contains

$y[5996]$, $y[5997]$, $y[5998]$, $y[5999]$,
$y[6000]$, $y[6001]$, $y[6002]$, $y[6003]$,

then it's conceivable that there might be false sharing of this cache line. However, thread 2 will access

$y[5996]$, $y[5997]$, $y[5998]$, $y[5999]$

at the *end* of its **for** i loop, while thread 3 will access

$y[6000]$, $y[6001]$, $y[6002]$, $y[6003]$

at the *beginning* of its iterations. So it's very likely that when thread 2 accesses, say, $y[5996]$, thread 3 will be long done with all four of

$y[6000]$, $y[6001]$, $y[6002]$, $y[6003]$.

Similarly, when thread 3 accesses, say, $y[6003]$, it's very likely that thread 2 won't be anywhere near starting to access

$y[5996]$, $y[5997]$, $y[5998]$, $y[5999]$.

It's therefore unlikely that false sharing of the elements of y will be a significant problem with the 8000×8000 input. Similar reasoning suggests that false sharing of y is unlikely to be a problem with the $8,000,000 \times 8$ input. Also note that we don't need to worry about false sharing of A or x, since their values are never updated by the matrix-vector multiplication code.

This brings up the question of how we might avoid false sharing in our matrix-vector multiplication program. One possible solution is to "pad" the y vector with dummy elements to ensure that any update by one thread won't affect another thread's cache line. Another alternative is to have each thread use its own private storage during the multiplication loop, and then update the shared storage when they're done. (See Exercise 5.15.)

5.10 Tasking

While many problems are straightforward to parallelize with OpenMP, they generally have a fixed or predetermined number of parallel blocks and loop iterations to schedule across participating threads. When this is not the case, the constructs we've seen thus far make it difficult (or even impossible) to effectively parallelize the problem at hand. Consider, for instance, parallelizing a web server; HTTP requests may arrive at irregular times, and the server itself should ideally be able to respond to a potentially infinite number of requests. This is easy to conceptualize using a **while** loop, but recall our discussion in Section 5.5.1: **while** and **do−while** loops cannot be parallelized with OpenMP, nor can **for** loops that have an unbounded number of iterations. This poses potential issues for dynamic problems, including recursive algorithms, such as graph traversals, or producer-consumer style programs like web servers. To address these issues, OpenMP 3.0 introduced *Tasking* functionality [47]. Tasking has been successfully applied to a number of problems that were previously difficult to parallelize with OpenMP [1].

Tasking allows developers to specify independent units of computation with the `task` directive:

#pragma omp task

When a thread reaches a block of code with this directive, a new task is generated by the OpenMP run-time that will be scheduled for execution. It is important to note that the task will not necessarily be executed immediately, since there may be other tasks already pending execution. Task blocks behave similarly to a standard `parallel` region, but can launch an arbitrary number of tasks instead of only `num_threads`. In fact, tasks must be launched from within a `parallel` region but generally by only one of the threads in the team. Therefore a majority of programs that use Tasking functionality will contain an outer region that looks somewhat like:

```
#    pragma omp parallel
#    pragma omp single
     {
         . . .
#        pragma omp task
         . . .
     }
```

where the `parallel` directive creates a team of threads and the `single` directive instructs the runtime to only launch tasks from a single thread. If the `single` directive is omitted, subsequent `task` instances will be launched multiple times, one for each thread in the team.

To demonstrate OpenMP tasking functionality, recall our discussion on parallelizing the calculation of the first *n* Fibonacci numbers in Section 5.5.2. Due to the loop-carried dependence, results were unpredictable and, more importantly, often incorrect. However, we *can* parallelize this algorithm with the `task` directive. First, let's take a look at a recursive serial implementation that stores the sequence in a global array called `fibs`:

```
int fib(int n) {
    int i = 0;
    int j = 0;
```

```
if (n <= 1) {
    // fibs is a global variable
    // It needs storage for n+1 ints
    fibs[n] = n;
    return n;
}

i = fib(n - 1);
j = fib(n - 2);
fibs[n] = i + j;
return fibs[n];
}
```

This chain of recursive calls will be time-consuming, so let's execute each as a separate task that can run in parallel. We can do this by adding a `parallel` and a `single` directive before the initial (nonrecursive) call that starts `fib`, and adding **#pragma** omp task before each of the two recursive calls in `fib`. However, after we make this change, the results are incorrect—more specifically, except for `fib[1]`, the sequence is all zeroes. This is because the default data scope for variables in tasks is private. So after completing each of the tasks

```
#   pragma omp task
i = fib(n - 1);
#   pragma omp task
j = fib(n - 2);
```

the results in `i` and `j` are lost: `i` and `j` retain their values from the initializations

```
int i = 0;
int j = 0;
```

at the beginning of the function. In other words, the memory locations that are assigned the results of `fib(n−1)` and `fib(n−2)` are not the same as the memory locations declared at the beginning of the function. So the values that are used to update `fibs[n]` are the zeroes assigned at the beginning of the function.

We can adjust the scope of `i` and `j` by declaring the variables to be `shared` in the tasks that execute the recursive call. However executing the program now will produce unpredictable results similar to our original attempt at parallelization. The problem here is that the order in which the various tasks execute isn't specified. In other words, our recursive function calls, `fib(n − 1)` and `fib(n − 2)` will be run eventually, but the thread executing the task that makes the recursive calls can continue to run and simply **return** the current value of `fibs[n]` early. We need to force this task to wait for its subtasks to complete with the `taskwait` directive, which operates as a `barrier` for tasks. We've put this all together in Program 5.6.

```
1   int fib(int n) {
2       int i = 0;
3       int j = 0;
4
5       if (n <= 1) {
6           fibs[n] = n;
7           return n;
8       }
9
10  #   pragma omp task shared(i)
11      i = fib(n - 1);
12
13  #   pragma omp task shared(j)
14      j = fib(n - 2);
15
16  #   pragma omp taskwait
17      fibs[n] = i + j;
18      return fibs[n];
19  }
```

Program 5.6: Computing the Fibonacci numbers using OpenMP tasks.

Our parallel Fibonacci program will now produce the correct results, but you may notice significant slowdowns with larger values of n; in fact, there is a good chance that the serial version of the program executes much faster! To gain an intuition as to why this occurs, recall our discussion of the overhead associated with forking and joining threads. Similarly, each task requires its own data environment to be generated upon creation, which takes time. There are a few options we can use to help reduce task creation overhead. The first option is to only create tasks in situations where n is large enough. We can do this with the **if** directive:

#pragma omp task shared(i) **if**(n > 20)

which in this case will restrict task creation to only occur when n is larger than 20 (chosen arbitrarily in this case based on some experimentation). Reviewing fib again, we can see that there will be a task executing fib itself, another executing fib($n - 1$), and a third executing fib($n - 2$) for each recursive call. This is inefficient, because the parent task executing fib only launches two subtasks and then simply waits for their results. We can eliminate a task by having the parent thread perform one of the recursive calls to fib instead before doing the final calculation after the taskwait directive. On our 64-core testbed, these two changes halved the execution time of the program with $n = 35$.

While using the Tasking API requires a bit more planning and care to use—especially with data scoping and limiting runaway task creation—it allows a much broader set of problems to be parallelized by OpenMP.

5.11 Thread-safety[8]

Let's look at another potential problem that occurs in shared-memory programming: *thread-safety*. A block of code is **thread-safe** if it can be simultaneously executed by multiple threads without causing problems.

As an example, suppose we want to use multiple threads to "tokenize" a file. Let's suppose that the file consists of ordinary English text, and that the tokens are just contiguous sequences of characters separated from the rest of the text by white space—a space, a tab, or a newline. A simple approach to this problem is to divide the input file into lines of text and assign the lines to the threads in a round-robin fashion: the first line goes to thread 0, the second goes to thread 1, ..., the tth goes to thread t, the $t + 1$st goes to thread 0, and so on.

We'll read the text into an array of strings, with one line of text per string. Then we can use a `parallel` **for** directive with a `schedule(`**static**`,1)` clause to divide the lines among the threads.

One way to tokenize a line is to use the `strtok` function in `string.h`. It has the following prototype:

```
char* strtok(
        char*         string       /* in/out */,
        const char*   separators    /* in     */);
```

Its usage is a little unusual: the first time it's called the `string` argument should be the text to be tokenized, so in our example it should be the line of input. For subsequent calls, the first argument should be `NULL`. The idea is that in the first call, `strtok` caches a pointer to `string`, and for subsequent calls it returns successive tokens taken from the cached copy. The characters that delimit tokens should be passed in `separators`, so we should pass in the string `" \t\n"` as the `separators` argument.

Given these assumptions, we can write the `Tokenize` function shown in Program 5.7. The main function has initialized the array `lines` so that it contains the input text, and `line_count` is the number of strings stored in `lines`. Although for our purposes, we only need the `lines` argument to be an input argument, the `strtok` function modifies its input. Thus when `Tokenize` returns, `lines` will be modified. When we run the program with a single thread, it correctly tokenizes the input stream. The first time we run it with two threads and the input

> *Pease porridge hot.*
> *Pease porridge cold.*
> *Pease porridge in the pot*
> *Nine days old.*

the output is also correct. However, the second time we run it with this input, we get the following output:

---

[8] This material is also covered in Chapter 4. So if you've already read that chapter, you may want to just skim this section.

```
1   void Tokenize(
2         char*    lines[]        /* in/out */,
3         int      line_count     /* in      */,
4         int      thread_count   /* in      */) {
5      int my_rank, i, j;
6      char *my_token;
7
8   #  pragma omp parallel num_threads(thread_count) \
9         default(none) private(my_rank, i, j, my_token) \
10        shared(lines, line_count)
11     {
12        my_rank = omp_get_thread_num();
13  #     pragma omp for schedule(static, 1)
14        for (i = 0; i < line_count; i++) {
15           printf("Thread %d > line %d = %s",
16                    my_rank, i, lines[i]);
17           j = 0;
18           my_token = strtok(lines[i], " \t\n");
19           while ( my_token != NULL ) {
20              printf("Thread %d > token %d = %s\n",
21                       my_rank, j, my_token);
22              my_token = strtok(NULL, " \t\n");
23              j++;
24           }
25        } /* for i */
26     } /* omp parallel */
27
28  } /* Tokenize */
```

Program 5.7: A first attempt at a multi-threaded tokenizer.

```
Thread 0 > line 0 = Pease porridge hot.
Thread 1 > line 1 = Pease porridge cold.
Thread 0 > token 0 = Pease
Thread 1 > token 0 = Pease
Thread 0 > token 1 = porridge
Thread 1 > token 1 = cold.
Thread 0 > line 2 = Pease porridge in the pot
Thread 1 > line 3 = Nine days old.
Thread 0 > token 0 = Pease
Thread 1 > token 0 = Nine
Thread 0 > token 1 = days
Thread 1 > token 1 = old.
```

What happened? Recall that strtok caches the input line. It does this by declaring a variable to have static storage class. This causes the value stored in this variable to persist from one call to the next. Unfortunately for us, this cached string is shared, not private. Thus it appears that thread 1's call to strtok with the second line has apparently overwritten the contents of thread 0's call with the first line. Even worse, thread 0 has found a token ("days") that should be in thread 1's output.

The strtok function is therefore *not* thread-safe: if multiple threads call it simultaneously, the output it produces may not be correct. Regrettably, it's not uncommon for C library functions to fail to be thread-safe. For example, neither the random number generator rand in stdlib.h nor the time conversion function localtime in time.h is guaranteed to be thread-safe. In some cases, the C standard specifies an alternate, thread-safe version of a function. In fact, there is a thread-safe version of strtok:

```
char* strtok_r(
        char*           string          /* in/out */,
        const char*     separators      /* in      */,
        char**          saveptr_p       /* in/out */);
```

The "_r" is supposed to suggest that the function is *re-entrant*, which is sometimes used as a synonym for thread-safe.[9] The first two arguments have the same purpose as the arguments to strtok. The saveptr argument is used by strtok_r for keeping track of where the function is in the input string; it serves the purpose of the cached pointer in strtok. We can correct our original Tokenize function by replacing the calls to strtok with calls to strtok_r. We simply need to declare a **char*** variable to pass in for the third argument, and replace the calls in line 18 and line 22 with the following calls:

```
my_token = strtok_r(lines[i], " \t\n", &saveptr);
. . .
my_token = strtok_r(NULL, " \t\n", &saveptr);
```

respectively.

5.11.1 Incorrect programs can produce correct output

Notice that our original version of the tokenizer program shows an especially insidious form of program error: The first time we ran it with two threads, the program produced correct output. It wasn't until a later run that we saw an error. This, unfortunately, is not a rare occurrence in parallel programs. It's especially common in shared-memory programs. Since, for the most part, the threads are running independently of each other, as we noted back at the beginning of the chapter, the exact

---

[9] However, the distinction is a bit more nuanced; being reentrant means a function can be interrupted and called again (reentered) in different parts of a program's control flow and still execute correctly. This can happen due to nested calls to the function or a trap/interrupt sent from the operating system. Since strtok uses a single static pointer to track its state while parsing, multiple calls to the function from different parts of a program's control flow will corrupt the string—therefore it is *not* reentrant. It's worth noting that although reentrant functions, such as strtok_r, can also be thread safe, there is no guarantee a reentrant function will *always* be thread safe—and vice versa. It's best to consult the documentation if there's any doubt.

sequence of statements executed is nondeterministic. For example, we can't say when thread 1 will first call `strtok`. If its first call takes place after thread 0 has tokenized its first line, then the tokens identified for the first line should be correct. However, if thread 1 calls `strtok` before thread 0 has finished tokenizing its first line, it's entirely possible that thread 0 may not identify all the tokens in the first line, so it's especially important in developing shared-memory programs to resist the temptation to assume that since a program produces correct output, it must be correct. We always need to be wary of race conditions.

5.12 Summary

OpenMP is a standard for programming shared-memory MIMD systems. It uses both special functions and preprocessor directives called **pragmas**, so unlike Pthreads and MPI, OpenMP requires compiler support. One of the most important features of OpenMP is that it was designed so that developers could *incrementally* parallelize existing serial programs, rather than having to write parallel programs from scratch.

OpenMP programs start multiple **threads** rather than multiple processes. Threads can be much lighter weight than processes; they can share almost all the resources of a process, except each thread must have its own stack and program counter.

To get OpenMP's function prototypes and macros, we include the `omp.h` header in OpenMP programs. There are several OpenMP directives that start multiple threads; the most general is the `parallel` directive:

```
#   pragma omp parallel
      structured block
```

This directive tells the run-time system to execute the following structured block of code in parallel. It may **fork** or start several threads to execute the structured block. A **structured block** is a block of code with a single entry point and a single exit point, although calls to the C library function `exit` are allowed within a structured block. The number of threads started is system dependent, but most systems will start one thread for each available core. The collection of threads executing `block of code` is called a **team**. One of the threads in the team is the thread that was executing the code before the `parallel` directive. This thread is called the **parent**. The additional threads started by the `parallel` directive are called **child** threads. When all of the threads are finished, the child threads are terminated or **joined**, and the parent thread continues executing the code beyond the structured block.

Many OpenMP directives can be modified by **clauses**. We made frequent use of the `num_threads` clause. When we use an OpenMP directive that starts a team of threads, we can modify it with the `num_threads` clause so that the directive will start the number of threads we desire.

When OpenMP starts a team of threads, each of the threads is assigned a rank or ID in the range $0, 1, \ldots,$ `thread_count` $- 1$. The OpenMP library function `omp_get_thread_num` the returns the calling thread's rank. The function `omp_get_num_threads` returns the number of threads in the current team.

A major problem in the development of shared-memory programs is the possibility of **race conditions**. A race condition occurs when multiple threads attempt to access a shared resource, at least one of the accesses is an update, and the accesses can result in an error. Code that is executed by multiple threads that update a shared resource that can only be updated by one thread at a time is called a **critical section**. Thus if multiple threads try to update a shared variable, the program has a race condition, and the code that updates the variable is a critical section. OpenMP provides several mechanisms for ensuring **mutual exclusion** in critical sections. We examined four of them:

1. `Critical` directives ensure that only one thread at a time can execute the structured block. If multiple threads try to execute the code in the critical section, all but one of them will block before the critical section. When one thread finishes the critical section, another thread will be unblocked and enter the code.
2. Named `critical` directives can be used in programs having different critical sections that can be executed concurrently. Multiple threads trying to execute code in critical section(s) with the same name will be handled in the same way as multiple threads trying to execute an unnamed critical section. However, threads entering critical sections with different names can execute concurrently.
3. An `atomic` directive can only be used when the critical section has the form `x <op> = <expression>`, `x++`, `++x`, `x--`, or `--x`. It's designed to exploit special hardware instructions, so it can be much faster than an ordinary critical section.
4. Simple locks are the most general form of mutual exclusion. They use function calls to restrict access to a critical section:

```
omp_set_lock(&lock);
critical section
omp_unset_lock(&lock);
```

When multiple threads call `omp_set_lock`, only one of them will proceed to the critical section. The others will block until the first thread calls `omp_unset_lock`. Then one of the blocked threads can proceed.

All of the mutual exclusion mechanisms can cause serious program problems, such as deadlock, so they need to be used with great care.

A **for** directive can be used to partition the iterations in a **for** loop among the threads. This directive doesn't start a team of threads; it divides the iterations in a **for** loop among the threads in an existing team. If we want also to start a team of threads, we can use the `parallel` **for** directive. There are a number of restrictions on the form of a **for** loop that can be parallelized; basically, the run-time system must be able to determine the total number of iterations through the loop body before the loop begins execution. For details, see Program 5.3.

It's not enough, however, to ensure that our **for** loop has one of the canonical forms. It must also not have any **loop-carried dependences**. A loop-carried dependence occurs when a memory location is read or written in one iteration and written in another iteration. OpenMP won't detect loop-carried dependences; it's up to us, the programmers, to detect and eliminate them. It may, however, be impossible to eliminate them, in which case the loop isn't a candidate for parallelization.

By default, most systems use a **block partitioning** of the iterations in a parallelized **for** loop. If there are n iterations, this means that roughly the first $n/thread\_count$ are assigned to thread 0, the next $n/thread\_count$ are assigned to thread 1, and so on. However, there are a variety of **scheduling** options provided by OpenMP. The schedule clause has the form

```
schedule(<type> [,<chunksize >])
```

The type can be static, dynamic, guided, auto, or runtime. In a static schedule, the iterations can be assigned to the threads before the loop starts execution. In dynamic and guided schedules, the iterations are assigned on the fly. When a thread finishes a chunk of iterations—a contiguous block of iterations—it requests another chunk. If **auto** is specified, the schedule is determined by the compiler or run-time system, and if runtime is specified, the schedule is determined at run-time by examining the environment variable OMP_SCHEDULE.

Only static, dynamic, and guided schedules can have a chunksize. In a static schedule, the chunks of chunksize iterations are assigned in round robin fashion to the threads. In a dynamic schedule, each thread is assigned chunksize iterations, and when a thread completes its chunk, it requests another chunk. In a guided schedule, the size of the chunks decreases as the iteration proceeds.

In OpenMP the **scope** of a variable is the collection of threads to which the variable is accessible. Typically, any variable that was defined before the OpenMP directive has **shared** scope within the construct. That is, all the threads have access to it. There are a couple of important exceptions to this. The first is that the loop variable in a **for** or parallel **for** construct is **private**, that is, each thread has its own copy of the variable. The second exception is that variables used in a task construct have private scope. Variables that are defined within an OpenMP construct have private scope, since they will be allocated from the executing thread's stack.

As a rule of thumb, it's a good idea to explicitly assign the scope of variables. This can be done by modifying a parallel or parallel **for** directive with the *scoping* clause:

default(none)

This tells the system that the scope of every variable that's used in the OpenMP construct must be explicitly specified. Most of the time this can be done with private or shared clauses.

The only exceptions we encountered were **reduction variables**. A **reduction operator** is an associative binary operation (such as addition or multiplication), and a **reduction** is a computation that repeatedly applies the same reduction operator to a sequence of operands to get a single result. Furthermore, all of the intermediate results of the operation should be stored in the same variable: the **reduction variable**. For example, if A is an array with n elements, then the code

```
int sum = 0;
for (i = 0; i < n; i++)
    sum += A[i];
```

is a reduction. The reduction operator is addition and the reduction variable is `sum`. If we try to parallelize this loop, the reduction variable should have properties of both private and shared variables. Initially we would like each thread to add its array elements into its own private `sum`, but when the threads are done, we want the private `sum`'s combined into a single, shared `sum`. OpenMP therefore provides the `reduction` clause for identifying reduction variables and operators.

A `barrier` directive will cause the threads in a team to block until all the threads have reached the directive. We've seen that the `parallel`, `parallel for`, and `for` directives have implicit barriers at the end of the structured block.

We recalled that modern microprocessor architectures use caches to reduce memory access times, so typical architectures have special hardware to ensure that the caches on the different chips are **coherent**. Since the unit of cache coherence, a **cache line** or **cache block**, is usually larger than a single word of memory, this can have the unfortunate side effect that two threads may be accessing different memory locations, but when the two locations belong to the same cache line, the cache-coherence hardware acts as if the threads were accessing the same memory location—if one of the threads updates its memory location, and then the other thread tries to read its memory location, it will have to retrieve the value from main memory. That is, the hardware is forcing the thread to act as if it were actually sharing the memory location. Hence, this is called **false sharing**, and it can seriously degrade the performance of a shared-memory program.

The `task` directive can be used to get parallel threads to execute different sequences of instructions. For example, if our code contains three functions, `f1`, `f2`, and `f3` that are independent of each other, then the serial code

```
f1(args1);
f2(args2);
f3(args3);
```

can be parallelized by using the following directives:

```
#   pragma omp parallel numthreads(3)
#       pragma omp single
        {
#           pragma omp task
            f1(args1);

#           pragma omp task
            f2(args2);

#           pragma omp task
            f3(args3);
        }
```

The `parallel` directive will start three threads. If we omitted the `single` directive, each function call would be executed by all three threads. The `single` directive ensures that only one of the threads calls each of the functions. If each of the functions returns a value and we want to (say) add these values, we can put a barrier before the addition by adding a `taskwait` directive:

```
#   pragma omp parallel numthreads(3)
#      pragma omp single
       {
#         pragma omp task shared(x1)
          x1 = f1(args1);

#         pragma omp task shared(x2)
          x2 = f2(args2);

#         pragma omp task shared(x3)
          x3 = f3(args3);

#         pragma omp taskwait
          x = x1 + x2 + x3;
       }
```

Some C functions cache data between calls by declaring variables to be `static`. This can cause errors when multiple threads call the function; since static storage is shared among the threads, one thread can overwrite another thread's data. Such a function is not **thread-safe**, and, unfortunately, there are several such functions in the C library. Sometimes, however, the library has a thread-safe variant of a function that isn't thread-safe.

In one of our programs, we saw a particularly insidious problem: when we ran the program with multiple threads and a fixed set of input, it sometimes produced correct output, even though it had an error. Producing correct output during testing doesn't guarantee that the program is in fact correct. It's up to us to identify possible race conditions.

5.13 Exercises

5.1 If it's defined, the `_OPENMP` macro is a decimal **int**. Write a program that prints its value. What is the significance of the value?

5.2 Download `omp_trap_1.c` from the book's website, and delete the `critical` directive. Now compile and run the program with more and more threads and larger and larger values of *n*. How many threads and how many trapezoids does it take before the result is incorrect?

5.3 Modify `omp_trap1.c` so that
   (i) it uses the first block of code on page 234, and
   (ii) the time used by the parallel block is timed using the OpenMP function `omp_get_wtime()`. The syntax is

```
double  omp_get_wtime ( void )
```

It returns the number of seconds that have elapsed since some time in the past. For details on taking timings, see Section 2.6.4. Also recall that OpenMP has a `barrier` directive:

```
#   pragma omp barrier
```

Now on a system with at least two cores, time the program with
  a. one thread and a large value of n, and
  b. two threads and the same value of n.
What happens? Download `omp_trap_2.c` from the book's website. How does its performance compare? Explain your answers.

5.4 Recall that OpenMP creates private variables for reduction variables, and these private variables are initialized to the identity value for the reduction operator. For example, if the operator is addition, the private variables are initialized to 0, while if the operator is multiplication, the private variables are initialized to 1. What are the identity values for the operators `&&`, `||`, `&`, `|`, `^`?

5.5 Suppose that on the amazing Bleeblon computer, variables with type **float** can store three decimal digits. Also suppose that the Bleeblon's floating point registers can store four decimal digits, and that after any floating point operation, the result is rounded to three decimal digits before being stored. Now suppose a C program declares an array `a` as follows:

```
float  a[] = {4.0, 3.0, 3.0, 1000.0};
```

  a. What is the output of the following block of code if it's run on the Bleeblon?

```
int  i;
float  sum = 0.0;
for  (i = 0; i < 4; i++)
    sum += a[i];
printf("sum = %4.1f\n", sum);
```

  b. Now consider the following code:

```
    int  i;
    float  sum = 0.0;
#   pragma omp parallel for num_threads(2) \
        reduction(+:sum)
    for  (i = 0; i < 4; i++)
        sum += a[i];
    printf("sum = %4.1f\n", sum);
```

Suppose that the run-time system assigns iterations $i = 0$, 1 to thread 0 and $i = 2$, 3 to thread 1. What is the output of this code on the Bleeblon?

5.6 Write an OpenMP program that determines the default scheduling of parallel **for** loops. Its input should be the number of iterations, and its output should be which iterations of a parallelized **for** loop are executed by which thread. For example, if there are two threads and four iterations, the output might be the following:

```
Thread 0: Iterations 0 — 1
Thread 1: Iterations 2 — 3
```

5.7 In our first attempt to parallelize the program for estimating π, our program was incorrect. In fact, we used the result of the program when it was run with one thread as evidence that the program run with two threads was incorrect. Explain why we could "trust" the result of the program when it was run with one thread.

5.8 Consider the loop

```
a[0] = 0;
for (i = 1; i < n; i++)
    a[i] = a[i−1] + i;
```

There's clearly a loop-carried dependence, as the value of a[i] can't be computed without the value of a[i−1]. Can you see a way to eliminate this dependence and parallelize the loop?

5.9 Modify the trapezoidal rule program that uses a `parallel` **for** directive (`omp_trap_3.c`) so that the `parallel` **for** is modified by a `schedule(runtime)` clause. Run the program with various assignments to the environment variable `OMP_SCHEDULE` and determine which iterations are assigned to which thread. This can be done by allocating an array `iterations` of n **int**s and in the Trap function assigning `omp_get_thread_num()` to `iterations[i]` in the ith iteration of the **for** loop. What is the default assignment of iterations on your system? How are `guided` schedules determined?

5.10 Recall that all structured blocks modified by an unnamed `critical` directive form a single critical section. What happens if we have a number of `atomic` directives in which different variables are being modified? Are they all treated as a single critical section?

We can write a small program that tries to determine this. The idea is to have all the threads simultaneously execute something like the following code:

```
        int i;
        double my_sum = 0.0;
        for (i = 0; i < n; i++)
#           pragma omp atomic
            my_sum += sin(i);
```

We can do this by modifying the code by a `parallel` directive:

```
#   pragma omp parallel num_threads(thread_count)
    {
        int i;
        double my_sum = 0.0;
        for (i = 0; i < n; i++)
```

```
#           pragma omp atomic
            my_sum += sin(i);
}
```

Note that since `my_sum` and `i` are declared in the `parallel` block, each thread has its own private copy. Now if we time this code for large n when `thread_count = 1` and we also time it when `thread_count > 1`, then as long as `thread_count` is less than the number of available cores, the run-time for the single-threaded run should be roughly the same as the time for the multi-threaded run if the different threads' executions of `my_sum += sin(i)` are treated as different critical sections. On the other hand, if the different executions of `my_sum += sin(i)` are all treated as a single critical section, the multithreaded run should be much slower than the single-threaded run. Write an OpenMP program that implements this test. Does your implementation of OpenMP allow simultaneous execution of updates to different variables when the updates are protected by `atomic` directives?

5.11 Recall that in C, a function that takes a two-dimensional array argument must specify the number of columns in the argument list, so it is quite common for C programmers to only use one-dimensional arrays, and to write explicit code for converting pairs of subscripts into a single dimension. (See Exercise 3.14.) Modify the OpenMP matrix-vector multiplication so that it uses a one-dimensional array for the matrix.

5.12 Download the source file `omp_mat_vect_rand_split.c` from the book's website. Find a program that does cache profiling (e.g., Valgrind [51]) and compile the program according to the instructions in the cache profiler documentation. (For example, with Valgrind you will want a symbol table and full optimization. With `gcc` use `gcc -g -02 .. ..`) Now run the program according to the instructions in the cache profiler documentation, using input $k \times (k \cdot 10^6)$, $(k \cdot 10^3) \times (k \cdot 10^3)$, and $(k \cdot 10^6) \times k$. Choose k so large that the number of level 2 cache misses is of the order 10^6 for at least one of the input sets of data.

a. How many level 1 cache write-misses occur with each of the three inputs?
b. How many level 2 cache write-misses occur with each of the three inputs?
c. Where do most of the write-misses occur? For which input data does the program have the most write-misses? Can you explain why?
d. How many level 1 cache read-misses occur with each of the three inputs?
e. How many level 2 cache read-misses occur with each of the three inputs?
f. Where do most of the read-misses occur? For which input data does the program have the most read-misses? Can you explain why?
g. Run the program with each of the three inputs but without using the cache profiler. With which input is the program the fastest? With which input is the program the slowest? Can your observations about cache misses help explain the differences? How?

5.13 Recall the matrix-vector multiplication example with the 8000×8000 input. Suppose that thread 0 and thread 2 are assigned to different processors. If a cache line contains 64 bytes or 8 **doubles**, is it possible for false sharing between threads 0 and 2 to occur for any part of the vector y? Why? What about if thread 0 and thread 3 are assigned to different processors; is it possible for false sharing to occur between them for any part of y?

5.14 Recall the matrix-vector multiplication example with an $8 \times 8,000,000$ matrix. Suppose that **doubles** use 8 bytes of memory and that a cache line is 64 bytes. Also suppose that our system consists of two dual-core processors.

   a. What is the minimum number of cache lines that are needed to store the vector y?

   b. What is the maximum number of cache lines that are needed to store the vector y?

   c. If the boundaries of cache lines always coincide with the boundaries of 8-byte **doubles**, in how many different ways can the components of y be assigned to cache lines?

   d. If we only consider which pairs of threads share a processor, in how many different ways can four threads be assigned to the processors in our computer? Here we're assuming that cores on the same processor share cache.

   e. Is there an assignment of components to cache lines and threads to processors that will result in no false-sharing in our example? In other words, is it possible that the threads assigned to one processor will have their components of y in one cache line and the threads assigned to the other processor will have their components in a different cache line?

   f. How many assignments of components to cache lines and threads to processors are there?

   g. Of these assignments, how many will result in no false sharing?

5.15 a. Modify the matrix-vector multiplication program so that it pads the vector y when there's a possibility of false sharing. The padding should be done so that if the threads execute in lock-step, there's no possibility that a single cache line containing an element of y will be shared by two or more threads. Suppose, for example, that a cache line stores eight **doubles** and we run the program with four threads. If we allocate storage for at least 48 **doubles** in y, then, on each pass through the **for** i loop, there's no possibility that two threads will simultaneously access the same cache line.

   b. Modify the matrix-vector multiplication program so that each thread uses private storage for its part of y during the **for** i loop. When a thread is done computing its part of y, it should copy its private storage into the shared variable.

   c. How does the performance of these two alternatives compare to the original program? How do they compare to each other?

5.16 The following code can be used as the basis of an implementation of merge sort:

```
/* Sort elements of list from list[lo] to list[hi] */
void Mergesort(int list[], int lo, int hi) {
    if (lo < hi) {
        int mid = (lo + hi)/2;
        Mergesort(list, lo, mid);
        Mergesort(list, mid+1, hi);
        Merge(list, lo, mid, hi);
    }
}   /* Mergesort */
```

The Mergesort function can be called from the following main function:

```
int main(int argc, char* argv[])
    int *list, n;

    Get_args(argc, argv, &n);
    list = malloc(n*sizeof(int));
    Get_list(list, n);
    Mergesort(list, 0, n-1);
    Print_list(list, n);
    free(list);
    return 0;
}   /* main */
```

Use the task directive (and any other OpenMP directives) to implement a parallel merge sort program.

5.17 Although strtok_r is thread-safe, it has the rather unfortunate property that it gratuitously modifies the input string. Write a tokenizer that is thread-safe and doesn't modify the input string.

5.14 Programming assignments

5.1 Use OpenMP to implement the parallel histogram program discussed in Chapter 2.

5.2 Suppose we toss darts randomly at a square dartboard whose bullseye is at the origin and whose sides are two feet in length. Suppose also that there's a circle inscribed in the square dartboard. The radius of the circle is 1 foot, and its area is π square feet. If the points that are hit by the darts are uniformly distributed (and we always hit the square), then the number of darts that hit inside the circle should approximately satisfy the equation

$$\frac{\text{number in circle}}{\text{total number of tosses}} = \frac{\pi}{4},$$

since the ratio of the area of the circle to the area of the square is $\pi/4$.

We can use this formula to estimate the value of π with a random number generator:

```
number_in_circle = 0;
for (toss = 0; toss < number_of_tosses; toss++) {
    x = random double between −1 and 1;
    y = random double between −1 and 1;
    distance_squared = x*x + y*y;
    if (distance_squared <= 1) number_in_circle++;
}

pi_estimate
    = 4*number_in_circle / ((double) number_of_tosses);
```

This is called a "Monte Carlo" method, since it uses randomness (the dart tosses). Write an OpenMP program that uses a Monte Carlo method to estimate π. Read in the total number of tosses before forking any threads. Use a reduction clause to find the total number of darts hitting inside the circle. Print the result after joining all the threads. You may want to use **long long int**s for the number of hits in the circle and the number of tosses, since both may have to be very large to get a reasonable estimate of π.

5.3 Count sort is a simple serial sorting algorithm that can be implemented as follows:

```
void Count_sort(int a[], int n) {
    int i, j, count;
    int* temp = malloc(n*sizeof(int));

    for (i = 0; i < n; i++) {
        count = 0;
        for (j = 0; j < n; j++)
            if (a[j] < a[i])
                count++;
            else if (a[j] == a[i] && j < i)
                count++;
        temp[count] = a[i];
    }

    memcpy(a, temp, n*sizeof(int));
    free(temp);
}  /* Count_sort */
```

The basic idea is that for each element a[i] in the list a, we count the number of elements in the list that are less than a[i]. Then we insert a[i] into a temporary list using the subscript determined by the count. There's a slight problem with this approach when the list contains equal elements, since they could get assigned to the same slot in the temporary list. The code deals with this by incrementing the count for equal elements on the basis of the subscripts. If both a[i] == a[j] *and* j < i, then we count a[j] as being "less than" a[i].

After the algorithm has completed, we overwrite the original array with the temporary array using the string library function `memcpy`.

a. If we try to parallelize the **for** i loop (the outer loop), which variables should be private and which should be shared?

b. If we parallelize the **for** i loop using the scoping you specified in the previous part, are there any loop-carried dependences? Explain your answer.

c. Can we parallelize the call to `memcpy`? Can we modify the code so that this part of the function will be parallelizable?

d. Write a C program that includes a parallel implementation of `Count_sort`.

e. How does the performance of your parallelization of `Count_sort` compare to serial `Count_sort`? How does it compare to the serial `qsort` library function?

5.4 Recall that when we solve a large linear system, we often use Gaussian elimination followed by *backward substitution*. Gaussian elimination converts an $n \times n$ linear system into an *upper triangular* linear system by using the "row operations":

- Add a multiple of one row to another row
- Swap two rows
- Multiply one row by a nonzero constant

An upper triangular system has zeroes below the "diagonal" extending from the upper left-hand corner to the lower right-hand corner.
For example, the linear system

$$
\begin{aligned}
2x_0 &- 3x_1 & & = 3 \\
4x_0 &- 5x_1 &+ x_2 &= 7 \\
2x_0 &- x_1 &- 3x_2 &= 5
\end{aligned}
$$

can be reduced to the upper triangular form

$$
\begin{aligned}
2x_0 &- 3x_1 & & = 3 \\
& x_1 &+ x_2 &= 1 \;, \\
& & - 5x_2 &= 0
\end{aligned}
$$

and this system can be easily solved by first finding x_2 using the last equation, then finding x_1 using the second equation, and finally finding x_0 using the first equation.

We can devise a couple of serial algorithms for back substitution. The "row-oriented" version is

```
for (row = n-1; row >= 0; row--) {
    x[row] = b[row];
    for (col = row+1; col < n; col++)
        x[row] -= A[row][col]*x[col];
    x[row] /= A[row][row];
}
```

Here the "right-hand side" of the system is stored in the array b, the two-dimensional array of coefficients is stored in the array A, and the solutions are stored in the array x. An alternative is the following "column-oriented" algorithm:

```
for (row = 0; row < n; row++)
    x[row] = b[row];

for (col = n−1; col >= 0; col−−) {
    x[col] /= A[col][col];
    for (row = 0; row < col; row++)
        x[row] −= A[row][col]*x[col];
}
```

   a. Determine whether the outer loop of the row-oriented algorithm can be parallelized.
   b. Determine whether the inner loop of the row-oriented algorithm can be parallelized.
   c. Determine whether the (second) outer loop of the column-oriented algorithm can be parallelized.
   d. Determine whether the inner loop of the column-oriented algorithm can be parallelized.
   e. Write one OpenMP program for each of the loops that you determined could be parallelized. You may find the single directive useful—when a block of code is being executed in parallel and a sub-block should be executed by only one thread, the sub-block can be modified by a **#pragma** omp single directive. The threads in the executing team will block at the end of the directive until all of the threads have completed it.
   f. Modify your parallel loop with a schedule(runtime) clause and test the program with various schedules. If your upper triangular system has 10,000 variables, which schedule gives the best performance?

5.5 Use OpenMP to implement a program that does Gaussian elimination. (See the preceding problem.) You can assume that the input system doesn't need any row-swapping.

5.6 Use OpenMP to implement a producer consumer program in which some of the threads are producers and others are consumers. The producers read text from a collection of files, one per producer. They insert lines of text into a single shared queue. The consumers take the lines of text and tokenize them. Tokens are "words" separated by white space. When a consumer finds a token, it writes it to stdout.

5.7 Use your answer to Exercise 5.16 to implement a parallel merge sort program. Get the number of threads from the command line. Either get the input list from stdin or use the C library random function to generate the list. How does the performance of the parallel program compare to a serial merge sort?

GPU programming with CUDA

6.1 GPUs and GPGPU

In the late 1990s and early 2000s, the computer industry responded to the demand for highly realistic computer video games and video animations by developing extremely powerful **graphics processing units** or **GPUs**. These processors, as their name suggests, are designed to improve the performance of programs that need to render many detailed images.

The existence of this computational power was a temptation to programmers who didn't specialize in computer graphics, and by the early 2000s they were trying to apply the power of GPUs to solving general computational problems, problems such as searching and sorting, rather than graphics. This became known as **General Purpose computing on GPUs** or **GPGPU**.

One of the biggest difficulties faced by the early developers of GPGPU was that the GPUs of the time could only be programmed using computer graphics APIs, such as Direct3D and OpenGL. So programmers needed to reformulate algorithms for general computational problems so that they used graphics concepts, such as vertices, triangles, and pixels. This added considerable complexity to the development of early GPGPU programs, and it wasn't long before several groups started work on developing languages and compilers that allowed programmers to implement general algorithms for GPUs in APIs that more closely resembled conventional, high-level languages for CPUs.

These efforts led to the development of several APIs for general purpose programming on GPUs. Currently the most widely used APIs are CUDA and OpenCL. CUDA was developed for use on Nvidia GPUs. OpenCL, on the other hand, was designed to be highly portable. It was designed for use on arbitrary GPUs *and* other processors—processors such as field programmable gate arrays (FPGAs) and digital signal processors (DSPs). To ensure this portability, an OpenCL program must include a good deal of code providing information about which systems it can be run on and information about how it should be run. Since CUDA was developed to run only on Nvidia GPUs, it requires relatively modest setup, and, as a consequence, we'll use it instead of OpenCL.

An Introduction to Parallel Programming. https://doi.org/10.1016/B978-0-12-804605-0.00013-0

Table 6.1 Execution of branch on a SIMD system.

Time	Datapaths with $x[i] >= 0$	Datapaths with $x[i] < 0$
1	Test $x[i] >= 0$	Test $x[i] >= 0$
2	$x[i] += 1$	Idle
3	Idle	$x[i] -= 2$

6.2 GPU architectures

As we've seen (see Chapter 2), CPU architectures can be extremely complex. However, we often think of a conventional CPU as a SISD device in Flynn's Taxonomy (see Section 2.3): the processor fetches an instruction from memory and executes the instruction on a small number of data items. The instruction is an element of the *Single Instruction stream*—the "SI" in SISD—and the data items are elements of the *Single Data stream*—the "SD" in SISD. GPUs, however, are composed of SIMD or *Single Instruction stream, Multiple Data stream* processors. So, to understand how to program them, we need to first look at their architecture.

Recall (from Section 2.3) that we can think of a SIMD processor as being composed of a single control unit and multiple datapaths. The control unit fetches an instruction from memory and broadcasts it to the datapaths. Each datapath either executes the instruction on *its* data or is idle.

For example, suppose there are n datapaths that share an n-element array x. Also suppose that the ith datapath will work with $x[i]$. Now suppose we want to add 1 to the nonnegative elements of x and subtract 2 from the negative elements of x. We might implement this with the following code:

```
/* Datapath i executes the following code */
if (x[i] >= 0)
    x[i] += 1;
else
    x[i] -= 2;
```

In a typical SIMD system, each datapath carries out the test $x[i] >= 0$. Then the datapaths for which the test is true execute $x[i] += 1$, while those for which $x[i] < 0$ are *idle*. Then the roles of the datapaths are reversed: those for which $x[i] >= 0$ are idle while the other datapaths execute $x[i] -= 2$. See Table 6.1.

A typical GPU can be thought of as being composed of one or more SIMD processors. Nvidia GPUs are composed of **Streaming Multiprocessors** or **SMs**.[1] One SM can have several control units and many more datapaths. So an SM can be thought of as consisting of one or more SIMD processors. The SMs, however, operate asynchronously: there is no penalty if one branch of an **if**–**else** executes on one SM, and

---

[1] The abbreviation that Nvidia uses for a streaming multiprocessor depends on the particular GPU microarchitecture. For example, Tesla and Fermi multiprocessors have SMs, Kepler multiprocessors have SMXs, and Maxwell multiprocessors have SMMs. More recent GPUs have SMs. We'll use SM, regardless of the microarchitecture.

Table 6.2 Execution of branch on a system with multiple SMs.

Time	Datapaths with x[i] >= 0 (on SM A)	Datapaths with x[i] < 0 (on SM B)
1	Test x[i] >= 0	Test x[i] >= 0
2	x[i] += 1	x[i] -= 2

the other executes on another SM. So in our preceding example, if all the threads with x[i] >= 0 were executing on one SM, and all the threads with x[i] < 0 were executing on another, the execution of our **if−else** example would require only two stages. (See Table 6.2.)

In Nvidia parlance, the datapaths are called cores, **Streaming Processors**, or **SPs**. Currently,[2] one of the most powerful Nvidia processor has 82 SMs, and each SM has 128 SPs for a total of 10,496 SPs. Since we use the term "core" to mean something else when we're discussing MIMD architectures, we'll use SP to denote an Nvidia datapath. Also note that Nvidia uses the term **SIMT** instead of SIMD. SIMT stands for Single Instruction Multiple Thread, and the term is used because threads on an SM that are executing the same instruction may not execute simultaneously: to hide memory access latency, some threads may block while memory is accessed and other threads, that have already accessed the data, may proceed with execution.

Each SM has a relatively small block of memory that is shared among its SPs. As we'll see, this memory can be accessed very quickly by the SPs. All of the SMs on a single chip also have access to a much larger block of memory that is shared among all the SPs. Accessing this memory is relatively slow. (See Fig. 6.1.)

The GPU and its associated memory are usually physically separate from the CPU and its associated memory. In Nvidia documentation, the CPU together with its associated memory is often called the **host**, and the GPU together with its memory is called the **device**. In earlier systems the physical separation of host and device memories required that data was usually explicitly transferred between CPU memory and GPU memory. That is, a function was called that would transfer a block of data from host memory to device memory or vice versa. So, for example, data read from a file by the CPU or output data generated by the GPU would have to be transferred between the host and device with an explicit function call. However, in more recent Nvidia systems (those with compute capability ≥ 3.0), the explicit transfers in the source code aren't needed for correctness, although they may be able to improve overall performance. (See Fig. 6.2.)

6.3 Heterogeneous computing

Up to now we've implicitly assumed that our parallel programs will be run on systems in which the individual processors have identical architectures. Writing a program

---

[2] Spring 2021.

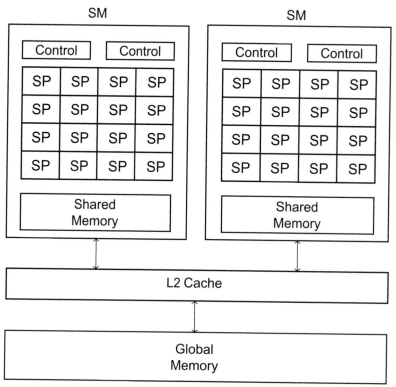

FIGURE 6.1

Simplified block diagram of a GPU.

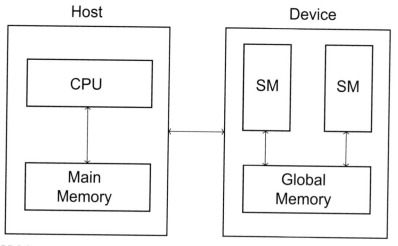

FIGURE 6.2

Simplified block diagram of a CPU and a GPU.

that runs on a GPU is an example of **heterogeneous** computing. The reason is that the programs make use of both a host processor—a conventional CPU—and a device processor—a GPU—and, as we've just seen, the two processors have different architectures.

We'll still write a single program (using the SPMD approach—see Section 2.4.1), but now there will be functions that we write for conventional CPUs and functions that we write for GPUs. So, effectively, we'll be writing two programs.

Heterogeneous computing has become much more important in recent years. Recall from Chapter 1 that from about 1986 to 2003, the single-thread performance of conventional CPUs was increasing, on average, more than 50% per year, but since 2003, the improvement in single-thread performance has decreased to the point that from 2015 to 2017, it has been growing at less than 4% per year [28]. So programmers are leaving no stone unturned in their search for ways to bolster performance, and one possibility is to make use of other types of processors, processors other than CPUs. Our focus is GPUs, but other possibilities include **Field Programmable Gate Arrays** or **FPGAs**, and **Digital Signal Processors** or **DSPs**. FPGAs contain programmable logic blocks and interconnects that can be configured prior to program execution. DSPs contain special circuitry for manipulating (e.g., compressing, filtering) signals, especially "real-world" analog signals.

6.4 CUDA hello

So let's start talking about the CUDA API, the API we'll be using to program heterogeneous CPU–GPU systems.

CUDA is a software platform that can be used to write GPGPU programs for heterogeneous systems equipped with an Nvidia GPU. CUDA was originally an acronym for "Compute Unified Device Architecture," which was meant to suggest that it provided a single interface for programming both CPU and GPU. More recently, however, Nvidia has decided that CUDA is not an acronym; it's simply the name of an API for GPGPU programming.

There is a language-specific CUDA API for several languages; for example, there are CUDA APIs for C, C++, Fortran, Python, and Java. We'll be using CUDA C, but we need to be aware that sometimes we'll need to use some C++ constructs. This is because the CUDA C compiler can compile both C and C++ programs, and it can do this because it is a modified C++ compiler. So where the specifications for C and C++ differ the CUDA compiler sometimes uses C++. For example, since the C library function `malloc` returns a **void**∗ pointer, a C program doesn't need a cast in the instruction

```
float *x = malloc(n*sizeof(float));
```

However, in C++ a cast is required

```
float *x = (float*) malloc(n*sizeof(float));
```

As usual, we'll begin by implementing a version of the "hello, world" program. We'll write a CUDA C program in which each CUDA thread prints a greeting.[3] Since the program is heterogeneous, we will, effectively, be writing two programs: a host or CPU program and a device or GPU program.

Note that even though our programs are written in CUDA C, CUDA programs cannot be compiled with an ordinary C compiler. So unlike MPI and Pthreads, CUDA is not just a library that can be linked in to an ordinary C program: CUDA requires a special compiler. For example, an ordinary C compiler (such as gcc) generates a machine language executable for a single CPU (e.g., an x86 processor), but the CUDA compiler must generate machine language for two different processors: the host processor and the device processor.

6.4.1 The source code

The source code for a CUDA program that prints a greeting from each thread on the GPU is shown in Program 6.1.

As you might guess, there's a header file for CUDA programs, which we include in Line 2.

The Hello function follows the include directives and starts on Line 5. This function is run by each thread on the GPU. In CUDA parlance, it's called a **kernel**, a function that is started by the host but runs on the device. CUDA kernels are identified by the keyword __global__, and they always have return type **void**.

The main function follows the kernel on Line 12. Like ordinary C programs, CUDA C programs start execution in main, and the main function runs on the *host*. The function first gets the number of threads from the command line. It then starts the required number of copies of the kernel on Line 18. The call to cudaDeviceSynchronize will cause the main program to wait until all the threads have finished executing the kernel, and when this happens, the program terminates as usual with **return** 0.

6.4.2 Compiling and running the program

A CUDA program file that contains both host code and device code should be stored in a file with a ".cu" suffix. For example, our hello program is in a file called cuda_hello.cu. We can compile it using the CUDA compiler nvcc. The command should look something like this[4]:

```
$ nvcc -o cuda_hello cuda_hello.cu
```

If we want to run one thread on the GPU, we can type

```
$ ./cuda_hello 1
```

and the output will be

---

[3] This program requires an Nvidia GPU with compute capability ≥ 2.0.

[4] Recall that the dollar sign ($) denotes the shell prompt, and it should not be typed.

```
1   #include <stdio.h>
2   #include <cuda.h>    /* Header file for CUDA */
3
4   /* Device code:  runs on GPU */
5   __global__ void Hello(void) {
6
7       printf("Hello from thread %d!\n", threadIdx.x);
8   }  /* Hello */
9
10
11  /* Host code:  Runs on CPU */
12  int main(int argc, char* argv[]) {
13      int thread_count;        /* Number of threads to run on GPU */
14
15      thread_count = strtol(argv[1], NULL, 10);
16                       /* Get thread_count from command line */
17
18      Hello <<<1, thread_count >>>();
19                       /* Start thread_count threads on GPU, */
20
21      cudaDeviceSynchronize();      /* Wait for GPU to finish */
22
23      return 0;
24  }  /* main */
```

Program 6.1: CUDA program that prints greetings from the threads.

```
Hello from thread 0!
```

If we want to run ten threads on the GPU, we can type

```
$ ./cuda_hello 10
```

and the output of will be

```
Hello from thread 0!
Hello from thread 1!
Hello from thread 2!
Hello from thread 3!
Hello from thread 4!
Hello from thread 5!
Hello from thread 6!
Hello from thread 7!
Hello from thread 8!
Hello from thread 9!
```

6.5 A closer look

So what exactly happens when we run `cuda_hello`? Let's take a closer look.

As we noted earlier, execution begins on the host in the `main` function. It gets the number of threads from the command line by calling the C library `strtol` function.

Things get interesting in the call to the kernel in Line 18. Here we tell the system how many threads to start on the GPU by enclosing the pair

```
1, thread_count
```

in triple angle brackets. If there were any arguments to the `Hello` function, we would enclose them in the following parentheses.

The kernel specifies the code that each thread will execute. So each of our threads will print a message

```
"Hello from thread %d\n"
```

The decimal int format specifier (`%d`) refers to the variable `threadIdx.x`. The struct `threadIdx` is one of several variables defined by CUDA when a kernel is started. In our example, the field `x` gives the relative index or rank of the thread that is executing. So we use it to print a message containing the thread's rank.

After a thread has printed its message, it terminates execution.

Notice that our kernel code uses the Single-Program Multiple-Data or SPMD paradigm: each thread runs a copy of the same code on its own data. In this case, the only thread-specific data is the thread rank stored in `threadIdx.x`.

One very important difference between the execution of an ordinary C function and a CUDA kernel is that kernel execution is **asynchronous**. This means that the call to the kernel on the host *returns* as soon as the host has notified the system that it should start running the kernel, and even though the call in `main` has returned, the threads executing the kernel may not have finished executing. The call to `cudaDeviceSynchronize` in Line 21 forces the `main` function to wait until all the threads executing the kernel have completed. If we omitted the call to `cudaDeviceSynchronize`, our program could terminate before the threads produced any output, and it might appear that the kernel was never called.

When the host returns from the call to `cudaDeviceSynchronize`, the `main` function then terminates as usual with **return** 0.

To summarize, then:

- Execution begins in `main`, which is running on the host.
- The number of threads is taken from the command line.
- The call to `Hello` starts the kernel.
  - The `<<<1, thread_count>>>` in the call specifies that `thread_count` copies of the kernel should be started on the device.
  - When the kernel is started, the struct `threadIdx` is initialized by the system, and in our example the field `threadIdx.x` contains the thread's index or rank.
  - Each thread prints its message and terminates.
- The call to `cudaDeviceSynchronize` in `main` forces the host to wait until all of the threads have completed kernel execution before continuing and terminating.

6.6 Threads, blocks, and grids

You're probably wondering why we put a "1" in the angle brackets in our call to `Hello`:

```
Hello <<<1, thread_count >>>();
```

Recall that an Nvidia GPU consists of a collection of streaming multiprocessors (SMs), and each streaming multiprocessor consists of a collection of streaming processors (SPs). When a CUDA kernel runs, each individual thread will execute its code on an SP. With "1" as the first value in angle brackets, all of the threads that are started by the kernel call will run on a single SM. If our GPU has two SMs, we can try to use both of them with the kernel call

```
Hello <<<2, thread_count/2 >>>();
```

If `thread_count` is even, this kernel call will start a total of `thread_count` threads, and the threads will be divided between the two SMs: `thread_count`/2 threads will run on each SM. (What happens if `thread_count` is odd?)

    CUDA organizes threads into blocks and grids. A **thread block** (or just a **block** if the context makes it clear) is a collection of threads that run on a single SM. In a kernel call the first value in the angle brackets specifies the number of thread blocks. The second value is the number of threads in each thread block. So when we started the kernel with

```
Hello <<<1, thread_count >>>();
```

we were using one thread block, which consisted of `thread_count` threads, and, as a consequence, we only used one SM.

    We can modify our greetings program so that it uses a user-specified number of blocks, each consisting of a user-specified number of threads. (See Program 6.2.) In this program we get both the number of thread blocks and the number of threads in each block from the command line. Now the kernel call starts `blk_ct` thread blocks, each of which contains `th_per_blk` threads.

    When the kernel is started, each block is assigned to an SM, and the threads in the block are then run on that SM. The output is similar to the output from the original program, except that now we're using two system-defined variables: `threadIdx.x` and `blockIdx.x`. As you've probably guessed, `threadIdx.x` gives a thread's rank or index in its block, and `blockIdx.x` gives a block's rank in the *grid*.

    A **grid** is just the collection of thread blocks started by a kernel. So a thread block is composed of threads, and a grid is composed of thread blocks.

    There are several built-in variables that a thread can use to get information on the grid started by the kernel. The following four variables are structs that are initialized in each thread's memory when a kernel begins execution:

- `threadIdx`: the rank or index of the thread in its thread block.
- `blockDim`: the dimensions, shape, or size of the thread blocks.

```
1   #include <stdio.h>
2   #include <cuda.h>      /* Header file for CUDA */
3
4   /* Device code:  runs on GPU */
5   __global__ void Hello(void) {
6
7      printf("Hello from thread %d in block %d\n",
8              threadIdx.x, blockIdx.x);
9   }  /* Hello */
10
11
12  /* Host code:  Runs on CPU */
13  int main(int argc, char* argv[]) {
14     int blk_ct;              /* Number of thread blocks */
15     int th_per_blk;       /* Number of threads in each block */
16
17     blk_ct = strtol(argv[1], NULL, 10);
18                 /* Get number of blocks from command line */
19     th_per_blk = strtol(argv[2], NULL, 10);
20        /* Get number of threads per block from command line */
21
22     Hello <<<blk_ct, th_per_blk >>>();
23             /* Start blk_ct*th_per_blk threads on GPU, */
24
25     cudaDeviceSynchronize();        /* Wait for GPU to finish */
26
27     return 0;
28  }  /* main */
```

Program 6.2: CUDA program that prints greetings from threads in multiple blocks.

- `blockIdx`: the rank or index of the block within the grid.
- `gridDim`: the dimensions, shape, or size of the grid.

All of these structs have three fields, x, y, and z,[5] and the fields all have unsigned integer types. The fields are often convenient for applications. For example, an application that uses graphics may find it convenient to assign a thread to a point in two- or three-dimensional space, and the fields in `threadIdx` can be used to indicate the point's position. An application that makes extensive use of matrices may find it convenient to assign a thread to an element of a matrix, and the fields in `threadIdx` can be used to indicate the column and row of the element.

---

[5] Nvidia devices that have compute capability < 2 (see Section 6.7) only allow x- and y-dimensions in a grid.

When we call a kernel with something like

```
int blk_ct, th_per_blk;
...
Hello <<<blk_ct, th_per_blk >>();
```

the three-element structures `gridDim` and `blockDim` are initialized by assigning the values in angle brackets to the x fields. So, effectively, the following assignments are made:

```
gridDim.x = blk_ct;
blockDim.x = th_per_blk;
```

The y and z fields are initialized to 1. If we want to use values other than 1 for the y and z fields, we should declare two variables of type `dim3`, and pass them into the call to the kernel. For example,

```
dim3 grid_dims, block_dims;
grid_dims.x = 2;
grid_dims.y = 3;
grid_dims.z = 1;
block_dims.x = 4;
block_dims.y = 4;
block_dims.z = 4;
...
Kernel <<<grid_dims, block_dims >>> (...);
```

This should start a grid with $2 \times 3 \times 1 = 6$ blocks, each of which has $4^3 = 64$ threads.

Note that all the blocks must have the same dimensions. More importantly, CUDA requires that thread blocks be independent. So one thread block must be able to complete its execution, regardless of the states of the other thread blocks: the thread blocks can be executed sequentially in any order, or they can be executed in parallel. This ensures that the GPU can schedule a block to execute solely on the basis of the state of that block: it doesn't need to check on the state of any other block.[6]

6.7 Nvidia compute capabilities and device architectures[7]

There are limits on the number of threads and the number of blocks. The limits depend on what Nvidia calls the **compute capability** of the GPU. The compute capability is a number having the form $a.b$. Currently the a-value or major revision number can be 1, 2, 3, 5, 6, 7, 8. (There is no major revision number 4.) The possible b-values or minor revision numbers depend on the major revision value, but currently

---

[6] With the introduction of CUDA 9 and the Pascal processor, it became possible to synchronize threads in multiple blocks. See Subsection 7.1.13 and Exercise 7.6.

[7] The values in this section are current as of spring 2021, but some of them may change when Nvidia releases new GPUs and new versions of CUDA.

Table 6.3 GPU architectures and compute capabilities.

Name	Ampere	Tesla	Fermi	Kepler	Maxwell	Pascal	Volta	Turing
Compute capability	8.0	1.b	2.b	3.b	5.b	6.b	7.0	7.5

they fall in the range 0–7. CUDA no longer supports devices with compute capability < 3.

For devices with compute capability > 1, the maximum number of threads per block is 1024. For devices with compute capability 2.b, the maximum number of threads that can be assigned to a single SM is 1536, and for devices with compute capability > 2, the maximum is currently 2048. There are also limits on the sizes of the dimensions in both blocks and grids. For example, for compute capability > 1, the maximum x- or y-dimension is 1024, and the maximum z-dimension is 64. For further information, see the appendix on compute capabilities in the CUDA C++ Programming Guide [11].

Nvidia also has names for the microarchitectures of its GPUs. Table 6.3 shows the current list of architectures and some of their corresponding compute capabilities. Somewhat confusingly, Nvidia also uses Tesla as the name for their products targeting GPGPU.

We should note that Nvidia has a number of "product families" that can consist of anything from an Nvidia-based graphics card to a "system on a chip," which has the main hardware components of a system, such as a mobile phone in a single integrated circuit.

Finally, note that there are a number of versions of the CUDA API, and they do *not* correspond to the compute capabilities of the different GPUs.

6.8 Vector addition

GPUs and CUDA are designed to be especially effective when they run data-parallel programs. So let's write a very simple, data-parallel CUDA program that's embarrassingly parallel: a program that adds two vectors or arrays. We'll define three n-element arrays, x, y, and z. We'll initialize x and y on the host. Then a kernel can start at least n threads, and the ith thread will add

```
z[i] = x[i] + y[i];
```

Since GPUs tend to have more 32-bit than 64-bit floating point units, let's use arrays of `floats` rather than `doubles`:

```
float *x, *y, *z;
```

After allocating and initializing the arrays, we'll call the kernel, and after the kernel completes execution, the program checks the result, frees memory, and quits. See Program 6.3, which shows the kernel and the main function.

Let's take a closer look at the program.

```
1   __global__ void Vec_add(
2         const float x[]    /* in  */,
3         const float y[]    /* in  */,
4         float       z[]    /* out */,
5         const int   n      /* in  */) {
6      int my_elt = blockDim.x * blockIdx.x + threadIdx.x;
7
8      /* total threads = blk_ct*th_per_blk may be > n */
9      if (my_elt < n)
10        z[my_elt] = x[my_elt] + y[my_elt];
11  } /* Vec_add */
12
13  int main(int argc, char* argv[]) {
14     int n, th_per_blk, blk_ct;
15     char i_g; /* Are x and y user input or random? */
16     float *x, *y, *z, *cz;
17     double diff_norm;
18
19     /* Get the command line arguments, and set up vectors */
20     Get_args(argc, argv, &n, &blk_ct, &th_per_blk, &i_g);
21     Allocate_vectors(&x, &y, &z, &cz, n);
22     Init_vectors(x, y, n, i_g);
23
24     /* Invoke kernel and wait for it to complete           */
25     Vec_add <<<blk_ct, th_per_blk>>>(x, y, z, n);
26     cudaDeviceSynchronize();
27
28     /* Check for correctness */
29     Serial_vec_add(x, y, cz, n);
30     diff_norm = Two_norm_diff(z, cz, n);
31     printf("Two-norm of difference between host and ");
32     printf("device = %e\n", diff_norm);
33
34     /* Free storage and quit */
35     Free_vectors(x, y, z, cz);
36     return 0;
37  } /* main */
```

Program 6.3: Kernel and `main` function of a CUDA program that adds two vectors.

6.8.1 The kernel

In the kernel (Lines 1–11), we first determine which element of z the thread should compute. We've chosen to make this index the same as the *global* rank or index of the thread. Since we're only using the x fields of the `blockDim` and `threadIdx` structs, there are a total of

Table 6.4 Global thread ranks or indexes in a grid with 4 blocks and 5 threads per block.

blockIdx.x	threadIdx.x				
	0	**1**	**2**	**3**	**4**
0	0	1	2	3	4
1	5	6	7	8	9
2	10	11	12	13	14
3	15	16	17	18	19

```
gridDim.x * blockDim.x
```

threads. So we can assign a unique "global" rank or index to each thread by using the formula

```
rank = blockDim.x * blockIdx.x + threadIdx.x
```

For example, if we have four blocks and five threads in each block, then the global ranks or indexes are shown in Table 6.4. In the kernel, we assign this global rank to `my_elt` and use this as the subscript for accessing each thread's elements of the arrays x, y, and z.

Note that we've allowed for the possibility that the total number of threads may not be exactly the same as the number of components of the vectors. So before carrying out the addition,

```
z[my_elt] = x[my_elt] + y[my_elt];
```

we first check that `my_elt` < n. For example, if we have $n = 997$, and we want at least two blocks with at least two threads per block, then, since 997 is prime, we can't possibly have exactly 997 threads. Since this kernel needs to be executed by at least n threads, we must start more than 997. For example, we might use four blocks of 256 threads, and the last 27 threads in the last block would skip the line

```
z[my_elt] = x[my_elt] + y[my_elt];
```

Note that if we needed to run our program on a system that didn't support CUDA, we could replace the kernel with a serial vector addition function. (See Program 6.4.) So we can view the CUDA kernel as taking the serial **for** loop and assigning each iteration to a different thread. This is often how we start the design process when we want to parallelize a serial code for CUDA: assign the iterations of a loop to individual threads.

Also note that if we apply Foster's method to parallelizing the serial vector sum, and we make the tasks the additions of the individual components, then we don't need to do anything for the communication and aggregation phases, and the mapping phase simply assigns each addition to a thread.

```
1   void  Serial_vec_add(
2          const  float    x[]     /*  in   */,
3          const  float    y[]     /*  in   */,
4          float           cz[]    /*  out  */,
5          const  int      n       /*  in   */) {
6
7      for  (int  i = 0;  i < n;  i++)
8          cz[i] = x[i] + y[i];
9   }  /* Serial_vec_add */
```

Program 6.4: Serial vector addition function.

6.8.2 Get_args

After declaring the variables, the `main` function calls a `Get_args` function, which returns n, the number of elements in the arrays, `blk_ct`, the number of thread blocks, and `th_per_blk`, the number of threads in each block. It gets these from the command line. It also returns a **char** `i_g`. This tells the program whether the user will input x and y or whether it should generate them using a random number generator. If the user doesn't enter the correct number of command line arguments, the function prints a usage summary and terminates execution. Also if n is greater than the total number of threads, it prints a message and terminates. (See Program 6.5.) Note that `Get_args` is written in standard C, and it runs completely on the host.

6.8.3 Allocate_vectors **and managed memory**

After getting the command line arguments, the main function calls `Allocate_vectors`, which allocates storage for four n-element arrays of **float**:

```
x, y, z, cz
```

The first three arrays are used on both the host and the device. The fourth array, `cz`, is only used on the host: we use it to compute the vector sum with one core of the host. We do this so that we can check the result computed on the device. (See Program 6.6.)

First note that since `cz` is only used on the host, we allocate its storage using the standard C library function `malloc`. For the other three arrays, we allocate storage in Lines 9–11 using the CUDA function

```
__host__  cudaError_t cudaMallocManaged (
                void**    devPtr   /*  out  */,
                size_t    size     /*  in   */,
                unsigned  flags    /*  in   */);
```

The __host__ qualifier is a CUDA addition to C, and it indicates that the function should be called and run on the host. This is the default for functions in CUDA

```
1  void Get_args(
2        const int argc              /* in   */,
3        char*      argv[]           /* in   */,
4        int*       n_p              /* out  */,
5        int*       blk_ct_p         /* out  */,
6        int*       th_per_blk_p     /* out  */,
7        char*      i_g              /* out  */) {
8     if (argc != 5) {
9        /* Print an error message and exit */
10       ...
11    }
12
13    *n_p = strtol(argv[1], NULL, 10);
14    *blk_ct_p = strtol(argv[2], NULL, 10);
15    *th_per_blk_p = strtol(argv[3], NULL, 10);
16    *i_g = argv[4][0];
17
18    /* Is n > total thread count = blk_ct*th_per_blk? */
19    if (*n_p > (*blk_ct_p)*(*th_per_blk_p)) {
20       /* Print an error message and exit */
21       ...
22    }
23 }  /* Get_args */
```

Program 6.5: Get_args function from CUDA program that adds two vectors.

```
1  void Allocate_vectors(
2        float** x_p     /* out */,
3        float** y_p     /* out */,
4        float** z_p     /* out */,
5        float** cz_p    /* out */,
6        int     n       /* in  */) {
7
8     /* x, y, and z are used on host and device */
9     cudaMallocManaged(x_p, n*sizeof(float));
10    cudaMallocManaged(y_p, n*sizeof(float));
11    cudaMallocManaged(z_p, n*sizeof(float));
12
13    /* cz is only used on host */
14    *cz_p = (float*) malloc(n*sizeof(float));
15 }  /* Allocate_vectors */
```

Program 6.6: Array allocation function of CUDA program that adds two vectors.

programs, so it can be omitted when we're writing our own functions, and they'll only be run on the host.

The return value, which has type `cudaError_t`, allows the function to return an error. Most CUDA functions return a `cudaError_t` value, and if you're having problems with your code, it is a very good idea to check it. However, always checking it tends to clutter the code, and this can distract us from the main purpose of a program. So in the code we discuss we'll generally ignore `cudaError_t` return values.

The first argument is a pointer to a pointer: it refers to the pointer that's being allocated. The second argument specifies the number of bytes that should be allocated. The `flags` argument controls which kernels can access the allocated memory. It defaults to `cudaMemAttachGlobal` and can be omitted.

The function `cudaMallocManaged` is one of several CUDA memory allocation functions. It allocates memory that will be automatically managed by the "unified memory system." This is a relatively recent addition to CUDA,[8] and it allows a programmer to write CUDA programs as if the host and device shared a single memory: pointers referring to memory allocated with `cudaMallocManaged` can be used on both the device and the host, even when the host and the device have separate physical memories. As you can imagine this greatly simplifies programming, but there are some cautions. Here are a few:

1. Unified memory requires a device with compute capability ≥ 3.0, and a 64-bit host operating system.
2. On devices with compute capability < 6.0 memory allocated with `cudaMallocManaged` cannot be simultaneously accessed by both the device and the host. When a kernel is executing, it has exclusive access to memory allocated with `cudaMallocManaged`.
3. Kernels that use unified memory can be slower than kernels that treat device memory as separate from host memory.

The last caution has to do with the transfer of data between the host and the device. When a program uses unified memory, it is up to the system to decide when to transfer from the host to the device or vice versa. In programs that explicitly transfer data, it is up to the programmer to include code that implements the transfers, and she may be able to exploit her knowledge of the code to do things that reduce the cost of transfers, things such as omitting some transfers or overlapping data transfer with computation.

At the end of this section we'll briefly discuss the modifications required if you want to explicitly handle the transfers between host and device.

6.8.4 **Other functions called from** main

Except for the `Free_vectors` function, the other host functions that we call from main are just standard C.

---

[8] It first became available in CUDA 6.0.

The function `Init_vectors` either reads x and y from `stdin` using `scanf` or generates them using the C library function `random`. It uses the last command line argument `i_g` to decide which it should do.

The `Serial_vec_add` function (Program 6.4) just adds x and y on the host using a **for** loop. It stores the result in the host array `cz`.

The `Two_norm_diff` function computes the "distance" between the vector z computed by the kernel and the vector `cz` computed by `Serial_vec_add`. So it takes the difference between corresponding components of z and `cz`, squares them, adds the squares, and takes the square root:

$$\sqrt{(z[0] - cz[0])^2 + (z[1] - cz[1])^2 + \cdots + (z[n-1] - cz[n-1])^2}.$$

See Program 6.7.

```
1   double Two_norm_diff (
2          const float   z []    /* in */,
3          const float   cz []   /* in */,
4          const int     n       /* in */) {
5       double diff, sum = 0.0;
6
7       for (int i = 0; i < n; i++) {
8           diff = z[i] - cz[i];
9           sum += diff*diff;
10      }
11      return sqrt(sum);
12  }  /* Two_norm_diff */
```

Program 6.7: C function that finds the distance between two vectors.

The `Free_vectors` function just frees the arrays allocated by `Allocate_vectors`. The array `cz` is freed using the C library function `free`, but since the other arrays are allocated using `cudaMallocManaged`, they must be freed by calling `cudaFree`:

```
__host__ __device__ cudaError_t cudaFree ( void* ptr )
```

The qualifier `__device__` is a CUDA addition to C, and it indicates that the function can be called from the device. So `cudaFree` can be called from the host or the device. However, if a pointer is allocated on the device, it cannot be freed on the host, and vice versa.

It's important to note that unless memory allocated on the device is explicitly freed by the program, it won't be freed until the program terminates. So if a CUDA program calls two (or more) kernels, and the memory used by the first kernel isn't explicitly freed before the second is called, it will remain allocated, regardless of whether the second kernel actually uses it.

See Program 6.8.

```
1  void Free_vectors(
2        float* x      /* in/out */,
3        float* y      /* in/out */,
4        float* z      /* in/out */,
5        float* cz     /* in/out */) {
6
7     /* Allocated with cudaMallocManaged */
8     cudaFree(x);
9     cudaFree(y);
10    cudaFree(z);
11
12    /* Allocated with malloc */
13    free(cz);
14 }  /* Free_vectors */
```

Program 6.8: CUDA function that frees four arrays.

6.8.5 Explicit memory transfers[9]

Let's take a look at how to modify the vector addition program for a system that doesn't provide unified memory. Program 6.9 shows the kernel and main function for the modified program.

The first thing to notice is that the kernel is unchanged: the arguments are x, y, z, and n. It finds the thread's global index, my_elt, and if this is less than n, it adds the elements of x and y to get the corresponding element of z.

The basic structure of the main function is almost the same. However, since we're assuming unified memory is unavailable, pointers on the host aren't valid on the device, and vice versa: an address on the host may be illegal on the device, or, even worse, it might refer to memory that the device is using for some other purpose. Similar problems occur if we try to use a device address on the host. So instead of declaring and allocating storage for three arrays that are all valid on both the host and the device, we declare and allocate storage for three arrays that are valid on the host hx, hy, and hz, *and* we declare and allocate storage for three arrays that are valid on the device, dx, dy, and dz. The declarations are in Lines 15–16, and the allocations are in the Allocate_vectors function called in Line 20. The function itself is in Program 6.10. Since unified memory isn't available, instead of using cudaMallocManaged, we use the C library function malloc for the host arrays, and the CUDA function cudaMalloc for the device arrays:

```
__host__  __device__  cudaError_t cudaMalloc (
            void** dev_p    /* out */,
            size_t size     /* in  */);
```

---

[9] If your device has compute capability $\geq$ **3.0**, you can skip this section.

```
1   __global__ void Vec_add(
2        const float    x[]    /* in   */,
3        const float    y[]    /* in   */,
4        float          z[]    /* out  */,
5        const int      n      /* in   */) {
6      int my_elt = blockDim.x * blockIdx.x + threadIdx.x;
7
8      if (my_elt < n)
9         z[my_elt] = x[my_elt] + y[my_elt];
10  }  /* Vec_add */
11
12  int main(int argc, char* argv[]) {
13     int n, th_per_blk, blk_ct;
14     char i_g;  /* Are x and y user input or random? */
15     float *hx, *hy, *hz, *cz; /* Host arrays       */
16     float *dx, *dy, *dz;      /* Device arrays      */
17     double diff_norm;
18
19     Get_args(argc, argv, &n, &blk_ct, &th_per_blk, &i_g);
20     Allocate_vectors(&hx, &hy, &hz, &cz, &dx, &dy, &dz, n);
21     Init_vectors(hx, hy, n, i_g);
22
23     /* Copy vectors x and y from host to device */
24     cudaMemcpy(dx, hx, n*sizeof(float), cudaMemcpyHostToDevice);
25     cudaMemcpy(dy, hy, n*sizeof(float), cudaMemcpyHostToDevice);
26
27
28     Vec_add <<<blk_ct, th_per_blk>>>(dx, dy, dz, n);
29
30     /* Wait for kernel to complete and copy result to host */
31     cudaMemcpy(hz, dz, n*sizeof(float), cudaMemcpyDeviceToHost);
32
33     Serial_vec_add(hx, hy, cz, n);
34     diff_norm = Two_norm_diff(hz, cz, n);
35     printf("Two-norm of difference between host and ");
36     printf("device = %e\n", diff_norm);
37
38     Free_vectors(hx, hy, hz, cz, dx, dy, dz);
39
40     return 0;
41  }  /* main */
```

Program 6.9: Part of CUDA program that implements vector addition without unified memory.

```
1   void  Allocate_vectors (
2         float**  hx_p    /* out */,
3         float**  hy_p    /* out */,
4         float**  hz_p    /* out */,
5         float**  cz_p    /* out */,
6         float**  dx_p    /* out */,
7         float**  dy_p    /* out */,
8         float**  dz_p    /* out */,
9         int       n      /* in  */)  {
10
11        /* dx,  dy,  and dz  are used on device */
12        cudaMalloc(dx_p, n*sizeof(float));
13        cudaMalloc(dy_p, n*sizeof(float));
14        cudaMalloc(dz_p, n*sizeof(float));
15
16        /* hx,  hy,  hz,  cz  are used on host */
17        *hx_p = (float*) malloc(n*sizeof(float));
18        *hy_p = (float*) malloc(n*sizeof(float));
19        *hz_p = (float*) malloc(n*sizeof(float));
20        *cz_p = (float*) malloc(n*sizeof(float));
21   }  /* Allocate_vectors */
```

Program 6.10: Allocate_vectors function for CUDA vector addition program that doesn't use unified memory.

The first argument is a reference to a pointer that will be used *on the device*. The second argument specifies the number of bytes to allocate on the device.

After we've initialized hx and hy on the host, we copy their contents over to the device, storing the transferred contents in the memory allocated for dx and dy, respectively. The copying is done in Lines 24–26 using the CUDA function cudaMemcpy:

```
__host__  cudaError_t cudaMemcpy (
        void*           dest      /* out */,
        const void*     source    /* in  */,
        size_t          count     /* in  */,
        cudaMemcpyKind  kind      /* in  */);
```

This copies count bytes from the memory referred to by source into the memory referred to by dest. The type of the kind argument, cudaMemcpyKind, is an enumerated type defined by CUDA that specifies where the source and dest pointers are located. For our purposes the two values of interest are cudaMemcpyHostToDevice and cudaMemcpyDeviceToHost. The first indicates that we're copying from the host to the device, and the second indicates that we're copying from the device to the host.

The call to the kernel in Line 28 uses the pointers dx, dy, and dz, because these are addresses that are valid on the *device*.

After the call to the kernel, we copy the result of the vector addition from the device to the host in Line 31 using `cudaMemcpy` again. A call to `cudaMemcpy` is *synchronous*, so it waits for the kernel to finish executing before carrying out the transfer. So in this version of vector addition we do *not* need to use `cudaDeviceSynchronize` to ensure that the kernel has completed before proceeding.

After copying the result from the device back to the host, the program checks the result, frees the memory allocated on the host and the device, and terminates. So for this part of the program, the only difference from the original program is that we're freeing seven pointers instead of four. As before, the `Free_vectors` function frees the storage allocated on the host with the C library function `free`. It uses `cudaFree` to free the storage allocated on the device.

6.9 Returning results from CUDA kernels

There are several things that you should be aware of regarding CUDA kernels. First, they always have return type **void**, so they can't be used to return a value. They also can't return anything to the host through the standard C pass-by-reference. The reason for this is that addresses on the host are, in most systems, invalid on the device, and vice versa. For example, suppose we try something like this:

```
__global__ void Add(int x, int y, int* sum_p) {
   *sum_p = x + y;
}  /*  Add */
```

```
int main(void) {
   int sum = -5;
   Add <<<1, 1>>> (2, 3, &sum);
   cudaDeviceSynchronize();
   printf("The sum is %d\n", sum);

   return 0;
}
```

It's likely that either the host will print -5 or the device will hang. The reason is that the address `&sum` is probably invalid on the device. So the dereference

```
   *sum_p = x + y;
```

is attempting to assign `x + y` to an invalid memory location.

There are several possible approaches to "returning" a result to the host from a kernel. One is to declare pointer variables and allocate a single memory location. On a system that supports unified memory, the computed value will be automatically copied back to host memory:

```
__global__ void Add(int x, int y, int* sum_p) {
   *sum_p = x + y;
}  /*  Add */
```

```
int main(void) {
    int* sum_p;
    cudaMallocManaged(&sum_p, sizeof(int));
    *sum_p = -5;
    Add <<<1, 1>>> (2, 3, sum_p);
    cudaDeviceSynchronize();
    printf("The sum is %d\n", *sum_p);
    cudaFree(sum_p);

    return 0;
}
```

If your system doesn't support unified memory, the same idea will work, but the result will have to be explicitly copied from the device to the host:

```
__global__ void Add(int x, int y, int *sum_p) {
    *sum_p = x + y;
} /* Add */

int main(void) {
    int *hsum_p, *dsum_p;
    hsum_p = (int*) malloc(sizeof(int));
    cudaMalloc(&dsum_p, sizeof(int));
    *hsum_p = -5;
    Add <<<1, 1>>> (2, 3, dsum_p);
    cudaMemcpy(hsum_p, dsum_p, sizeof(int),
            cudaMemcpyDeviceToHost);
    printf("The sum is %d\n", *hsum_p);
    free(hsum_p);
    cudaFree(dsum_p);

    return 0;
}
```

Note that in both the unified and non-unified memory settings, we're returning a *single* value from the device to the host.

If unified memory is available, another option is to use a global managed variable for the sum:

```
__managed__ int sum;

__global__ void Add(int x, int y) {
    sum = x + y;
} /* Add */

int main(void) {
    sum = -5;
    Add <<<1, 1>>> (2, 3);
```

```
cudaDeviceSynchronize();
printf("After kernel:   The sum is %d\n", sum);

    return 0;
}
```

The qualifier __managed__ declares sum to be a managed **int** that is accessible to all the functions, regardless of whether they run on the host or the device. Since it's managed, the same restrictions apply to it that apply to managed variables allocated with cudaMallocManaged. So this option is unavailable on systems with compute capability < 3.0, and on systems with compute capability < 6.0, sum can't be accessed on the host while the kernel is running. So after the call to Add has started, the host can't access sum until after the call to cudaDeviceSynchronize has completed.

Since this last approach uses a global variable, it has the usual problem of reduced modularity associated with global variables.

6.10 CUDA trapezoidal rule I

6.10.1 The trapezoidal rule

Let's try to implement a CUDA version of the trapezoidal rule. Recall (see Section 3.2.1) that the trapezoidal rule estimates the area between an interval on the x-axis and the graph of a function by dividing the interval into subintervals and approximating the area between each subinterval and the graph by the area of a trapezoid. (See Fig. 3.3.) So if the interval is $[a, b]$ and there are n trapezoids, we'll divide $[a, b]$ into n equal subintervals, and the length of each subinterval will be

$$h = (b - a)/n.$$

Then if x_i is the left end point of the ith subinterval,

$$x_i = a + ih,$$

for $i = 0, 1, 2, \ldots, n - 1$. To simplify the notation, we'll also denote b, the right end point of the interval, as

$$b = x_n = a + nh.$$

Recall that if a trapezoid has height h and base lengths c and d, then its area is

$$\frac{h}{2}(c + d).$$

So if we think of the length of the subinterval $[x_i, x_{i+1}]$ as the height of the ith trapezoid, and $f(x_i)$ and $f(x_{i+1})$ as the two base lengths (see Fig. 3.4), then the area of the ith trapezoid is

$$\frac{h}{2}[f(x_i) + f(x_{i+1})].$$

This gives us a total approximation of the area between the graph and the x-axis as

$$\frac{h}{2}[f(x_0) + f(x_1)] + \frac{h}{2}[f(x_1) + f(x_2)] + \cdots + \frac{h}{2}[f(x_{n-1}) + f(x_n)],$$

and we can rewrite this as

$$h\left[\frac{1}{2}(f(a) + f(b)) + (f(x_1) + f(x_2) + \cdots + f(x_{n-1}))\right].$$

We can implement this with the serial function shown in Program 6.11.

```
1   float  Serial_trap(
2            const float    a    /* in */,
3            const float    b    /* in */,
4            const int      n    /* in */) {
5        float  x, h = (b−a)/n;
6        float  trap = 0.5*(f(a) + f(b));
7
8        for (int i = 1; i <= n−1; i++) {
9            x = a + i*h;
10           trap += f(x);
11       }
12       trap = trap*h;
13
14       return trap;
15  }  /* Serial_trap */
```

Program 6.11: A serial function implementing the trapezoidal rule for a single CPU.

6.10.2 A CUDA implementation

If n is large, the vast majority of the work in the serial implementation is done by the **for** loop. So when we apply Foster's method to the trapezoidal rule, we're mainly interested in two types of tasks: the first is the evaluation of the function f at x_i, and the second is the addition of $f(x_i)$ into trap. Here $i = 1, \ldots, n - 1$. The second type of task depends on the first. So we can aggregate these two tasks.

This suggests that each thread in our CUDA implementation might carry out one iteration of the serial **for** loop. We can assign a unique integer rank to each thread as we did with the vector addition program. Then we can compute an x-value, the function value, and add in the function value to the "running sum":

```
/* h and trap are formal arguments to the kernel */
int my_i = blockDim.x * blockIdx.x + threadIdx.x;
float my_x = a + my_i*h;
float my_trap = f(my_x);
float trap += my_trap;
```

However, it's immediately obvious that there are several problems here:

1. We haven't initialized h or trap.
2. The my_i value can be too large or too small: the serial loop ranges from 1 up to and including $n - 1$. The smallest value for my_i is 0 and the largest is the total number of threads minus 1.
3. The variable trap must be shared among the threads. So the addition of my_trap forms a race condition: when multiple threads try to update trap at roughly the same time, one thread can overwrite another thread's result, and the final value in trap may be wrong. (For a discussion of race conditions, see Section 2.4.3.)
4. The variable trap in the serial code is returned by the function, and, as we've seen, kernels must have **void** return type.
5. We see from the serial code that we need to multiply the total in trap by h after all of the threads have added their results.

Program 6.12 shows how we might deal with these problems. In the following sections, we'll look at the rationales for the various choices we've made.

6.10.3 Initialization, return value, and final update

To deal with the initialization and the final update (Items 1 and 5), we could try to select a single thread—say, thread 0 in block 0—to carry out the operations:

```
int my_i = blockDim.x * blockIdx.x + threadIdx.x;
if (my_i == 0) {
    h = (b-a)/n;
    trap = 0.5*(f(a) + f(b));
}
...
if (my_i == 0)
    trap = trap*h;
```

There are (at least) a couple of problems with these options: formal arguments to functions are private to the executing thread and **thread synchronization**.

Kernel and function arguments are private to the executing thread.

Like the threads started in Pthreads and OpenMP, each CUDA thread has its own stack and, and since formal arguments are allocated on the thread's stack, each thread has its own private variables h and trap. So any changes made to one of these variables by one thread won't be visible to the other threads. We could have each thread initialize h, but we could also just do the initialization once in the host. If we do this before the kernel is called, each thread will get a copy of the value of h.

Things are more complicated with trap. Since it's updated by multiple threads, it must be *shared* among the threads. We can achieve the *effect* of sharing trap by allocating storage for a memory location before the kernel is called. This allocated memory location will correspond to what we've been calling trap. Now we can pass

```
1    __global__ void Dev_trap(
2         const float    a        /* in      */,
3         const float    b        /* in      */,
4         const float    h        /* in      */,
5         const int      n        /* in      */,
6         float*         trap_p   /* in/out */) {
7      int my_i = blockDim.x * blockIdx.x + threadIdx.x;
8
9      /* f(x_0) and f(x_n) were computed on the host. So */
10     /* compute f(x_1), f(x_2), ..., f(x_(n-1))        */
11     if (0 < my_i && my_i < n) {
12        float my_x = a + my_i*h;
13        float my_trap = f(my_x);
14        atomicAdd(trap_p, my_trap);
15     }
16  }   /* Dev_trap */
17
18  /* Host code */
19  void Trap_wrapper(
20        const float    a          /* in  */,
21        const float    b          /* in  */,
22        const int      n          /* in  */,
23        float*         trap_p     /* out */,
24        const int      blk_ct     /* in  */,
25        const int      th_per_blk /* in  */) {
26
27     /* trap_p storage allocated in main with
28      * cudaMallocManaged */
29     *trap_p = 0.5*(f(a) + f(b));
30     float h = (b-a)/n;
31
32     Dev_trap<<<blk_ct, th_per_blk>>>(a, b, h, n, trap_p);
33     cudaDeviceSynchronize();
34
35     *trap_p = h*(*trap_p);
36  }   /* Trap_wrapper */
```

Program 6.12: CUDA kernel and wrapper implementing trapezoidal rule.

a *pointer* to the memory location to the kernel. That is, we can do something like this:

```
/* Host code */
float* trap_p;
cudaMallocManaged(&trap_p, sizeof(float));
    ...
```

```
*trap_p = 0.5*(f(a) + f(b));

/* Call kernel */
...

/* After return from kernel */
*trap_p = h*(*trap_p);
```

When we do this, each thread will get its own copy of `trap_p`, but all of the copies of `trap_p` will refer to the same memory location. So `*trap_p` will be shared.

Note that using a pointer instead of a simple float also solves the problem of returning the value of `trap` in Item 4.

A wrapper function

If you look at the code in Program 6.12, you'll see that we've placed most of the code we use before and after calling the kernel in a **wrapper function**, `Trap_wrapper`. A wrapper function is a function whose main purpose is to call another function. It can perform any preparation needed for the call. It can also perform any additional work needed after the call.

6.10.4 Using the correct threads

We assume that the number of threads, `blk_ct*th_per_blk`, is at least as large as the number of trapezoids. Since the serial **for** loop iterates from 1 up to $n-1$, thread 0 and any thread with `my_i` $> n - 1$, shouldn't execute the code in the body of the serial **for** loop. So we should include a test before the main part of the kernel code

```
if (0 < my_i && my_i < n) {
    /* Compute x, f(x), and add in to *trap_p */
    ...
}
```

See Line 11 in Program 6.12.

6.10.5 Updating the return value and `atomicAdd`

This leaves the problem of updating `*trap_p` (Item 3 in the list above). Since the memory location is shared, an update such as

```
*trap_p += my_trap;
```

forms a race condition, and the actual value ultimately stored in `*trap_p` will be unpredictable. We're solving this problem by using a special CUDA library function, `atomicAdd`, to carry out the addition.

An operation carried out by a thread is **atomic** if it appears to all the other threads as if it were "indivisible." So if another thread tries to access the result of the operation or an operand used in the operation, the access will occur either before the operation started or after the operation completed. Effectively, then, the operation appears to consist of a single, indivisible, machine instruction.

As we saw earlier (see Section 2.4.3), addition is not ordinarily an atomic operation: it consists of several machine instructions. So if one thread is executing an addition, it's possible for another thread to access the operands and the result while the addition is in progress. Because of this, the CUDA library defines several atomic addition functions. The one we're using has the following syntax:

```
__device__  float  atomicAdd(
       float*  float_p    /* in/out */,
       float   val        /* in     */);
```

This atomically adds the contents of `val` to the contents of the memory referred to by `float_p` and stores the result in the memory referred to by `float_p`. It returns the value of the memory referred to by `float_p` at the beginning of the call. See Line 14 of Program 6.12.

6.10.6 Performance of the CUDA trapezoidal rule

We can find the run-time of our trapezoidal rule by finding the execution time of the `Trap_wrapper` function. The execution of this function includes all of the computations carried out by the serial trapezoidal rule, including the initialization of `*trap_p` (Line 29) and `h` (Line 30), and the final update to `*trap_p` (Line 35). It also includes all of the calculations in the body of the serial **for** loop in the `Dev_trap` kernel. So we can effectively determine the run-time of the CUDA trapezoidal rule by timing a host function, and we only need to insert calls to our timing functions before and after the call to `Trap_wrapper`. We use the `GET_TIME` macro defined in the `timer.h` header file on the book's website:

```
double start, finish;
...
GET_TIME(start);
Trap_wrapper(a, b, n, trap_p, blk_ct, th_per_blk);
GET_TIME(finish);
printf("Elapsed time for cuda = %e seconds\n",
       finish-start);
```

The same approach can be used to time the serial trapezoidal rule:

```
GET_TIME(start)
trap = Serial_trap(a, b, n);
GET_TIME(finish);
printf("Elapsed time for cpu = %e seconds\n",
       finish-start);
```

Recall from the section on taking timings (Section 2.6.4) that we take a number of timings, and we ordinarily report the minimum elapsed time. However, if the vast majority of the times are much greater (e.g., 1% or 0.1% greater), then the minimum time may not be reproducible. So other users who run the program may get a time

Table 6.5 Mean run-times for serial and CUDA trapezoidal rule (times are in ms).

System	ARM Cortex-A15	Nvidia GK20A	Intel Core i7	Nvidia GeForce GTX Titan X
Clock	2.3 GHz	852 MHz	3.5 GHz	1.08 GHz
SMs, SPs		1, 192		24, 3072
Run-time	33.6	20.7	4.48	3.08

much larger than ours. When this happens, we report the mean or median of the elapsed times.

Now when we ran this program on our hardware, there *were* a number of times that were within 1% of the minimum time. However, we'll be comparing the run-times of this program with programs that had very few run-times within 1% of the minimum. So for our discussion of implementing the trapezoidal rule using CUDA (Sections 6.10–6.13), we'll use the *mean* run-time, and the means are taken over at least 50 executions.

When we run the serial trapezoidal and the CUDA trapezoidal rule functions many times and take the means of the elapsed times, we get the results shown in Table 6.5. These were taken using $n = 2^{20} = 1,048,576$ trapezoids with $f(x) = x^2 + 1$, $a = -3$, and $b = 3$. The GPUs use 1024 blocks with 1024 threads per block for a total of 1,048,576 threads. The 192 SPs of the GK20A are clearly much faster than a fairly slow conventional processor, an ARM Cortex-A15, but a single core of an Intel Core i7 is much faster than the GK20A. The 3072 SPs on a Titan X *were* 45% faster than the single core of the Intel, but it would seem that with 3072 SPs, we should be able to do better.

6.11 CUDA trapezoidal rule II: improving performance

If you've read the Pthreads or OpenMP chapter, you can probably make a good guess at how to make the CUDA program run faster. For a thread's call to `atomicAdd` to actually be atomic, no other thread can update `*trap_p` while the call is in progress. In other words, the updates to `*trap_p` can't take place simultaneously, and our program may not be very parallel at this point.

One way to improve the performance is to carry out a tree-structured global sum that's similar to the tree-structured global sum we introduced in the MPI chapter (Section 3.4.1). However, because of the differences between the GPU architecture and the distributed-memory CPU architecture, the details are somewhat different.

6.11.1 Tree-structured communication

We can visualize the execution of the "global sum" we implemented in the CUDA trapezoidal rule as a more or less random, linear ordering of the threads. For ex-

Table 6.6 Basic global sum with eight threads.

Time	Thread	my_trap	*trap_p
Start	—	—	9
t_0	5	11	20
t_1	2	5	25
t_2	3	7	32
t_3	7	15	47
t_4	4	9	56
t_5	6	13	69
t_6	0	1	70
t_7	1	3	73

ample, suppose we have only 8 threads and one thread block. Then our threads are $0, 1, \ldots, 7$, and one of the threads will be the first to succeed with the call to atomicAdd. Say it's thread 5. Then another thread will succeed. Say it's thread 2. Continuing in this fashion we get a sequence of atomicAdds, one per thread. Table 6.6 shows how this might proceed over time. Here, we're trying to keep the computations simple: we're assuming that $f(x) = 2x + 1$, $a = 0$, and $b = 8$. So $h = (8 - 0)/8 = 1$, and the value referenced by trap_p at the start of the global sum is

$$0.5 \times (f(a) + f(b)) = 0.5 \times (1 + 17) = 9.$$

What's important is that this approach may serialize the threads. So the computation may require a *sequence* of 8 calculations. Fig. 6.3 illustrates a possible computation.

So rather than have each thread wait for its turn to do an addition into *trap_p, we can pair up the threads so that half of the "active" threads add their partial sum to their partner's partial sum. This gives us a structure that resembles a tree (or, perhaps better, a shrub). See Fig. 6.4.

In our figures, we've gone from requiring a sequence of 8 consecutive additions to a sequence of 4. More generally, if we double the number of threads and values (e.g., increase from 8 to 16), we'll double the length of the sequence of additions using the basic approach, while we'll only add one using the second, tree-structured approach. For example, if we increase the number of threads and values from 8 to 16, the first approach requires a sequence of 16 additions, but the tree-structured approach only requires 5. In fact, if there are t threads and t values, the first approach requires a sequence of t additions, while the tree-structured approach requires $\lceil \log_2(t) \rceil + 1$. For example, if we have 1000 threads and values, we'll go from 1000 communications and sums using the basic approach to 11 using the tree-structured approach, and if we have 1,000,000, we'll go from 1,000,000 to 21!

There are two standard implementations of a tree-structured sum in CUDA. One implementation uses shared memory, and in devices with compute capability < 3 this

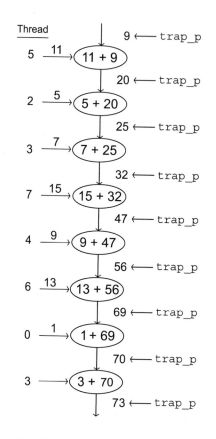

FIGURE 6.3

Basic sum.

is the best implementation. However, in devices with compute capability ≥ 3 there are several functions called **warp shuffles**, that allow a collection of threads within a *warp* to read variables stored by other threads in the warp.

6.11.2 Local variables, registers, shared and global memory

Before we explain the details of how warp shuffles work, let's digress for a moment and talk about memory in CUDA. In Section 6.2 we mentioned that SMs in an Nvidia processor have access to two collections of memory locations: each SM has access to its own "shared" memory, which is accessible only to the SPs belonging to the SM. More precisely, the shared memory allocated for a thread block is only accessible to the threads in that block. On the other hand, all of the SPs and all of the threads have access to "global" memory. The number of shared memory locations is relatively small, but they are quite fast, while the number of global memory locations is relatively large, but they are relatively slow. So we can think of the GPU memory

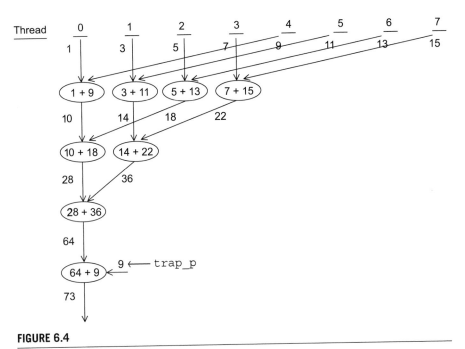

FIGURE 6.4

Tree-structured sum.

Table 6.7 Memory statistics for some Nvidia GPUs.

GPU	Compute Capability	Registers: Bytes per Thread	Shared Mem: Bytes per Block	Global Mem: Bytes per GPU
Quadro 600	2.1	504	48K	1G
GK20A (Jetson TK1)	3.2	504	48K	2G
GeForce GTX Titan X	5.2	504	48K	12G

as a hierarchy with three "levels." At the bottom, is the slowest, largest level: global memory. In the middle is a faster, smaller level: shared memory. At the top is the fastest, smallest level: the registers. For example, Table 6.7 gives some information on relative sizes. Access times also increase dramatically. It takes on the order of 1 cycle to copy a 4-byte int from one register to another. Depending on the system it can take up to an order of magnitude more time to copy from one shared memory location to another, and it can take from two to three orders of magnitude more time to copy from one global memory location to another.

An obvious question here: what about local variables? How much storage is available for them? And how fast is it? This depends on total available memory and program memory usage. If there is enough storage, local variables are stored in registers. However, if there isn't enough register storage, local variables are "spilled" to a region of global memory that's thread private, i.e., only the thread that owns the local variables can access them.

So as long as we have sufficient register storage, we expect the performance of a kernel to improve if we increase our use of registers and reduce our use of shared and/or global memory. The catch, of course, is that the storage available in registers is tiny compared to the storage available in shared and global memory.

6.11.3 Warps and warp shuffles

In particular, if we can implement a global sum in registers, we expect its performance to be superior to an implementation that uses shared or global memory, and the **warp shuffle** functions introduced in CUDA 3.0 allow us to do this.

In CUDA a **warp** is a set of threads with consecutive ranks belonging to a thread block. The number of threads in a warp is currently 32, although Nvidia has stated that this could change. There is a variable initialized by the system that stores the size of a warp:

```
int  warpSize
```

The threads in a warp operate in SIMD fashion. So threads in different warps can execute different statements with no penalty, while threads within the same warp must execute the same statement. When the threads within a warp attempt to execute different statements—e.g., they take different branches in an **if**−**else** statement—the threads are said to have **diverged**. When divergent threads finish executing different statements, and start executing the same statement, they are said to have **converged**.

The rank of a thread within a warp is called the thread's **lane**, and it can be computed using the formula

```
lane = threadIdx.x % warpSize;
```

The warp shuffle functions allow the threads in a warp to read from registers used by another thread in the same warp. Let's take a look at the one we'll use to implement a tree-structured sum of the values stored by the threads in a warp[10]:

```
__device__  float  __shfl_down_sync(
         unsigned   mask                    /* in */,
         float      var                     /* in */,
         unsigned   diff                    /* in */,
         int        width = warpSize /* in */);
```

The `mask` argument indicates which threads are participating in the call. A bit, representing the thread's lane, must be set for each participating thread to ensure that all of the threads in the call have converged—i.e., arrived at the call—before any thread begins executing the call to `__shfl_down_sync`. We'll ordinarily use all the threads in the warp. So we'll usually define

---

[10] Note that the syntax of the warp shuffles was changed in CUDA 9.0. So you may run across CUDA programs that use the older syntax.

```
mask = 0xffffffff;
```

Recall that $0x$ denotes a hexadecimal (base 16) value and $0xf$ is 15_10, which is 1111_2.[11] So this value of `mask` is 32 1's in binary, and it indicates that every thread in the warp participates in the call to __shfl_down_sync. If the thread with lane l calls __shfl_down_sync, then the value stored in `var` on the thread with

$$\text{lane} = l + \text{diff}$$

is returned on thread l. Since `diff` has type **unsigned**, it is ≥ 0. So the value that's returned is from a *higher*-ranked thread. Hence the name "shuffle *down*."

We'll only use `width` = `warpSize`, and since its default value is `warpSize`, we'll omit it from our calls.

There are several possible issues:

- What happens if thread l calls __shfl_down_sync but thread $l + \text{diff}$ doesn't? In this case, the value returned by the call on thread l is *undefined*.
- What happens if thread l calls __shfl_down_sync but $l + \text{diff} \geq \text{warpSize}$? In this case the call will return the value in `var` already stored on thread l.
- What happens if thread l calls __shfl_down_sync, and $l + \text{diff} < \text{warpSize}$, but $l + \text{diff} >$ largest lane in the warp. In other words, because the thread block size is not a multiple of `warpSize`, the last warp in the block has fewer than `warpSize` threads. Say there are w threads in the last warp, where $0 < w < \text{warpSize}$. Then if

$$l + \text{diff} \geq w,$$

the value returned by the call is also undefined.

So to avoid undefined results, it's best if

- All the threads in the warp call __shfl_down_sync, and
- All the warps have `warpSize` threads, or, equivalently, the thread block size (`blockDim.x`) is a multiple of `warpSize`.

6.11.4 Implementing tree-structured global sum with a warp shuffle

So we can implement a tree-structured global sum using the following code:

```
__device__ float Warp_sum(float var) {
   unsigned mask = 0xffffffff;

   for (int diff = warpSize/2; diff > 0; diff = diff/2)
      var += __shfl_down_sync_sync(mask, var, diff);
   return var;
}  /* Warp_sum */
```

---

[11] The subscripts indicate the base.

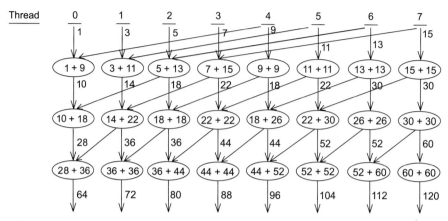

FIGURE 6.5

Tree-structured sum using warp shuffle.

Fig. 6.5 shows how the function would operate if `warpSize` were 8. (The diagram would be illegible if we used a `warpSize` of 32.) Perhaps the most confusing point in the behavior of `__shfl_down_sync` is that when the lane ID

$$l + \texttt{diff} \geq \texttt{warpSize},$$

the call returns the value in the *caller's* `var`. In the diagram this is shown by having only one arrow entering the oval with the sum, and it's labeled with the value just calculated by the thread carrying out the sum. In the row corresponding to `diff = 4` (the first row of sums), the threads with lane IDs $l = 4, 5, 6$, and 7 all have $l + 4 \geq 8$. So the call to `__shfl_down_sync` returns their current `var` values, 9, 11, 13, and 15, respectively, and these values are doubled, because the return value of the call is added into the calling thread's variable `var`. Similar behavior occurs in the row corresponding to the sums for `diff = 2` and lane IDs $l = 6$ and 7, and in the last row when `diff = 1` for the thread with lane ID $l = 7$.

From a practical standpoint, it's important to remember that this implementation will only return the correct sum on the thread with lane ID 0. If all of the threads need the result, we can use an alternative warp shuffle function, `__shfl_xor`. See Exercise 6.6.

6.11.5 Shared memory and an alternative to the warp shuffle

If your GPU has compute capability < 3.0, you won't be able to use the warp shuffle functions in your code, and a thread won't be able to directly access the registers of other threads. However, your code *can* use shared memory, and threads in the same thread block can all access the same shared memory locations. In fact, although shared memory access is slower than register access, we'll see that the shared memory implementation can be just as fast as the warp shuffle implementation.

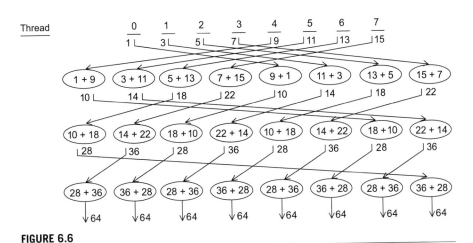

FIGURE 6.6

Dissemination sum using shared memory.

Since the threads belonging to a single warp operate synchronously, we can implement something very similar to a warp shuffle using shared memory instead of registers.

```
__device__ float Shared_mem_sum(float shared_vals[]) {
   int my_lane = threadIdx.x % warpSize;

   for (int diff = warpSize/2; diff > 0; diff = diff/2) {
      /* Make sure 0 <= source < warpSize */
      int source = (my_lane + diff) % warpSize;
      shared_vals[my_lane] += shared_vals[source];
   }
   return shared_vals[my_lane];
}
```

This should be called by all the threads in a warp, and the array shared_vals should be stored in the shared memory of the SM that's running the warp. Since the threads in the warp are operating in SIMD fashion, they effectively execute the code of the function in lockstep. So there's no race condition in the updates to shared_vals: all the threads read the values in shared_vals[source] before any thread updates shared_vals[my_lane].

Technically speaking, this isn't a tree-structured sum. It's sometimes called a **dissemination sum** or **dissemination reduction**. Fig. 6.6 illustrates the copying and additions that take place. Unlike the earlier figures, this figure doesn't show the direct contributions that a thread makes to its sums: including these lines would have made the figure too difficult to read. Also note that every thread reads a value from another thread in each pass through the **for** statement. After all these values have been added in, every thread has the correct sum—not just thread 0. Although we won't need this for the trapezoidal rule, this can be useful in other applications. Furthermore, in any cycle in which the threads in a warp are working, each thread either executes the current instruction or it is idle. So the cost of having every thread execute the same instruction shouldn't be any greater than having some of the threads execute one instruction and the others idle.

An obvious question here is: how does `Shared_mem_sum` make use of Nvidia's shared memory? The answer is that it's not required to use shared memory. The function's argument, the array `shared_vals`, could reside in either global memory or shared memory. In either case, the function would return the sum of the elements of `shared_vals`.

However, to get the best performance, the argument `shared_vals` should be defined to be __shared__ in a kernel. For example, if we know that `shared_vals` will need to store at most 32 floats in each thread block, we can add this definition to our kernel:

```
__shared__ float shared_vals[32];
```

For each thread block this sets aside storage for a collection of 32 floats in the shared memory of the SM to which the block is assigned.

Alternatively, if it isn't known at compile time how much shared memory is needed, it can be declared as

```
extern __shared__ float shared_vals[];
```

and when the kernel is called, a third argument can be included in the triple angle brackets specifying the size *in bytes* of the block of shared memory. For example, if we were using `Shared_mem_sum` in a trapezoidal rule program, we might call the kernel `Dev_trap` with

```
Dev_trap <<<blk_ct, th_per_blk, th_per_blk*sizeof(float)>>>
         (... args to Dev_trap ...);
```

This would allocate storage for `th_per_blk` floats in the `shared_vals` array in each thread block.

6.12 Implementation of trapezoidal rule with `warpSize` thread blocks

Let's put together what we've learned about more efficient sums, warps, warp shuffles, and shared memory to create a couple of new implementations of the trapezoidal rule.

For both versions we'll assume that the thread blocks consist of `warpSize` threads, and we'll use one of our "tree-structured" sums to add the results of the threads in the warp. After computing the function values and adding the results within a warp, the thread with lane ID 0 in the warp will add the warp sum into the total using `Atomic_add`.

6.12.1 Host code

For both the warp shuffle and the shared memory versions, the host code is virtually identical to the code for our first CUDA version. The only substantive difference is that there is no th_per_blk variable in the new versions, since we're assuming that each thread block has warpSize threads.

6.12.2 Kernel with warp shuffle

Our kernel is shown in Program 6.13. Initialization of my_trap is the same as it was in our original implementation (Program 6.12). However, instead of adding each

```
1   __global__ void Dev_trap(
2         const float    a        /*    in    */,
3         const float    b        /*    in    */,
4         const float    h        /*    in    */,
5         const int      n        /*    in    */,
6         float*         trap_p   /* in/out */) {
7      int my_i = blockDim.x * blockIdx.x + threadIdx.x;
8
9      float my_trap = 0.0f;
10     if (0 < my_i && my_i < n) {
11        float my_x = a + my_i*h;
12        my_trap = f(my_x);
13     }
14
15     float result = Warp_sum(my_trap);
16
17     /* result is correct only on thread 0 */
18     if (threadIdx.x == 0) atomicAdd(trap_p, result);
19  } /* Dev_trap */
```

Program 6.13: CUDA kernel implementing trapezoidal rule and using Warp_sum.

thread's calculation directly into *trap_p, each warp (or, in this case, thread block) calls the Warp_sum function (Fig. 6.5) to add the values computed by the threads in the warp. Then, when the warp returns, thread (or lane) 0 adds the warp sum for its thread block (result) into the global total. Since, in general, this version will use multiple thread blocks, there will be multiple warp sums that need to be added to *trap_p. So if we didn't use atomicAdd, the addition of result to *trap_p would form a race condition.

6.12.3 Kernel with shared memory

The kernel that uses shared memory is shown in Program 6.14. It is almost identical to the version that uses the warp shuffle. The main differences are that it declares an

```
1   __global__ void Dev_trap(
2         const float    a         /* in  */,
3         const float    b         /* in  */,
4         const float    h         /* in  */,
5         const int      n         /* in  */,
6         float*         trap_p    /* out */) {
7      __shared__ float shared_vals[WARPSZ];
8      int my_i = blockDim.x * blockIdx.x + threadIdx.x;
9      int my_lane = threadIdx.x % warpSize;
10
11     shared_vals[my_lane] = 0.0f;
12     if (0 < my_i && my_i < n) {
13        float my_x = a + my_i*h;
14        shared_vals[my_lane] = f(my_x);
15     }
16
17     float result = Shared_mem_sum(shared_vals);
18
19     /* result is the same on all threads in a block. */
20     if (threadIdx.x == 0) atomicAdd(trap_p, result);
21  } /* Dev_trap */
```

Program 6.14: CUDA kernel implementing trapezoidal rule and using shared memory.

array of shared memory in Line 7; it initializes this array in Lines 11 and 14; and, of course, the call to Shared_mem_sum is passed this array rather than a scalar register.

Since we know at compile time how much storage we'll need in shared_vals, we can define the array by simply preceding the ordinary C definition with the CUDA qualifier __shared__:

```
__shared__ float shared_vals[WARPSZ];
```

Note that the CUDA defined variable warpSize is *not* defined at compile-time. So our program defines a preprocessor macro

```
#define WARPSZ 32
```

6.12.4 Performance

Of course, we want to see how the various implementations perform. (See Table 6.8.) The problem is the same as the problem we ran earlier (see Table 6.5): we're integrating $f(x) = x^2 + 1$ on the interval $[-3, 3]$, and there are $2^{20} = 1,048,576$ trapezoids. However, since the thread block size is 32, we're using 32,768 thread blocks ($32 \times 32,768 = 1,048,576$).

Table 6.8 Mean run-times for trapezoidal rule using block size of 32 threads (times in ms).

System	ARM Cortex-A15	Nvidia GK20A	Intel Core i7	Nvidia GeForce GTX Titan X
Clock	2.3 GHz	852 MHz	3.5 GHz	1.08 GHz
SMs, SPs		1, 192		24, 3072
Original	33.6	20.7	4.48	3.08
Warp Shuffle		14.4		0.210
Shared Memory		15.0		0.206

We see that on both systems and with both sum implementations, the new programs do significantly better than the original. For the GK20A, the warp shuffle version runs in about 70% of the time of the original, and the shared memory version runs in about 72% of the time of the original. For the Titan X, the improvements are much more impressive: both versions run in less than 7% of the time of the original. Perhaps most striking is the fact that on the Titan X, the warp shuffle is, on average, slightly slower than the shared memory version.

6.13 CUDA trapezoidal rule III: blocks with more than one warp

Limiting ourselves to thread blocks with only 32 threads reduces the power and flexibility of our CUDA programs. For example, devices with compute capability ≥ 2.0 can have blocks with as many as 1024 threads or 32 warps, and CUDA provides a fast barrier that can be used to synchronize *all* the threads in a block. So if we limited ourselves to only 32 threads in a block, we wouldn't be using one of the most useful features of CUDA: the ability to efficiently synchronize large numbers of threads.

So what would a "block" sum look like if we allowed ourselves to use blocks with up to 1024 threads? We could use one of our existing warp sums to add the values computed by the threads in each warp. Then we would have as many as $1024/32 = 32$ warp sums, and we could use one warp in the thread block to add the warp sums.

Since two threads belong to the same warp if their ranks in the block have the same quotient when divided by `warpSize`, to add the warp sums, we can use warp 0, the threads with ranks $0, 1, \ldots, 31$ in the block.

6.13.1 __syncthreads

We might try to use the following pseudocode for finding the sum of the values computed by all the threads in a block:

```
Each thread computes its contribution;
Each warp adds its threads' contributions;
Warp 0 in block adds warp sums;
```

However, there's a race condition. Do you see it? When warp 0 tries to compute the total of the warp sums in the block, it doesn't know whether all the warps in the block have completed their sums. For example, suppose we have two warps, warp 0 and warp 1, each of which has 32 threads. Recall that the threads in a warp operate in SIMD fashion: no thread in the warp proceeds to a new instruction until all the threads in the warp have completed (or skipped) the current instruction. But the threads in warp 0 can operate independently of the threads in warp 1. So if warp 0 finishes computing its sum before warp 1 computes its sum, warp 0 could try to add warp 1's sum to its sum before warp 1 has finished, and, in this case, the block sum could be incorrect.

So we must make sure that warp 0 doesn't start adding up the warp sums until all of the warps in the block are done. We can do this by using CUDA's fast barrier:

```
__device__ void __syncthreads(void);
```

This will cause the threads in the thread block to wait in the call until all of the threads have started the call. Using __syncthreads, we can modify our pseudocode so that the race condition is avoided:

```
Each thread computes its contribution;
Each warp adds its threads' contributions;
__syncthreads();
Warp 0 in block adds warp sums;
```

Now warp 0 won't be able to add the warp sums until every warp in the block has completed its sum.

There are a couple of important caveats when we use __syncthreads. First, it's critical that *all* of the threads in the block execute the call. For example, if the block contains at least two threads, and our code includes something like this:

```
int my_x = threadIdx.x;
if (my_x < blockDim.x/2)
    __syncthreads();
my_x++;
```

then only half the threads in the block will call __syncthreads, and these threads can't proceed until *all* the threads in the block have called __syncthreads. So they will wait forever for the other threads to call __syncthreads.

The second caveat is that __syncthreads only synchronizes the threads in a block. If a grid contains at least two blocks, and if all the threads in the grid call __syncthreads then the threads in different blocks will continue to operate independently of each other. So we can't synchronize the threads in a general grid with __syncthreads.[12]

---

[12] CUDA 9 includes an API that allows programs to define barriers across more general collections of threads than thread blocks, but defining a barrier across multiple thread blocks requires hardware support that's not available in processors with compute capability < 6.

6.13.2 More shared memory

If we try to implement the pseudocode in CUDA, we'll see that there's an important detail that the pseudocode doesn't show: after the call to __syncthreads, how does warp 0 obtain access to the sums computed by the other warps? It can't use a warp shuffle and registers: the warp shuffles only allow a thread to read a register belonging to another thread when that thread belongs to the same warp, and, for the final warp sum we would like the threads in warp 0 to read registers belonging to threads in *other* warps.

You may have guessed that the solution is to use shared memory. If we use warp shuffles to compute the warp sums, we can just declare a shared array that can store up to 32 floats, and the thread with lane 0 in warp w can store its warp sum in element w of the array:

```
__shared__ float warp_sum_arr[WARPSZ];
int my_warp = threadIdx.x / warpSize;
int my_lane = threadIdx.x % warpSize;
// Threads calculate their contributions;
...
float my_result = Warp_sum(my_trap);
if (my_lane == 0) warp_sum_arr[my_warp] = my_result;
__syncthreads();
// Warp 0 adds the sums in warp_sum_arr
...
```

6.13.3 Shared memory warp sums

If we're using shared memory instead of warp shuffles to compute the warp sums, we'll need enough shared memory for each warp in a thread block. Since shared variables are shared by *all* the threads in a thread block, we need an array large enough to hold the contributions of all of the threads to the sum. So we can declare an array with 1024 elements—the largest possible block size—and partition it among the warps:

```
// Make max thread block size available at compile time
#define MAX_BLKSZ 1024
...
___shared__ float thread_calcs[MAX_BLKSZ];
```

Now each warp will store its threads' calculations in a subarray of thread_calcs:

```
float* shared_vals = thread_calcs + my_warp*warpSize;
```

In this setting a thread stores its contribution in the subarray referred to by shared_vals:

```
shared_vals[my_lane] = f(my_x);
```

Now each warp can compute the sum of its threads' contributions by using our shared memory implementation that uses blocks with 32 threads:

```
float my_result = Shared_mem_sum(shared_vals);
```

To continue we need to store the warp sums in locations that can be accessed by the threads in warp 0 in the block, and it might be tempting to try to make a subarray of `thread_calcs` do "double duty." For example, we might try to use the first 32 elements for both the contributions of the threads in warp 0, and the warp sums computed by the warps in the block. So if we have a block with 32 warps of 32 threads, warp w might store its sum in `thread_calcs[w]` for $w = 0, 1, 2, \ldots, 31$.

The problem with this approach is that we'll get another race condition. When can the other warps safely overwrite the elements in warp 0's block? After a warp has completed its call to `Shared_mem_sum`, it would need to wait until warp 0 has finished its call to `Shared_mem_sum` before writing to `thread_calcs`:

```
float my_result = Shared_mem_sum(shared_vals);
__syncthreads();
if (my_lane == 0) thread_calcs[my_warp] = my_result.
```

This is all well and good, but warp 0 still can't proceed with the final call to `Shared_mem_sum`: it must wait until all the warps have written to `thread_calcs`. So we would need a *second* call to `__syncthreads` before warp 0 could proceed:

```
if (my_lane == 0) thread_calcs[my_warp] = my_result.
__syncthreads();
// It's safe for warp 0 to proceed ...
if (my_warp == 0)
    my_result = Shared_mem_sum(thread_calcs);
```

Calls to `__syncthreads` are fast, but they're not free: every thread in the thread block will have to wait until all the threads in the block have called `__syncthreads`. So this can be costly. For example, if there are more threads in the block than there are SPs in an SM, the threads in the block won't all be able to execute simultaneously. So some threads will be delayed reaching the second call to `__syncthreads`, and all of the threads in the block will be delayed until the last thread is able to call `__syncthreads`. So we should only call `__syncthreads()` when we have to.

Alternatively, each warp could store its warp sum in the "first" element of its subarray:

```
float my_result = Shared_mem_sum(shared_vals);
if (my_lane == 0) shared_vals[0] = my_result;
__syncthreads();
...
```

It might at first appear that this would result in a race condition when the thread with lane 0 attempts to update `shared_vals`, but the update is OK. Can you explain why?

6.13.4 Shared memory banks

However, this implementation may not be as fast as possible. The reason has to do with details of the design of shared memory: Nvidia divides the shared memory on

Table 6.9 Shared memory banks: Columns are memory banks. The entries in the body of the table show subscripts of elements of `thread_calcs`.

	Bank					
	0	1	2	...	30	31
Subscripts	0	1	2	...	30	31
	32	33	34	...	62	63
	64	65	66	...	94	95
	96	97	98	...	126	127
	⋮	⋮	⋮	⋱	⋮	⋮
	992	993	994	...	1022	1023

an SM into 32 "banks" (16 for GPUs with compute capability < 2.0). This is done so that the 32 threads in a warp can simultaneously access shared memory: the threads in a warp can simultaneously access shared memory when each thread accesses a different bank.

Table 6.9 illustrates the organization of `thread_calcs`. In the table, the columns are banks, and the rows show the subscripts of consecutive elements of `thread_calcs`. So the 32 threads in a warp can simultaneously access the 32 elements in any one of the rows, or, more generally, if each thread access is to a different column.

When two or more threads access different elements in a single bank (or column in the table), then those accesses must be serialized. So the problem with our approach to saving the warp sums in elements 0, 32, 64, ..., 992 is that these are all in the same bank. So when we try to execute them, the GPU will serialize access, e.g., element 0 will be written, then element 32, then element 64, etc. So the writes will take something like 32 times as long as it would if the 32 elements were stored in different banks, e.g., a row of the table.

The details of bank access are a little complicated and some of the details depend on the compute capability, but the main points are

- If each thread in a warp accesses a different bank, the accesses can happen simultaneously.
- If multiple threads access different memory locations in a single bank, the accesses must be serialized.
- If multiple threads read the same memory location in a bank, the value read is broadcast to the reading threads, and the reads are simultaneous.

The CUDA programming Guide [11] provides full details.

Thus we could exploit the use of the shared memory banks if we stored the results in a contiguous subarray of shared memory. Since each thread block can use at least 16 Kbytes of shared memory, and our "current" definition of `shared_vals` only uses at most 1024 floats or 4 Kbytes of shared memory, there is plenty of shared memory available for storing 32 more floats.

So if we're using shared memory warp sums, a simple solution is to declare *two* arrays of shared memory: one for storing the computations made by each thread, and another for storing the warp sums.

```
__shared__ float thread_calcs[MAX_BLKSZ];
__shared__ float warp_sum_arr[WARPSZ];
float* shared_vals = thread_calcs + my_warp*warpSize;
...
float my_result = Shared_mem_sum(shared_vals);
if (my_lane == 0) warp_sum_arr[my_warp] = my_result;
__syncthreads();
...
```

6.13.5 Finishing up

The remaining codes for the warp sum kernel and the shared memory sum kernel are very similar. First warp 0 computes the sum of the elements in warp_sum_arr. Then thread 0 in the block adds the block sum into the total across all the threads in the grid using atomicAdd. Here's the code for the shared memory sum:

```
if (my_warp == 0) {
    if (threadIdx.x >= blockDim.x/warpSize)
        warp_sum_arr[threadIdx.x] = 0.0;
    blk_result = Shared_mem_sum(warp_sum_arr);
}

if (threadIdx.x == 0) atomicAdd(trap_p, blk_result);
```

In the test threadIdx.x > blockDim.x/warpSize we're checking to see if there are fewer than 32 warps in the block. If there are, then the final elements in warp_sum_arr won't have been initialized. For example, if there are 256 warps in the block, then

```
blockDim.x/warpSize = 256/32 = 8
```

So there are only 8 warps in a block and we'll have only initialized elements $0, 1, \ldots, 7$ of warp_sum_arr. But the warp sum function expects 32 values. So for the threads with threadIdx.x >= 8, we assign

```
warp_sum_arr[threadIdx.x] = 0.0;
```

For the sake of completeness, Program 6.15 shows the kernel that uses shared memory. The main differences between this kernel and the kernel that uses warp shuffles are that the declaration of the first shared array isn't needed in the warp shuffle version, and, of course, the warp shuffle version calls Warp_sum instead of Shared_mem_sum.

6.13.6 Performance

Before moving on, let's take a final look at the run-times for our various versions of the trapezoidal rule. (See Table 6.10.) The problem is the same: find the area under

```
1    __global__ void Dev_trap(
2          const float    a        /* in  */,
3          const float    b        /* in  */,
4          const float    h        /* in  */,
5          const int      n        /* in  */,
6          float*         trap_p   /* out */) {
7       __shared__ float thread_calcs[MAX_BLKSZ];
8       __shared__ float warp_sum_arr[WARPSZ];
9       int my_i = blockDim.x * blockIdx.x + threadIdx.x;
10      int my_warp = threadIdx.x / warpSize;
11      int my_lane = threadIdx.x % warpSize;
12      float* shared_vals = thread_calcs + my_warp*warpSize;
13      float blk_result = 0.0;
14
15      shared_vals[my_lane] = 0.0f;
16      if (0 < my_i && my_i < n) {
17         float my_x = a + my_i*h;
18         shared_vals[my_lane] = f(my_x);
19      }
20
21      float my_result = Shared_mem_sum(shared_vals);
22      if (my_lane == 0) warp_sum_arr[my_warp] = my_result;
23      __syncthreads();
24
25      if (my_warp == 0) {
26         if (threadIdx.x >= blockDim.x/warpSize)
27            warp_sum_arr[threadIdx.x] = 0.0;
28         blk_result = Shared_mem_sum(warp_sum_arr);
29      }
30
31      if (threadIdx.x == 0) atomicAdd(trap_p, blk_result);
32   }  /* Dev_trap */
```

Program 6.15: CUDA kernel implementing trapezoidal rule and using shared memory. This version can use large thread blocks.

the graph of $y = x^2 + 1$ between $x = -3$ and $x = 3$ using $2^{20} = 1,048,576$ trapezoids. However, instead of using a block size of 32 threads, this version uses a block size of 1024 threads. The larger blocks provide a significant advantage on the GK20A: the warp shuffle version is more than 10% faster than the version that uses 32 threads per block, and the shared memory version is about 5% faster. On the Titan X, the performance improvement is huge: the warp shuffle version is more than 30% faster, and the shared memory version is more than 25% faster. So on the faster GPU, reducing the number of threads calling atomicAdd was well worth the additional programming effort.

Table 6.10 Mean run-times for trapezoidal rule using arbitrary block size (times in ms).

System	ARM Cortex-A15	Nvidia GK20A	Intel Core i7	Nvidia GeForce GTX Titan X
Clock	2.3 GHz	852 MHz	3.5 GHz	1.08 GHz
SMs, SPs		1, 192		24, 3072
Original	33.6	20.7	4.48	3.08
Warp Shuffle, 32 ths/blk		14.4		0.210
Shared Memory, 32 ths/blk		15.0		0.206
Warp Shuffle		12.8		0.141
Shared Memory		14.3		0.150

6.14 Bitonic sort

Bitonic sort is a somewhat unusual sorting algorithm, but it has the virtue that it can be parallelized fairly easily. Even better, it can be parallelized so that the threads operate independently of each other for clearly defined segments of code. On the down side, it's not a very intuitive sorting algorithm, and our implementation will require frequent use of barriers. In spite of this, our parallelization with CUDA will be considerably faster than a single core implementation on a CPU.

The algorithm proceeds by building subsequences of keys that form *bitonic se-quences*. A bitonic sequence is a sequence that first increases and then decreases.[13] The butterfly structure (see Section 3.4.4) is at the heart of bitonic sort. However, now, instead of defining the structure by communicating between pairs of processes, we define it by the use of compare-swap operations.

6.14.1 Serial bitonic sort

To see how this works suppose that n is a positive integer that is a power of 2, and we have a list of n integer keys. Any pair of consecutive elements can be turned into either an increasing or a decreasing sequence by a compare-swap: these are two-element butterflies. If we have a four-element list, we can first create a bitonic sequence with a couple of compare swaps or two-element butterflies. We then use a four-element butterfly to create a sorted list.

As an example, suppose our input list is $\{40, 20, 10, 30\}$. Then the first two ele-ments can be turned into an increasing sequence by a compare-swap:

```
if (list[0] > list[1]) Swap(&list[0], &list[1]);
```

---

[13] Technically, a bitonic sequence is either a sequence that first increases and then decreases, or it is a sequence that can be converted to such a sequence by one or more circular shifts. For example, 3, 5, 4, 2, 1 is a bitonic sequence, since it increases and then decreases, but 5, 4, 2, 1, 3 is also a bitonic sequence, since it can be converted to the first sequence by a circular shift.

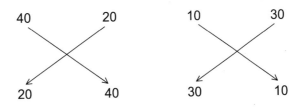

FIGURE 6.7

Two two-element butterflies.

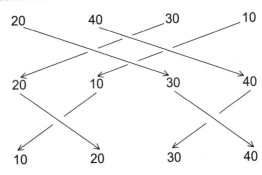

FIGURE 6.8

A four-element butterfly.

To get a bitonic sequence, we want the last two elements to form a decreasing sequence:

```
if (list[2] < list[3]) Swap(&list[2], &list[3]);
```

This gives us a list that's a bitonic sequence: {20, 40, 30, 10}. The two compare-swaps can be thought of as two two-element butterfly operations. (See Fig. 6.7.) The term "butterfly" comes from the fact that the diagram resembles two bowties or butterflies.

Now the bitonic sequence can be turned into a sorted, increasing list by a four-element butterfly. (See Fig. 6.8.) A four-element butterfly starts with two "linked" compare-swaps: the first and third elements and the second and fourth elements. The first pair of compare-swaps moves the smaller elements in the list into the first half, and the larger elements into the second half.

After the first pair of compare-swaps, we carry out another pair of compare-swaps: one on the first half of the list and another on the second half. Each of these compare-swaps ensures that each of the two halves is sorted into increasing order. Table 6.11 shows the details of the pairings in going from the two two-element butterflies to the four-element butterflies.

To summarize, then: to sort a four-element list, we first carry out two two-element butterflies: one on the first two elements and one on the last two. Then we carry out a four-element butterfly on the entire list. The purpose of the first two two-element butterflies is to build a bitonic sequence (a sequence that first increases and then

Table 6.11 Bitonic sort of a four-element list: first two two-element butterflies, then a four-element butterfly.

	Subscripts			
	0	**1**	**2**	**3**
List at start	40	20	10	30
Desired 2 element order	Incr		Decr	
2 element butterfly pairing	1	0	3	2
List after compare swaps	20	40	30	10
Desired 4 element order	Incr			
4 element butterfly: Stage A pairing	2	3	0	1
Stage A: List after compare swaps	20	10	30	40
Stage B pairing	1	0	3	2
Stage B: List after compare swaps	10	20	30	40

decreases). The four-element butterfly builds an increasing (sorted) sequence from the bitonic sequence.

The same idea can be extended to an eight-element list. Now we'll need to carry out the following:

1. Four two-element butterflies. This will build one bitonic sequence from the first four elements, and another bitonic sequence from the last four elements.
2. Two four-element butterflies. This will build an eight-element bitonic sequence: the first four elements will increase, and the last four will decrease.
3. One eight-element butterfly. This will turn the eight-element bitonic sequence into an increasing sequence.

An example that sorts the list $\{15, 77, 83, 86, 35, 85, 92, 93\}$ is shown in Fig. 6.9. The details of the element pairings are shown in Table 6.12.

From these examples, we see that the *serial* algorithm for bitonic sort chooses "pairings" for the compare-swaps by using the butterfly structure. We also see that the algorithm alternates between increasing and decreasing sequences until the final n-element butterfly is completed. This final butterfly results in an increasing, sorted sequence.

Using these observations and our examples as a guideline, we can write high-level pseudocode for a serial bitonic sort. (See Program 6.16.) The outer loop, **for** bf_sz, is iterating over the sizes of the butterflies. In our eight-element example, we start with bf_sz = 2, then bf_sz = 4, and finally bf_sz = 8.

A butterfly with size m requires $\log_2(m)$ stages to sort. In our example, the 2-element butterflies have $1 = \log_2(2)$ stage; the 4-element butterflies have $2 = \log_2(4)$; and the 8-element butterfly has $3 = \log_2(8)$. For reasons that will become apparent shortly, it's convenient to identify the stages by the integers

```
bf_sz/2, bf_sz/4,  ...,  4, 2, 1.
```

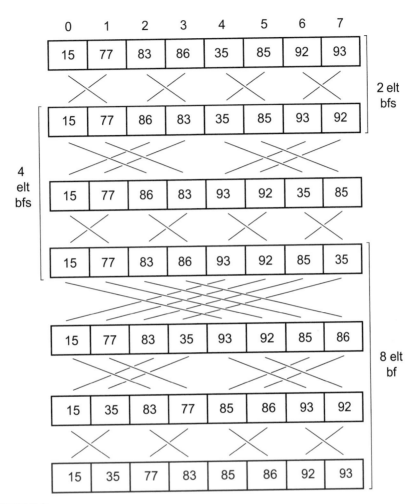

FIGURE 6.9

Sorting an eight-element list using bitonic sort.

```
1  for (bf_sz = 2; bf_sz <= n; bf_sz = 2*bf_sz)
2     for (stage = bf_sz/2; stage > 0; stage = stage/2)
3        for (th = 0; th < n/2; th++) {
4           Get_pair(th, stage, &my_elt1, &my_elt2);
5           if (Increasing_seq(my_elt1, bf_sz))
6              Compare_swap(list, my_elt1, my_elt2, INC);
7           else
8              Compare_swap(list, my_elt1, my_elt2, DEC);
9        }
```

Program 6.16: Pseudocode for serial bitonic sort.

Table 6.12 Bitonic sort of an eight-element list.

	Subscripts							
	0	**1**	**2**	**3**	**4**	**5**	**6**	**7**
List at start	15	77	83	86	35	85	92	93
Desired 2 element order	Incr		Decr		Incr		Decr	
2 element pairing	1	0	3	2	5	4	7	6
List after compare swaps	15	77	86	83	35	85	93	92
Desired 4 element order	Incr				Decr			
4 element: Stage A pairing	2	3	0	1	6	7	4	5
Stage A: List after compare swaps	15	77	86	83	93	92	35	85
4 element: Stage B pairing	0	1	3	2	5	4	7	6
Stage B: List after compare swaps	15	77	83	86	93	92	85	35
Desired 8 element order	Incr							
8 element: Stage A pairing	4	5	6	7	0	1	2	3
Stage A: List after compare swaps	15	77	83	35	93	92	85	86
8 element: Stage B pairing	2	3	0	1	6	7	4	5
Stage B: List after compare swaps	15	35	83	77	85	86	93	92
8 element: Stage C pairing	1	0	3	2	5	4	7	6
Stage C: List after compare swaps	15	35	77	83	85	86	92	93

The innermost loop is iterating sequentially over the pairs of elements that are being compared and possibly swapped: if we have n elements in the list, we'll have $n/2$ pairs. As we'll see, in the parallel implementation each of the $n/2$ pairs will be the responsibility of a single thread. So we chose to call the innermost loop variable th.

The function Get_pair determines which pair of elements a "thread" is responsible for, and the function Increasing_seq determines whether a pair of elements in the list should be in increasing or decreasing order.

6.14.2 Butterflies and binary representations

It's often easier to understand the structure of a butterfly if we look at the binary representation of the values involved. In our case, the values are the element subscripts, the thread ranks (or loop variable th), and the loop variables bf_sz and stage. An eight-element example is shown in Fig. 6.10. This shows the three individual stages in an eight element butterfly: stage $4 = 100_2$, stage $2 = 010_2$, and stage $= 1 = 001_2$.[14]

Each stage also shows which elements are paired. Since there are 8 elements, there are $8/2 = 4$ threads, and we see that we can determine which elements are assigned to a thread by looking at the binary representations of the stages, the element subscripts, and the thread ranks. (See Table 6.13.) The examples in the table show that

---

[14] The subscript of 2 indicates that the value is in base 2 or binary.

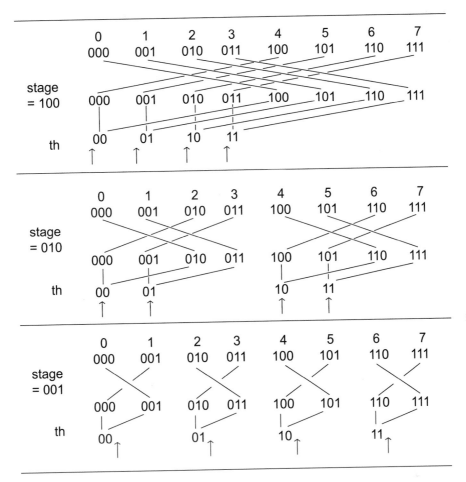

FIGURE 6.10

Binary representation of various values used in bitonic sort.

the elements a thread is responsible for can be determined by inserting an additional bit in the thread rank. For example, when the stage is 1, the elements are obtained by adding a bit to the right of the right-most bit (bit 0) in the thread rank. When the stage is 2, the elements are obtained by adding a bit between the least significant bit (bit 0) and the next bit to the left (bit 1). When the stage is 4, the bit is added between bit 1 and bit 2. The location for the additional bit is indicated in the figure by a small up arrow in Fig. 6.10.

If we introduce a variable `which_bit` representing $\log_2(\text{stage})$, the additional bit is inserted immediately to the right of bit `which_bit`, so we add a function `Insert_zero` that will add a zero bit to the binary representation of the thread rank. The subscript

Table 6.13 Mapping of threads to elements in bitonic sort.

Stage	Thread	1st Elt	2nd Elt
100 = 4	00 = 0	000 = 0	100 = 4
	01 = 1	001 = 1	101 = 5
	10 = 2	010 = 2	110 = 6
	11 = 3	011 = 3	111 = 7
010 = 2	00 = 0	000 = 0	010 = 2
	01 = 1	001 = 1	011 = 3
	10 = 2	100 = 4	110 = 6
	11 = 3	101 = 5	111 = 7
001 = 1	00 = 0	000 = 0	001 = 1
	01 = 1	010 = 2	011 = 3
	10 = 2	100 = 4	101 = 5
	11 = 3	110 = 6	111 = 7

Table 6.14 Is an element part of an increasing (Inc) or decreasing (Dec) sequence?

	bf_sz		
Element	0010 = 2	0100 = 4	1000 = 8
0000 = 0	Inc	Inc	Inc
0001 = 1	Inc	Inc	Inc
0010 = 2	Dec	Inc	Inc
0011 = 3	Dec	Inc	Inc
0100 = 4	Inc	Dec	Inc
0101 = 5	Inc	Dec	Inc
0110 = 6	Dec	Dec	Inc
0111 = 7	Dec	Dec	Inc

of the other element can be obtained by converting the zero bit to a one, and this can be done using a bitwise exclusive or.

Finally observe that in any one butterfly, an element in the list will always be part of an increasing or a decreasing sequence. That is, whether a pair of elements should swap so that the pair is increasing or decreasing depends only on the elements and the size of the butterfly. For example, in the 4-element butterfly, list[3] should always swap so that it's part of an increasing sequence, while list[4] should always swap so that it's part of a decreasing sequence. To see how to determine whether the sequence is increasing or decreasing, it helps to look at the binary representations of bf_sz and the element subscript. (See Table 6.14.) If we look at the *bitwise* and of the element subscript and bf_sz, we see that when the bitwise and is 0 (e.g., 2 & 4 = 0010 & 0100 = 0000), then the element should be part of an increasing sequence, while if the bitwise and is nonzero (e.g., 5 & 4 = 0101 & 0100 = 0100 = 4), then the element should be part of a decreasing sequence.

We can use these observations to write more detailed pseudocode. (See Program 6.17.) Here, we've replaced the two calls to `Compare_swap` in the original

```
1   for (bf_sz = 2; bf_sz <= n; bf_sz = 2*bf_sz)
2      for (stage = bf_sz/2; stage > 0; stage = stage/2) {
3         which_bit = Which_bit(stage);
4         for (th = 0; th < n/2; th++) {
5            my_elt1 = Insert_zero(th, which_bit);
6            my_elt2 = my_elt1 ^ stage;
7            Compare_swap(list, my_elt1, my_elt2, my_elt1 & bf_sz);
8         }
9      }
```

Program 6.17: Pseudocode for a second serial implementation of bitonic sort.

pseudocode with a single call, which will examine the value of `my_elt1 & bf_sz` to determine whether the pair should increase or decrease.

We leave writing the functions `Which_bit`, `Insert_zero` and `Compare_swap` to the exercises.

6.14.3 Parallel bitonic sort I

We designed our serial implementation with the idea that it would be converted into a multithreaded implementation by parallelizing the innermost loop in Program 6.17 with $n/2$ threads. So an aggregated task is choosing a pair of elements and putting them in the correct order. However, there's a race condition when the threads go from one stage to the next.

```
for (bf_sz = 2; bf_sz <= n; bf_sz = 2*bf_sz) {
   for (stage = bf_sz/2; stage > 0; stage = stage/2,
         which_bit++) {
      which_bit = Which_bit(stage);
      my_elt1 = Insert_zero(th, which_bit);
      my_elt2 = my_elt1 ^ stage;
      Compare_swap(list, my_elt1, my_elt2, my_elt1 & bf_sz);
   }
}
```

For example, suppose $n = 4$, and we've already completed the 2-element butterflies. So suppose we're executing a 4-element butterfly on the list $20, 40, 30, 10$, and suppose the sequence of events in Table 6.15 occurs. Also suppose for the moment that threads 0 and 1 are in different warps, so that they are not automatically synchronized. We see that Thread 0's second `Compare_swap` should happen *after* Thread 1's first `Compare_swap`, and because the order in which they were executed was reversed, the final list isn't sorted.

Of course, in most cases of interest, Threads 0 and 1 *will* belong to the same warp, and the second `Compare_swap` won't be executed until after all the threads in the

Table 6.15 Sequence of events in a four-element butterfly.

| Time | Thread 0 | | Thread 1 | | List After Stage |
	Stage	Execute	Stage	Execute	
0	—	—	—	—	20, 40, 30, 10
1	2	Compare_swap(elt 0, elt 2)	—	Idle	20, 40, 30, 10
2	1	Compare_swap(elt 0, elt 1)	—	Idle	20, 40, 30, 10
3	—	Idle	2	Compare_swap(elt 1, elt 3)	20, 10, 30, 40
4	—	Idle	1	Compare_swap(elt 2, elt 3)	20, 10, 30, 40

warp have completed the first. However, it's not difficult to see that this problem can happen if we have more than one warp. For example, suppose we have 128 elements in the list, 64 threads, and two warps of 32 threads. Then much the same situation can occur with (for example) threads 0 and 32. When $bf\_sz = 128$ and $stage = 64$, then thread 32 will be swapping elements 32 and 96, and when $stage = 32$, thread 0 will be swapping elements 0 and 32. So if thread 0 proceeds to $stage = 32$ before thread 32 has completed the compare-swap of elements 32 and 96, thread 0 can swap the wrong elements: it can swap the *old* element 32 with element 0.

So we should ensure that none of the threads will proceed to the next stage until all the threads have completed the current stage. As we have already seen, the threads in a single thread block can do this by calling __syncthreads.

So for our first parallel implementation, let's just use a single thread block, and call __syncthreads at the end of the body of the inner **for** loop. This gives us the code shown in Program 6.18. Comparing this to the serial bitonic sort function shown

```
1    unsigned th = threadIdx.x;
2
3    for (bf_sz = 2; bf_sz <= n; bf_sz = 2*bf_sz) {
4       for (stage = bf_sz/2; stage > 0;  stage = stage/2) {
5          which_bit = Which_bit(stage);
6          my_elt1 = Insert_zero(th, which_bit);
7          my_elt2 = my_elt1 ^ stage;
8          Compare_swap(list, my_elt1, my_elt2, my_elt1 & bf_sz);
9          __syncthreads();
10      }
11   }
```

Program 6.18: CUDA bitonic sort kernel that uses a single thread block.

in Program 6.17, we see that there are basically two differences. First, the innermost loop **for** th of the serial program has been eliminated, since the parallel program uses one thread per iteration of the serial code. Second, in the parallel program, we're calling __syncthreads at the end of each iteration of the **for** stage loop, and we're doing this to eliminate the race condition we just discussed.

6.14.4 Parallel bitonic sort II

Our work in implementing serial bitonic sort really paid off: we only needed to make two relatively straightforward changes to the serial code to parallelize it. However, the sizes of the problems that can be solved by this parallel implementation are *very* limited. Since current-generation CUDA thread blocks can consist of at most 1024 threads, and our implementation requires that the number of elements be twice the number of threads, the largest list that we can sort will consist of 2048 elements.

To increase the size of the lists that we can sort, there are (at least) a couple of options we might consider. One is to increase the amount of work done by each thread: instead of assigning two elements of the list to each thread, we might try assigning $2m$ elements to each thread for some integer $m > 1$. For example, instead of comparing and swapping two elements in the list, we might *merge* two m-element sublists of the list. If the algorithm is working on an increasing sublist, we would assign the lower half of the merged $2m$-element sublist to the lower-ranked thread and the upper half to the higher-ranked thread. We could reverse the assignments when the algorithm was working on a decreasing sublist. See programming assignment 6.4.

The alternative we'll look at now uses the same basic algorithm we used for a single thread block, except that we will synchronize multiple thread blocks by returning from the executing kernel and calling the same or another kernel.[15]

So suppose that we have a grid with at least two thread blocks and

$$n = 2 \times \mathtt{blk\_ct} \times \mathtt{th\_per\_blk}.$$

Also suppose that `blk_ct` and `th_per_blk` are powers of 2. (So n is also a power of 2.) We can start by using bitonic sort to sort each sublist of length $2 \times \mathtt{th\_per\_blk}$ into an increasing or decreasing list, and when we take pairs of consecutive sorted sublists, we'll have bitonic sequences of length $4 \times \mathtt{th\_per\_blk}$. This initial sorting will take place in a single kernel.

Now to continue with the bitonic sort, we need to get groups of thread blocks to cooperate in forming sorted sublists of length $4 \times \mathtt{th\_per\_blk}$, then $8 \times \mathtt{th\_per\_blk}$, etc. We can do this in two steps. We execute a stage of the current butterfly in a single kernel, and we synchronize the threads in the different blocks by returning from the kernel and then calling the kernel again for the next stage. Note that between kernel calls we don't need to call `cudaDeviceSynchronize`: the second kernel call will wait for the first to complete before it starts executing on the device.

After repeatedly executing the "one stage" kernel, we will eventually have "sub-butterflies" that are being applied to sublists with length

$$\leq 2 \times \mathtt{th\_per\_blk},$$

and the thread blocks can work independently of each other. So we can return to simply synchronizing across each thread block with calls to `__syncthreads`.

---

[15] As we noted earlier, CUDA 9 running on a processor with compute capability ≥ 6 provides support for synchronizing all of the threads in a grid.

This gives us the host function shown in Program 6.19.

A point to note here is that in our implementation there's a difference between "sublists" and "stages." A `stage` is a loop variable. In a butterfly on an n-element list, the first `stage` is $n/2$. Subsequent stages are $n/4, n/8$, etc. The first sublist in an n-element butterfly is the entire list. Subsequent sublists have sizes $n/2, n/4$, etc.

```
1   /* Sort sublists with 2*th_per_blk elements */
2   Pbitonic_start <<<blk_ct, th_per_blk>>> (list, n);
3
4   for (bf_sz = 4*th_per_blk; bf_sz <= n; bf_sz = bf_sz/2) {
5      for (stage = bf_sz/2; stage >= 2*th_per_blk;
6             stage = stage/2) {
7         /* Do a single stage of a butterfly */
8         Pbutterfly_one_stage <<<blk_ct, th_per_blk>>> (list, n,
9              bf_sz, stage);
10     }
11     /* Finish the current butterfly by working with "short"
12      * sublists */
13     Pbutterfly_finish <<<blk_ct, th_per_blk>>> (list, n,
14          bf_sz);
15  }
```

Program 6.19: Host function that implements a general CUDA bitonic sort.

Let's look at how this code would work if we have 8 thread blocks consisting of 1024 threads. So we can sort a list with

$$2 \times 8 \times 1024 = 16,384 \text{ elements.}$$

Then in `Pbitonic_start` each thread block will iterate through butterflies of sizes $2, 4, \ldots, 2048$. So when `Pbitonic_start` completes execution, the sublists consisting of the elements shown in Table 6.16 will be sorted. Then we'll proceed to the host **for**-loop and iterate through the values $bf\_sz = 4096, 8192, 16,384$. For each of these sizes, we'll iterate in the inner loop through the stages shown in Table 6.17. Of course, we're not done with a butterfly when we finish the inner loop. We still need to finish off the butterfly for stages $1024, 512, \ldots, 4, 2, 1$, and in each iteration of the outer loop we do this in the call to `Pbutterfly_finish`.

6.14.5 Performance of CUDA bitonic sort

Before moving on, let's take a look at the performance of our CUDA implementations of bitonic sort. These are run on a system with a Pascal GPU (compute capability 6.1) and an Intel Xeon host. Table 6.18 shows the run-times of serial bitonic sort on the host, the C-library `qsort` function on the host, and our first parallel implementation of bitonic sort, which uses one thread block with 1024 threads. All three run-times are

Table 6.16 Elements in sorted sublists after call to `Pbitonic_start`.

Block	Sorted Elements			
0	0	1	. . .	2047
1	2048	2049	. . .	4095
⋮	⋮	⋮		⋮
8	14,436	14,437	. . .	16,383

Table 6.17 Stages executed by `Pbutterfly_one_stage`.

bf_sz	stages		
4096	2048		
8192	4096	2048	
16,384	8192	4096	2048

Table 6.18 Mean run-times for various sorts of a list of 2048 ints (times in ms).

System	Intel Xeon 4116	Intel Xeon 4116	Nvidia Quadro P5000
Clock	2.1 GHz	2.1 GHz	1.73 GHz
SMs, SPs			20, 1280
Sort	qsort	bitonic	parallel bitonic
Blks, Threads			1, 1024
Run-time	0.116	1.19	0.128

taken by sorting a list with 2048 elements. Not surprisingly, for such a small list, the `qsort` library function is significantly faster than parallel bitonic sort with one thread block.

Table 6.19 shows the times for sorting a list with 2,097,152 (2^{21}) elements on the same systems, but in this case the parallel sort uses the multiple block implementation of bitonic sort with 1024 blocks of 1024 threads. Now the parallel bitonic sort is more than 40 times faster than the serial qsort.

6.15 Summary

In the early years of this century, programmers tried to exploit the computational power of **Graphics Processing Units** or **GPUs** for general purpose computing. This became known as **General Purpose Programming on Graphics Processing Units** or **GPGPU**, and nowadays GPGPU is one of the most important types of parallel computing.

Table 6.19 Mean run-times for various sorts of a list of 2,097,152 ints (times in ms).

System	Intel Xeon 4116	Intel Xeon 4116	Nvidia Quadro P5000
Clock	2.1 GHz	2.1 GHz	1.73 GHz
SMs, SPs			20, 1280
Sort	qsort	bitonic	parallel bitonic
Blks, Threads			1024, 1024
Run-time	539	3710	13.1

For our purposes, the most basic GPU is a **single instruction multiple data** or **SIMD** processor. So a collection of datapaths will all execute the same instruction, but each datapath in the collection can have its own data. In an Nvidia GPU the datapaths are called **streaming processors** or **SPs**, and the SPs are grouped into **streaming multiprocessors** or **SMs**. In Nvidia systems, the software analogs of SPs and SMs are **threads** and **thread blocks**, respectively. However, the analogy isn't precise: while a thread block will run on a single SM, multiple, independent thread blocks can run on one SM. Furthermore, the threads in a thread block aren't constrained to execute in lockstep: a **warp** is a subset of the threads in a thread block that executes in lockstep. Currently, Nvidia warps consist of 32 consecutive threads.

We looked at two main types of memory in an Nvidia GPU: each SM has a small block of fast **shared memory**, which is shared among the threads in a thread block. There is also a large block of slower **global memory** that is shared among all the threads.

The API we used for GPGPU programming is called **CUDA**, and it is an extension of C/C++. It assumes that the system has both a CPU and a GPU: the `main` function runs on the CPU, and a **kernel** is a CUDA function that is called by the host but that runs on the GPU or **device**. So a CUDA program ordinarily runs on both a CPU and a GPU, and since these have different architectures, writing a CUDA program is sometimes called **heterogeneous programming**.

The code for a CUDA kernel is identified by the special identifier `__global__`. A kernel has return type **void**, but it can take arguments of any types. The call to a kernel must specify the number of thread blocks and the number of threads in each block. For example, suppose the kernel `My_kernel` has the prototype

```
__global__ void My_kernel(float x[], float y[], int n);
```

Then if we want to start `My_kernel` with four thread blocks, each of which has 256 threads, we could use the call

```
My_kernel <<< 4, 256 >>> (x, y, n);
```

The collection of thread blocks started by a kernel is called a **grid**. When a kernel is called, the threads can get information on the grid by examining the fields of several structs:

- `threadIdx`
- `blockIdx`
- `blockDim`
- `gridDim`

All of these have three fields: x, y, and z, but we only used the x field. When the kernel `My_kernel` is started by the call in the last paragraph

> `gridDim.x` = 4 and `blockDim.x` = 256.

(The y and z fields are set to 1.) Also `blockIdx.x` will give the rank of the block to which the thread belongs, and `threadIdx.x` will give its thread rank. (The y and z fields are set to 0.)

From the point of view of the host, a kernel call is ordinarily **asynchronous**: after the kernel is started on the device, the device and the host can execute simultaneously. If the host wants to use a result from the device, it can call

```
__host__ cudaError_t cudaDeviceSynchronize( void );
```

The host will block in the call until the device has finished. The __host__ identifier indicates that this function runs on the host, and `cudaError_t` is a CUDA defined type for specifying error conditions.

Note that by default only one kernel at a time runs on the device, and if the host first starts `Kernel_a` and then starts `Kernel_b`, then by default `Kernel_b` will not start executing until `Kernel_a` completes.

CUDA source files usually have a suffix of ".cu"; for example, `hello.cu`. Since CUDA is an extension of C/C++, it has its own compiler, `nvcc`, and we can compile programs using a shell command, such as

```
$ nvcc -arch=sm_52 -o hello hello.cu
```

There are many command line options. Some are the same as the options for `gcc`. For example, −g will create a symbol table for the host code. There are also many options that are not in `gcc`. For example, the command line option −arch=... is frequently used, since it can tell `nvcc` what the target GPU is. In our example, `sm_52` is indicating that the target is a maxwell processor.

Unlike MPI, Pthreads, and OpenMP, CUDA is not specified by a standards organization—Nvidia develops and maintains it. A consequence of this is that it tends to change more often than APIs that are standards. There are (at least) two reasons for this. When Nvidia introduces a new hardware feature, it wants to be sure that the feature is readily accessible to programmers, and as Nvidia's developers discover or create new software implementations, Nvidia wants to make these new features available to programmers. A downside to this is that Nvidia may discontinue support for old hardware and software much sooner than a standards organization might.

To guarantee that a CUDA program will run as desired, it may be necessary to consider both the *compute capability* of the GPU and the version of the CUDA software. For example, CUDA 9 included software that can implement barriers across any collection of threads. However, unless the compute capability of the GPU is ≥ 6, barriers can only be used across collections of threads that belong to a single thread block.

In most Nvidia systems, the host memory and the device memory are physically separate, and in older systems running CUDA, it was necessary to explicitly copy data from the host to the device and vice versa. However, since the introduction of the Kepler processor, Nvidia has provided **unified memory**. This is memory that is automatically replicated on the host and the device. For example, when a kernel is launched, any unified memory locations that have been updated on the host will be copied to the device, and when a kernel completes executing, unified memory locations that have been updated on the device will be copied to the host. The most commonly used method for allocating unified memory is to call

```
__host__ cudaError_t cudaMallocManaged ( void** ptr,
       size_t size );
```

This will allocate `size` bytes of unified memory, and when it returns, the pointer `*ptr` will refer to the allocated memory. The address stored in `*ptr` is valid on both the host and the device. So this address can be passed as an argument from the host to the device, and if the memory it refers to is updated by a kernel, a host pointer that refers to the memory will also refer to the updated memory.

Unified memory that's allocated with `cudaMallocManaged` should be freed with a call to

```
__host__ cudaError_t cudaFree ( void* ptr );
```

Note that if a pointer refers to a memory location on the host that is on the stack, then we cannot pass the pointer to a kernel: the address won't be valid. Consequently, we can't simulate pass-by-reference to a kernel by simply passing the address of a host stack variable. The simplest way to obtain the effect of pass-by-reference is to declare a pointer on the host, and use `cudaMallocManaged` to make the pointer refer to a memory location in unified memory.

Recall that a **warp** is a subset of the threads of a thread block, and the threads in a warp execute in SIMD fashion. So when all the threads in a warp are executing the same statements, then the threads will effectively execute in lockstep: if all the threads are executing

```
statement 1;
statement 2;
```

then no thread in the warp will start executing statement 2 until all the threads in the warp have finished executing statement 1. However, if some of the threads are executing one statement, and others are executing a different statement, then we'll have what's called **thread divergence**. For example, suppose we have the following code:

```
if (threadIdx.x % warpSize < 16)
    statement 1;
else
    statement 2;
```

Then all the threads in the first half of the warp will execute statement 1, but all the threads in the second half of the warp will execute statement 2. However, the threads in a warp execute in SIMD fashion. So while the threads in the first half of the warp execute statement 1, the threads in the second half are idle. After the threads in the first half have finished executing statement 1, the threads in the first half of the warp will be idle while the threads in the second half of the warp execute statement 2. So thread divergence can cause a large degradation in performance.

Recall that a **barrier** is a function that can be called by a collection of threads, and no thread will return from the call until all the threads in the collection have started the call. Since the threads in a warp execute in lockstep, there is, effectively, a barrier between each pair of consecutive statements that is executed by all the threads in a warp. CUDA also provides a fast barrier for the threads in a thread block:

```
__device__ void __syncthreads(void);
```

The __device__ identifier indicates that the function should be called on the device (but not the host). This particular function must be called by *all* the threads in the thread block. Recently (CUDA 9) Nvidia added barriers that can be defined across user-defined collections of threads. However, barriers across threads in different blocks require a Pascal or more recent processor. For older processors, programmers can form a barrier across all the threads in the grid by returning from a kernel and starting another (or the same) kernel. This does *not* require an intervening call to cudaDeviceSynchronize.

Recall that an operation is **atomic** if when it is executed by one thread, it appears indivisible to the other threads. CUDA provides a number of atomic operations that can be safely used by any thread. We used

```
__device__ float atomicAdd(float* x_p, float y);
```

This adds y to the memory location referenced by x_p. It returns the original value stored in the memory referenced by x_p.

In several chapters we've seen that using tree-structured pairings of threads (or processes) can greatly reduce the cost of operations such as a global sum. CUDA provides two alternative means of implementing fast tree-structured operations: warp shuffles and shared memory. Warp shuffles allow a thread to read a register belonging to another thread in the same warp. We used __shfl_down_sync to implement a tree-structured sum when each thread in the warp is storing a **float** my_val:

```
unsigned mask = 0xffffffff;
float result = my_val;
for (unsigned diff = warpSize/2; diff > 0; diff = diff/2)
    result += __shfl_down_sync(mask, result, diff);
```

This assumes that all the threads in the warp are executing the code. The function

```
__device__ float __shfl_down_sync (unsigned mask,
    float result, unsigned diff,
    int width=warpSize );
```

returns the value in result on the thread with warp rank or **lane**

```
(caller lane) + diff
```

When this value is ≥ 32, the function returns the value stored in the caller's result variable. So the thread with lane 0 will be the only thread with the correct result. Note that there isn't a race condition in the code, because no thread will update its result until all the threads have returned from the call to __shfl_down_sync. The argument mask specifies which threads in the warp are participating in the call: if the thread with lane l is participating, then bit l should be one. So setting mask to 0xffffffff will ensure that every thread is participating. The width argument specifies the number of threads, and it is optional; since we used a full 32 threads, we omitted it.

Shared memory provides an alternative to the use of warp shuffles. The simplest way to define a block of shared memory is to statically specify its size when it's declared in the kernel. For example,

```
__global__ void My_kernel (...) {
  ...
  // Number of elements must be available at compile time
  __shared__ float sum_arr[32];
  ...
  // Initialize sum_arr[my_lane]
  int my_lane = ...;
  ...
  for (unsigned diff = 16; diff > 0; diff >>= 1) {
    int source = (my_lane + diff) % warpSize;
    sum_arr[my_lane] += sum_arr[source];
  }
  ...
}
```

Technically, this is not a tree-structured sum: it's sometimes called a **dissemination** sum. Now, because of the remainder (%) in the calculation of source, all of the threads in the warp are getting correct addends at each stage. So this version has the virtue that all the threads in the warp will have the correct result. Note also that this function is executed by the threads in a warp. So there is no race condition in the addition into sum_arr[my_lane], since the load of sum_arr[source] will happen before the addition.

A final point to note is that shared memory is divided into **banks**. This is done so that all the threads in a warp can simultaneously execute a load or store, provided each thread accesses a different bank. Ordinarily consecutive 4-byte words are stored in successive banks.[16] So, for example, in the shared memory sum of the preceding

---

[16] This can be changed to 8-byte words with cudaDeviceSetSharedMemConfig(). See the CUDA Programming Guide [11] for details.

paragraph, the shared memory accesses in sum_arr[my_lane] are to consecutive 4-byte words. Also, in the accesses in sum_arr[my_lane], each thread is accessing a distinct 4-byte word among the 32 array elements. So neither of the accesses causes bank conflicts.

6.16 Exercises

6.1 Run the cuda_hello program (Program 6.1) without the call to cudaDeviceSynchronize. What happens?

6.2 When we ran cuda_hello (Program 6.1) with 10 threads, the output was ordered by the threads' ranks: the output from thread 0 was first, then the output from thread 1, then the output from thread 2, etc. Is this always the case? If not, what is the smallest number of threads you can start for which the output is *not* always ordered by the threads' ranks? Can you explain why?

6.3 When you run the cuda_hello program (Program 6.1), what is the largest number of threads you can start? What happens if you exceed this number?

6.4 Modify our first implementation of the trapezoidal rule (Program 6.12) so that it can be used on a system that doesn't support unified memory.

6.5 The __shfl_sync function

```
__device__  float  __shfl_sync(
              unsigned    mask            /* in */,
              float       var             /* in */,
              int         srcLane         /* in */,
              int         width = warpSize  /* in */);
```

returns the value in var from the thread with lane ID srcLane on the calling thread. Unlike __shfl_down_sync, if srcLane < 0 or srcLane > warpSize, then the return value is taken from the thread in the current warp with remainder srcLane % warpSize.

Use the __shfl_sync function to implement a broadcast across a warp, i.e., every thread in the warp gets a value stored in a designated thread of the warp. Write a driver kernel that calls your broadcast function, and a driver main program that initializes data structures, calls the kernel, and prints the results.

6.6 Our Warp_sum function (Section 6.11.4) implements a tree-structured sum of the values stored on the threads in a warp, and it only returns the correct sum on the thread with lane ID 0. If all of the threads in the warp need the correct sum, an alternative is to use a *butterfly*. (See Fig. 3.9.) In CUDA this is most easily implemented with another warp shuffle, __shfl_xor_sync:

```
__device__  float  __shfl_xor_sync(
              unsigned    mask            /* in */,
              float       var             /* in */,
              int         lanemask        /* in */,
              int         width = warpSize  /* in */);
```

This function computes the lane ID of the source by forming a bitwise exclusive or of the caller's lane ID with a `lanemask`.

For example, if warps consisted of 8 threads, we would start with a `lanemask` of $4_{10} = 100_2$ and `lanemask` would iterate through $2_{10} = 010_2$ and $1_{10} = 001_2$. The following table shows the lane ID of the source thread for each thread and for each value of the lanemask:

Caller Lane ID	Source Lane ID = lanemask XOR Caller Lane ID		
	lanemask 100_2	lanemask 010_2	lanemask 001_2
$0 = 000_2$	$4 = 100_2$	$2 = 010_2$	$1 = 001_2$
$1 = 001_2$	$5 = 101_2$	$3 = 011_2$	$0 = 000_2$
$2 = 010_2$	$6 = 110_2$	$0 = 000_2$	$3 = 011_2$
$3 = 011_2$	$7 = 111_2$	$1 = 001_2$	$2 = 010_2$
$4 = 100_2$	$0 = 000_2$	$6 = 110_2$	$5 = 101_2$
$5 = 101_2$	$1 = 001_2$	$7 = 111_2$	$4 = 100_2$
$6 = 110_2$	$2 = 010_2$	$4 = 100_2$	$7 = 111_2$
$7 = 111_2$	$3 = 011_2$	$5 = 101_2$	$6 = 110_2$

For a 32-lane warp, the bitmask should start at 16 and go through 8, 4, 2, and 1. Use `__shfl_xor_sync` to implement a butterfly-structured warp sum. Write a kernel that calls your function, and a main function that initializes data structures, calls the kernel, and prints the results.

6.7 An alternative implementation of the trapezoidal rule for large numbers of trapezoids can repeatedly reduce the number of threads involved in computing sums. For example, with 2^{30} trapezoids, we could start with 2^{20} blocks of 2^{10} threads. Each block can use our two-stage process to compute the sum of its threads' values. Then there are 2^{20} sums, which can be divided among 1024 blocks of 1024 threads, and we can use the two-stage process to find the sum of each block's sums. Finally, we'll have one block of 1024 sums of sums, and we can use the two-stage process to find a single sum. Use our two-stage sum as the heart of a program that finds the sum of up to $1024 \times (2^{31} - 1)$ trapezoids.

6.8 In our implementation of the trapezoidal rule that uses more than 32 threads per thread block and shared memory sums (Section 6.13.3), we suggested using the following code:

```
. . .
float result = Shared_mem_sum(shared_vals +
    warpSize*my_warp);
if (my_lane == 0) shared_vals[warpSize*my_warp] =
    result;
__syncthreads();
. . .
```

Explain why the assignment

```
shared_vals[warpSize*my_warp] = result;
```

doesn't cause a race condition.

6.9 There are several techniques we can use to try to improve the performance of the warp sum functions in our trapezoidal rule programs:

a. In the **for** statement

```
for (int diff = warpSize/2; diff > 0; diff = diff/2)
```

the identifier warpSize is a variable, and the compiler doesn't know its value. So at run-time, the system will probably execute the loop as written. If we replace warpSize/2 with a macro, e.g.,

#define WARPSZ_DIV_2 16

...

```
for (int diff = WARPSZ_DIV_2; diff > 0; diff = diff/2)
```

the compiler can determine the values of diff, and it may be able to optimize the execution of the loop. Try making this change to the warp sum and shared memory versions of the program. How much of an improvement is there (if any) in the program's performance?

b. Integer division (as in diff = diff/2) is an expensive operation compared to a right shift. If you make diff **unsigned** and replace diff = diff/2 with a right shift, how much of an improvement was there (if any) in the program's performance?

c. We, the programmers, know exactly how many times the body of the **for** statement will be executed. So we might try "unrolling" the **for** statement by replacing it with the body of the loop when diff = 16, diff = 8, diff = 4, diff = 2, diff = 1. If you do this, how much of an improvement is there in the program's performance?

6.10 This question refers to the implementation of the pseudocode in Program 6.17.

a. Implement the Which_bit function.

b. Use bitwise operators to implement the Insert_zero function in the serial bitonic sort.
Note that technically C right shift is undefined on signed ints, because the bits that are vacated by the shift can be filled with either a 0 or a sign-bit (0 or 1). There is no ambiguity with unsigned ints: since there's no sign-bit, the vacated bits are always filled with 0's. So it's sometimes a good idea to use **unsigned** ints when you're using bitwise operations.

c. Implement the Compare_swap function in the serial bitonic sort pseudocode in Program 6.17.

6.11 Write a serial bitonic sort that implements the following pseudocode:

```
for (bf_sz = 2; bf_sz <= n; bf_sz *= 2) {
    order = INC;   /* INC = 0, DEC = 1 */
    for (start = 0; start < n; start += bf_sz) {
```

```
            if (order = INC)
                Bitonic_incr(bf_sz, list+start);
            else
                Bitonic_decr(bf_sz, list+start);
            /* REVERSE "flips" the bit in order */
            order = REVERSE(order);
        }

        . . .

    void Bitonic_incr(int bf_sz, int sublist[]) {
        halfway = bf_sz/2;
        for (i = 0; i < half_way; i++)
            if (sublist[i] > sublist[halfway + i])
                Swap(sublist[i], sublist[halfway+1]);
        if (bf_sz > 2) {
            Bitonic_incr(bf_sz/2, sublist);
            Bitonic_incr(bf_sz/2, sublist+halfway);
        }
    }
```

6.12 Recall that in a **for** loop such as

```
        int incr = ...;
        for (int i = 0; i < n; i += incr)
            Do_stuff();
```

the value in `incr` is sometimes called the *stride* of the loop. We can modify our single-block implementation of bitonic sort (Program 6.18) so that the body of the inner loop (the **for** stage loop) is contained in a **grid-stride loop**. A grid-stride loop is a loop whose stride is equal to the number of threads in the grid. Grid-stride loops are very useful when we need more iterations than the number of available threads. (See [26].) In our first bitonic sort, the grid is just a single thread block. So the grid-stride is just the number of threads in the block.

Modify Program 6.18 so that it uses a grid-stride loop. If there is no limit to the number of elements in the list, what restrictions must be placed on n, the number of elements in the list? What restrictions must be placed on the number of threads relative to the number of elements in the list and the limit on the number of threads in a single block?

6.17 Programming assignments

6.1 In the trapezoidal rule programs, we always assigned at most one trapezoid to each thread. If there are a total of t threads, n trapezoids, and n is evenly divisible by t, an alternative is to assign n/t trapezoids to each thread. Modify the various trapezoid rule programs we wrote so that they will correctly implement the trapezoid rule when the number of trapezoids is a multiple of the number of threads. How does the performance of the modified programs compare to the performance of the original programs?

6.2 Computing a dot product is another type of global sum:

```
int n = ...;
float x[n], y[n];
/* Initialize x and y */

...
float dot = 0.0;
for (int i = 0; i < n; i++)
    dot += x[i]*y[i];
```

However, a possibly important difference between the dot product and the trapezoid rule is the fact that the dot product can require the use of a large amount of memory.

Read the material in the CUDA Programming Guide [11] on accessing global memory. If there are *t* threads (altogether) in a CUDA dot product program, which of the following implementations of the dot product should give the best performance?

```
int my_rank = blockIdx.x*blockDim.x + threadIdx.x;
int my_n = n/t;

/* "Block" Partition (nothing to do with "thread */
 *   block") */
int my_first = my_rank*my_n;
int my_last = my_first + my_n;
float my_dot = 0;
for (int i = my_first; i < my_last; i++)
    my_dot += x[i]*y[i];

/* "Cyclic" Partition */
float my_dot = 0;
for (int i = my_rank; i < n; i += t)
    my_dot + = x[i]*y[i];

/* "Block Cyclic" Partition (nothing to do with
 *   "thread block") */
/*   my_n = b*k for some unsigned ints b and k */
float my_dot = 0;
for (int i = my_rank*b; i < n; i += t*b)
    for (j = i; j < i+b; j++)
        my_dot += x[j]*y[j];

/* Finish all implementations */
atomicAdd(dot_p, my_dot);
```

Write a CUDA program with three kernels: one implementing a block partition of x and y, one implementing a cyclic partition, and one implementing a block cyclic partition. Does the performance of the actual program confirm your predictions?

Can you improve the performance of any of the partitions by using warp sums or shared memory sums?

6.3 Modify the second warp sum implementation of the trapezoidal rule program and the second bitonic sort program so that they run with different size inputs and different configurations of grids and thread blocks. Can you make any generalizations about how to predict input values that will optimize their performance?

6.4 Implement a bitonic sort in which each thread is responsible for two "chunks" of elements instead of two elements. If the array has n elements and there are blk_ct thread blocks and th_per_blk threads per block, assume that the total number of threads is a power of two, and n is evenly divisible by the number of threads. So

$$\text{chunk\_sz} = \frac{n}{\text{blk\_ct} \times \text{th\_per\_blk}}$$

is an integer. Now the basic program uses the same structure as the bitonic sort we developed in this chapter, except that the basic operation is a "merge-split" instead of a compare-swap.

So each thread is responsible for a contiguous sublist of chunk_sz elements, and each thread will initially sort its sublist in increasing order. Then if threads t and u are paired for a merge-split, $t < u$, and t and u are working on an increasing sequence, they will merge their sublists into an increasing sequence with t keeping the lower half and u keeping the upper half. If they're working on a decreasing sequence, t will keep the upper half and u will keep the lower half. So after each merge-split, each thread will always have an increasing sublist.

First implement the bitonic sort using a single thread block. Then modify the program so that it can handle an arbitrary number of thread blocks.

Parallel program development

In the last four chapters, we haven't just learned about parallel APIs; we've also developed a number of small parallel programs, and each of these programs has involved the implementation of a parallel algorithm. In this chapter, we'll look at a couple of larger examples: solving n-body problems and implementing a sorting algorithm called sample sort. For each problem, we'll start by looking at a serial solution and examining modifications to the serial solution. As we apply Foster's methodology, we'll see that there are some striking similarities between developing shared-memory, distributed-memory, and CUDA programs. We'll also see that in parallel programming there are problems that we need to solve for which there is no serial analog. We'll see that there are instances in which, as parallel programmers, we'll have to start "from scratch."

7.1 Two n-body solvers

In an n-body problem, we need to find the positions and velocities of a collection of interacting particles over a period of time. For example, an astrophysicist might want to know the positions and velocities of a collection of stars, while a chemist might want to know the positions and velocities of a collection of molecules or atoms. An n-body solver is a program that finds the solution to an n-body problem by simulating the behavior of the particles. The input to the problem is the mass, position, and velocity of each particle at the start of the simulation, and the output is typically the position and velocity of each particle at a sequence of user-specified times, or simply the position and velocity of each particle at the end of a user-specified time period.

Let's first develop a serial n-body solver. Then we'll try to parallelize it for both shared-memory systems, distributed-memory systems, and GPUs.

7.1.1 The problem

For the sake of explicitness, let's write an n-body solver that simulates the motions of planets or stars. We'll use Newton's second law of motion and his law of universal gravitation to determine the positions and velocities. Thus, if particle q has position $\mathbf{s}_q(t)$ at time t, and particle k has position $\mathbf{s}_k(t)$, then the force on particle q exerted

An Introduction to Parallel Programming. https://doi.org/10.1016/B978-0-12-804605-0.00014-2

by particle k is given by

$$\mathbf{f}_{qk}(t) = -\frac{Gm_q m_k}{\left|\mathbf{s}_q(t) - \mathbf{s}_k(t)\right|^3} \left[\mathbf{s}_q(t) - \mathbf{s}_k(t)\right]. \tag{7.1}$$

Here, G is the gravitational constant (6.673×10^{-11} m^3/(kg $\cdot$ s^2)), and m_q and m_k are the masses of particles q and k, respectively. Also, the notation $\left|\mathbf{s}_q(t) - \mathbf{s}_k(t)\right|$ represents the distance from particle k to particle q. Note that in general the positions, the velocities, the accelerations, and the forces are vectors, so we're using boldface to represent these variables. We'll use an italic font to represent the other, scalar variables, such as the time t and the gravitational constant G.

We can use Formula (7.1) to find the total force on any particle by adding the forces due to all the particles. If our n particles are numbered $0, 1, 2, \ldots, n-1$, then the total force on particle q is given by

$$\mathbf{F}_q(t) = \sum_{\substack{k=0 \\ k \neq q}}^{n-1} \mathbf{f}_{qk} = -Gm_q \sum_{\substack{k=0 \\ k \neq q}}^{n-1} \frac{m_k}{\left|\mathbf{s}_q(t) - \mathbf{s}_k(t)\right|^3} \left[\mathbf{s}_q(t) - \mathbf{s}_k(t)\right]. \tag{7.2}$$

Recall that the acceleration of an object is given by the second derivative of its position and that Newton's second law of motion states that the force on an object is given by its mass multiplied by its acceleration, so if the acceleration of particle q is $\mathbf{a}_q(t)$, then $\mathbf{F}_q(t) = m_q \mathbf{a}_q(t) = m_q \mathbf{s}_q''(t)$, where $\mathbf{s}_q''(t)$ is the second derivative of the position $\mathbf{s}_q(t)$. Thus, we can use Formula (7.2) to find the acceleration of particle q:

$$\mathbf{s}_q''(t) = -G \sum_{\substack{j=0 \\ j \neq q}}^{n-1} \frac{m_j}{\left|\mathbf{s}_q(t) - \mathbf{s}_j(t)\right|^3} \left[\mathbf{s}_q(t) - \mathbf{s}_j(t)\right]. \tag{7.3}$$

Thus Newton's laws give us a system of *differential* equations—equations involving derivatives—and our job is to find at each time t of interest the position $\mathbf{s}_q(t)$ and velocity $\mathbf{v}_q(t) = \mathbf{s}_q'(t)$.

We'll suppose that we either want to find the positions and velocities at the times

$$t = 0, \Delta t, 2\Delta t, \ldots, T\Delta t,$$

or, more often, simply the positions and velocities at the final time $T\Delta t$. Here, Δt and T are specified by the user, so the input to the program will be n, the number of particles, Δt, T, and, for each particle, its mass, its initial position, and its initial velocity. In a fully general solver, the positions and velocities would be three-dimensional vectors, but to keep things simple, we'll assume that the particles will move in a plane, and we'll use two-dimensional vectors instead.

The output of the program will be the positions and velocities of the n particles at the timesteps $0, \Delta t, 2\Delta t, \ldots$, or just the positions and velocities at $T\Delta t$. To get the output at only the final time, we can add an input option in which the user specifies whether she only wants the final positions and velocities.

7.1.2 **Two serial programs**

In outline, a serial *n*-body solver can be based on the following pseudocode:

```
1    Get input data;
2    for each timestep {
3        if (timestep output)
4            Print positions and velocities of particles;
5        for each particle q
6            Compute total force on q;
7        for each particle q
8            Compute position and velocity of q;
9    }
10   Print positions and velocities of particles;
```

We can use our formula for the total force on a particle (Formula (7.2)) to refine our pseudocode for the computation of the forces in Lines 5–6:

```
for each particle q {
    for each particle k != q {
        x_diff = pos[q][X] - pos[k][X];
        y_diff = pos[q][Y] - pos[k][Y];
        dist = sqrt(x_diff*x_diff + y_diff*y_diff);
        dist_cubed = dist*dist*dist;
        forces[q][X]
                -= G*masses[q]*masses[k]/dist_cubed * x_diff;
        forces[q][Y]
                -= G*masses[q]*masses[k]/dist_cubed * y_diff;
    }
}
```

Here, we're assuming that the forces and the positions of the particles are stored as two-dimensional arrays, forces and pos, respectively. We're also assuming we've defined constants X = 0 and Y = 1. So the *x*-component of the force on particle *q* is forces[q][X] and the *y*-component is forces[q][Y]. Similarly, the components of the position are pos[q][X] and pos[q][Y]. (We'll take a closer look at data structures shortly.)

We can use Newton's third law of motion, that is, for every action there is an equal and opposite reaction, to halve the total number of calculations required for the forces. If the force on particle *q* due to particle *k* is $\mathbf{f}_{qk}$, then the force on *k* due to *q* is $-\mathbf{f}_{qk}$. Using this simplification we can modify our code to compute forces, as shown in Program 7.1. To better understand this pseudocode, imagine the individual forces

```
for each particle q
    forces[q] = 0;
for each particle q {
    for each particle k > q {
        x_diff = pos[q][X] - pos[k][X];
        y_diff = pos[q][Y] - pos[k][Y];
        dist = sqrt(x_diff*x_diff + y_diff*y_diff);
        dist_cubed = dist*dist*dist;
        force_qk[X] = G*masses[q]*masses[k]/dist_cubed * x_diff;
        force_qk[Y] = G*masses[q]*masses[k]/dist_cubed * y_diff

        forces[q][X] += force_qk[X];
        forces[q][Y] += force_qk[Y];
        forces[k][X] -= force_qk[X];
        forces[k][Y] -= force_qk[Y];
    }
}
```

Program 7.1: A reduced algorithm for computing n-body forces.

as a two-dimensional array:

$$
\begin{bmatrix}
0 & \mathbf{f}_{01} & \mathbf{f}_{02} & \cdots & \mathbf{f}_{0,n-1} \\
-\mathbf{f}_{01} & 0 & \mathbf{f}_{12} & \cdots & \mathbf{f}_{1,n-1} \\
-\mathbf{f}_{02} & -\mathbf{f}_{12} & 0 & \cdots & \mathbf{f}_{2,n-1} \\
\vdots & \vdots & \vdots & \ddots & \vdots \\
-\mathbf{f}_{0,n-1} & -\mathbf{f}_{1,n-1} & -\mathbf{f}_{2,n-1} & \cdots & 0
\end{bmatrix}.
$$

(Why are the diagonal entries 0?) Our original solver simply adds all of the entries in row q to get forces[q]. In our modified solver, when $q = 0$, the body of the loop **for** each particle q will add the entries in row 0 into forces[0]. It will also add the kth entry in column 0 into forces[k], for $k = 1, 2, \ldots, n - 1$. In general, the qth iteration will add the entries to the right of the diagonal (that is, to the right of the 0) in row q into forces[q], and the entries below the diagonal in column q will be added into their respective forces, that is, the kth entry will be added into forces[k].

Note that in using this modified solver, it's necessary to initialize the forces array in a separate loop, since the qth iteration of the loop that calculates the forces will, in general, add the values it computes into forces[k] for $k = q + 1, q + 2, \ldots,$ $n - 1$, not just forces[q].

To distinguish between the two algorithms, we'll call the n-body solver with the original force calculation, the *basic* algorithm, and the solver with the number of calculations reduced, the *reduced* algorithm.

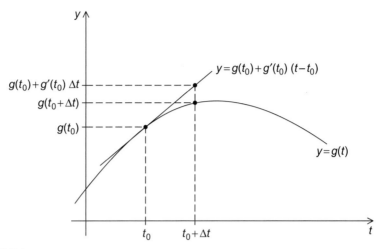

FIGURE 7.1

Using the tangent line to approximate a function.

The position and the velocity remain to be found. We know that the acceleration of particle q is given by

$$\mathbf{a}_q(t) = \mathbf{s}_q''(t) = \mathbf{F}_q(t)/m_q,$$

where $\mathbf{s}_q''(t)$ is the second derivative of the position $\mathbf{s}_q(t)$ and $\mathbf{F}_q(t)$ is the force on particle q. We also know that the velocity $\mathbf{v}_q(t)$ is the first derivative of the position $\mathbf{s}_q'(t)$, so we need to integrate the acceleration to get the velocity, and we need to integrate the velocity to get the position.

We might at first think that we can simply find an antiderivative of the function in Formula (7.3). However, a second look shows us that this approach has problems: the right-hand side contains unknown functions $\mathbf{s}_q$ and $\mathbf{s}_k$—not just the variable t—so we'll instead use a **numerical** method for *estimating* the position and the velocity. This means that rather than trying to find simple closed formulas, we'll approximate the values of the position and velocity at the times of interest. There are *many* possible choices for numerical methods, but we'll use the simplest one: Euler's method, which is named after the famous Swiss mathematician Leonhard Euler (1707–1783). In Euler's method, we use the tangent line to approximate a function. The basic idea is that if we know the value of a function $g(t_0)$ at time t_0 and we also know its derivative $g'(t_0)$ at time t_0, then we can approximate its value at time $t_0 + \Delta t$ by using the tangent line to the graph of $g(t_0)$. See Fig. 7.1 for an example. Now if we know a point $(t_0, g(t_0))$ on a line, and we know the slope of the line $g'(t_0)$, then an equation for the line is given by

$$y = g(t_0) + g'(t_0)(t - t_0).$$

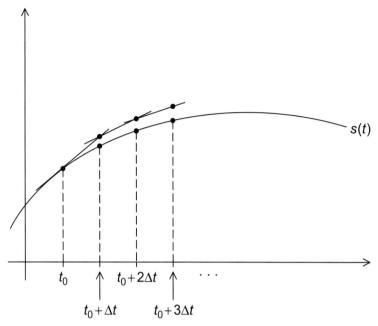

FIGURE 7.2

Euler's method.

Since we're interested in the time $t = t_0 + \Delta t$, we get

$$g(t + \Delta t) \approx g(t_0) + g'(t_0)(t + \Delta t - t) = g(t_0) + \Delta t g'(t_0).$$

Note that this formula will work even when $g(t)$ and y are vectors: when this is the case, $g'(t)$ is also a vector and the formula just adds a vector to a vector multiplied by a scalar, Δt.

Now we know the value of $\mathbf{s}_q(t)$ and $\mathbf{s}'_q(t)$ at time 0, so we can use the tangent line and our formula for the acceleration to compute $\mathbf{s}_q(\Delta t)$ and $\mathbf{v}_q(\Delta t)$:

$$\mathbf{s}_q(\Delta t) \approx \mathbf{s}_q(0) + \Delta t\, \mathbf{s}'_q(0) = \mathbf{s}_q(0) + \Delta t\, \mathbf{v}_q(0),$$

$$\mathbf{v}_q(\Delta t) \approx \mathbf{v}_q(0) + \Delta t\, \mathbf{v}'_q(0) = \mathbf{v}_q(0) + \Delta t\, \mathbf{a}_q(0) = \mathbf{v}_q(0) + \Delta t \frac{1}{m_q} \mathbf{F}_q(0).$$

When we try to extend this approach to the computation of $\mathbf{s}_q(2\Delta t)$ and $\mathbf{s}'_q(2\Delta t)$, we see that things are a little bit different, since we don't know the exact value of $\mathbf{s}_q(\Delta t)$ and $\mathbf{s}'_q(\Delta t)$. However, if our approximations to $\mathbf{s}_q(\Delta t)$ and $\mathbf{s}'_q(\Delta t)$ are good, then we should be able to get a reasonably good approximation to $\mathbf{s}_q(2\Delta t)$ and $\mathbf{s}'_q(2\Delta t)$ using the same idea. This is what Euler's method does (see Fig. 7.2).

Now we can complete our pseudocode for the two n-body solvers by adding in the code for computing position and velocity:

```
pos[q][X] += delta_t*vel[q][X];
pos[q][Y] += delta_t*vel[q][Y];
vel[q][X] += delta_t/masses[q]*forces[q][X];
vel[q][Y] += delta_t/masses[q]*forces[q][Y];
```

Here, we're using `pos[q]`, `vel[q]`, and `forces[q]` to store the position, the velocity, and the force, respectively, of particle q.

Before moving on to parallelizing our serial program, let's take a moment to look at data structures. We've been using an array type to store our vectors:

```
#define DIM 2

typedef double vect_t[DIM];
```

A struct is also an option. However, if we're using arrays and we decide to change our program so that it solves three-dimensional problems, in principle we only need to change the macro `DIM`. If we try to do this with structs, we'll need to rewrite the code that accesses individual components of the vector.

For each particle, we need to know the values of

- its mass,
- its position,
- its velocity,
- its acceleration, and
- the total force acting on it.

Since we're using Newtonian physics, the mass of each particle is constant, but the other values will, in general, change as the program proceeds. If we examine our code, we'll see that once we've computed a new value for one of these variables for a given timestep, we never need the old value again. For example, we don't need to do anything like this:

```
new_pos_q = f(old_pos_q);
new_vel_q = g(old_pos_q, new_pos_q);
```

Also, the acceleration is only used to compute the velocity, and its value can be computed in one arithmetic operation from the total force, so we only need to use a local, temporary variable for the acceleration.

For each particle it suffices to store its mass and the current value of its position, velocity, and force. We could store these four variables as a struct and use an array of structs to store the data for all the particles. Of course, there's no reason that all of the variables associated with a particle need to be grouped together in a struct. We can split the data into separate arrays in a variety of different ways. We've chosen to group the mass, position, and velocity into a single struct and store the forces in a separate array. With the forces stored in contiguous memory, we can use a fast function, such as `memset`, to quickly assign zeroes to all of the elements at the beginning of each iteration:

```
#include <string.h>    /* For memset */
.  .  .
vect_t* forces = malloc(n*sizeof(vect_t));
.  .  .
for (step = 1; step <= n_steps; step++) {
    .  .  .
    /* Assign 0 to each element of the forces array */
    forces = memset(forces, 0, n*sizeof(vect_t));
    for (part = 0; part < n-1; part++)
        Compute_force(part, forces, . . .)
    .  .  .
}
```

If the force on each particle were a member of a struct, the force members wouldn't occupy contiguous memory in an array of structs, and we'd have to use a relatively slow **for** loop to assign zero to each element.

7.1.3 Parallelizing the *n*-body solvers

Let's try to apply Foster's methodology to the *n*-body solver. Since we initially want *lots* of tasks, we can start by making our tasks the computations of the positions, the velocities, and the total forces at each timestep. In the basic algorithm, the algorithm in which the total force on each particle is calculated directly from Formula (7.2), the computation of $\mathbf{F}_q(t)$, the total force on particle q at time t, requires the positions of each of the particles $\mathbf{s}_r(t)$, for each r. The computation of $\mathbf{v}_q(t + \Delta t)$ requires the velocity at the previous timestep, $\mathbf{v}_q(t)$, and the force, $\mathbf{F}_q(t)$, at the previous timestep. Finally, the computation of $\mathbf{s}_q(t + \Delta t)$ requires $\mathbf{s}_q(t)$ and $\mathbf{v}_q(t)$. The communications among the tasks can be illustrated as shown in Fig. 7.3. The figure makes it clear that most of the communication among the tasks occurs among the tasks associated with an individual particle, so if we agglomerate the computations of $\mathbf{s}_q(t)$, $\mathbf{v}_q(t)$,

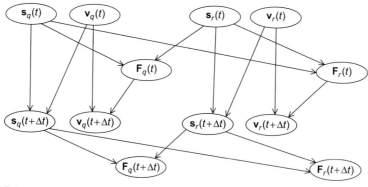

FIGURE 7.3

Communications among tasks in the basic *n*-body solver.

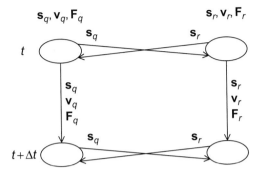

FIGURE 7.4

Communications among agglomerated tasks in the basic *n*-body solver.

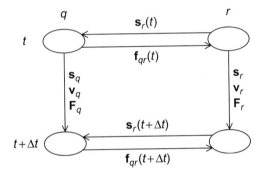

FIGURE 7.5

Communications among agglomerated tasks in the reduced *n*-body solver ($q < r$).

and $\mathbf{F}_q(t)$, our intertask communication is greatly simplified (see Fig. 7.4). Now the (agglomerated) tasks correspond to the particles and, in the figure, we've labeled the communications with the data that's being communicated. For example, the arrow from particle q at timestep t to particle r at timestep t is labeled with $\mathbf{s}_q$, the position of particle q.

For the reduced algorithm, the "intra-particle" communications are the same. That is, to compute $\mathbf{s}_q(t + \Delta t)$, we'll need $\mathbf{s}_q(t)$ and $\mathbf{v}_q(t)$, and to compute $\mathbf{v}_q(t + \Delta t)$, we'll need $\mathbf{v}_q(t)$ and $\mathbf{F}_q(t)$. Therefore, once again it makes sense to agglomerate the computations associated with a single particle into a composite task.

Recall that in the reduced algorithm, we make use of the fact that the force $\mathbf{f}_{rq} = -\mathbf{f}_{qr}$. So if $q < r$, then the communication *from* task r to task q is the same as in the basic algorithm—to compute $\mathbf{F}_q(t)$, task/particle q will need $\mathbf{s}_r(t)$ from task/particle r. However, the communication from task q to task r is no longer $\mathbf{s}_q(t)$, it's the force on particle q due to particle r, that is, $\mathbf{f}_{qr}(t)$. (See Fig. 7.5.)

The final stage in Foster's methodology is mapping. If we have n particles and T timesteps, then there will be nT tasks in both the basic and the reduced algorithm.

Astrophysical n-body problems typically involve thousands or even millions of particles, so n may be several orders of magnitude greater than the number of available cores. However, T may also be much larger than the number of available cores. In principle, then, we have two "dimensions" to work with when we map tasks to cores. However, if we consider the nature of Euler's method, we'll see that attempting to assign tasks associated with a single particle at different timesteps to different cores won't work very well. Before estimating $s_q(t + \Delta t)$ and $v_q(t + \Delta t)$, Euler's method must "know" $s_q(t)$, $v_q(t)$, and $a_q(t)$. Thus, if we assign particle q at time t to core c_0, and we assign particle q at time $t + \Delta t$ to core $c_1 \neq c_0$, then we'll have to communicate $s_q(t)$, $v_q(t)$, and $F_q(t)$ from c_0 to c_1. Of course, if particle q at time t and particle q at time $t + \Delta t$ are mapped to the same core, this communication won't be necessary, so once we've mapped the task consisting of the calculations for particle q at the first timestep to core c_0, we may as well map the subsequent computations for particle q to the same cores, since we can't simultaneously execute the computations for particle q at two different timesteps. Thus, mapping tasks to cores will, in effect, be an assignment of particles to cores.

At first glance, it might seem that any assignment of particles to cores that assigns roughly n/thread_count particles to each core will do a good job of balancing the workload among the cores, and for the basic algorithm this is the case. In the basic algorithm, the work required to compute the position, velocity, and force is the same for every particle. However, in the reduced algorithm, the work required in the forces computation loop is much greater for lower-numbered iterations than the work required for higher-numbered iterations. To see this, recall the pseudocode that computes the total force on particle q in the reduced algorithm:

```
for each particle k > q {
    x_diff = pos[q][X] - pos[k][X];
    y_diff = pos[q][Y] - pos[k][Y];
    dist = sqrt(x_diff*x_diff + y_diff*y_diff);
    dist_cubed = dist*dist*dist;
    force_qk[X] = G*masses[q]*masses[k]/dist_cubed * x_diff;
    force_qk[Y] = G*masses[q]*masses[k]/dist_cubed * y_diff;

    forces[q][X] += force_qk[X];
    forces[q][Y] += force_qk[Y];
    forces[k][X] -= force_qk[X];
    forces[k][Y] -= force_qk[Y];
}
```

Then, for example, when $q = 0$, we'll make $n - 1$ passes through the **for** each particle k > q loop, while when $q = n - 1$, we won't make any passes through the loop. Thus, for the reduced algorithm we would expect that a cyclic partition of the particles would do a better job than a block partition of evenly distributing the *computation*.

However, in a shared-memory setting, a cyclic partition of the particles among the cores is almost certain to result in a much higher number of cache misses than a block partition, and in a distributed-memory setting, the overhead involved in communicating data that has a cyclic distribution will probably be greater than the overhead involved in communicating data that has a block distribution (see Exercises 7.8 and 7.10).

Therefore, with a composite task consisting of all of the computations associated with a single particle throughout the simulation, we conclude the following:

1. A block distribution will give the best performance for the basic *n*-body solver.
2. For the reduced *n*-body solver, a cyclic distribution will best distribute the workload in the computation of the forces. However, this improved performance *may* be offset by the cost of reduced cache performance in a shared-memory setting and additional communication overhead in a distributed-memory setting.

To make a final determination of the optimal mapping of tasks to cores, we'll need to do some experimentation.

7.1.4 A word about I/O

You may have noticed that our discussion of parallelizing the *n*-body solver hasn't touched on the issue of I/O, even though I/O can figure prominently in both of our serial algorithms. We've discussed the problem of I/O several times in earlier chapters. Recall that different parallel systems vary widely in their I/O capabilities, and with the very basic I/O that is commonly available it is very difficult to obtain high performance. This basic I/O was designed for use by single-process, single-threaded programs, and when multiple processes or multiple threads attempt to access the I/O buffers, the system makes no attempt to schedule their access. For example, if multiple threads attempt to execute

```
printf("Hello from thread %d of %d\n", my_rank, thread_count);
```

more or less simultaneously, the order in which the output appears will be unpredictable. Even worse, one thread's output may not even appear as a single line. It can happen that the output from one thread appears as multiple segments, and the individual segments are separated by output from other threads.

Thus, as we've noted earlier, except for debug output, we generally assume that one process/thread does all the I/O, and when we're timing program execution, we'll use the option to only print output for the final timestep. Furthermore, we won't include this output in the reported run-times.

Of course, even if we're ignoring the cost of I/O, we can't ignore its existence. We'll briefly discuss its implementation when we discuss the details of our parallel implementations.

7.1.5 Parallelizing the basic solver using OpenMP

How can we use OpenMP to map tasks/particles to cores in the basic version of our *n*-body solver? Let's take a look at the pseudocode for the serial program:

```
for each timestep {
   if (timestep output)
      Print positions and velocities of particles;
   for each particle q
      Compute total force on q;
   for each particle q
      Compute position and velocity of q;
}
```

The two inner loops are both iterating over particles. So, in principle, parallelizing the two inner **for** loops will map tasks/particles to cores, and we might try something like this:

```
          for each timestep {
             if (timestep output)
                Print positions and velocities of particles;
#           pragma omp parallel for
             for each particle q
                Compute total force on q;
#           pragma omp parallel for
             for each particle q
                Compute position and velocity of q;
          }
```

We may not like the fact that this code could do a lot of forking and joining of threads, but before dealing with that, let's take a look at the loops themselves: we need to see if there are any race conditions caused by loop-carried dependences.

In the basic version, the first loop has the following form:

```
#   pragma omp parallel for
    for each particle q {
        forces[q][X] = forces[q][Y] = 0;
        for each particle k != q {
            x_diff = pos[q][X] − pos[k][X];
            y_diff = pos[q][Y] − pos[k][Y];
            dist = sqrt(x_diff*x_diff + y_diff*y_diff);
            dist_cubed = dist*dist*dist;
            forces[q][X]
                 −= G*masses[q]*masses[k]/dist_cubed*x_diff;
            forces[q][Y]
                 −= G*masses[q]*masses[k]/dist_cubed*y_diff;
        }
    }
```

Since the iterations of the **for** each particle q loop are partitioned among the threads, only one thread will access forces[q] for any q. Different threads do access the same elements of the pos array and the masses array. However, these arrays are only *read* in the loop. The remaining variables are used for temporary storage in a single iteration of the inner loop, and they can be private. Thus, the parallelization of the first loop in the basic algorithm won't introduce any race conditions.

The second loop has the form:

```
#   pragma omp parallel for
    for each particle q {
        pos[q][X] += delta_t*vel[q][X];
        pos[q][Y] += delta_t*vel[q][Y];
        vel[q][X] += delta_t/masses[q]*forces[q][X];
        vel[q][Y] += delta_t/masses[q]*forces[q][Y];
    }
```

Here, a single thread accesses pos[q], vel[q], masses[q], and forces[q], for any particle *q*, and the scalar variables are only read, so parallelizing this loop also won't introduce any race conditions.

Let's return to the issue of repeated forking and joining of threads. In our pseudocode, we have

```
    for each timestep {
        if (timestep output)
            Print positions and velocities of particles;
#       pragma omp parallel for
        for each particle q
            Compute total force on q;
#       pragma omp parallel for
        for each particle q
            Compute position and velocity of q;
    }
```

We encountered a similar issue when we parallelized odd-even transposition sort (see Section 5.6.2). In that case, we put a parallel directive before the outermost loop and used OpenMP **for** directives for the inner loops. Will a similar strategy work here? That is, can we do something like this?

```
#   pragma omp parallel
    for each timestep {
        if (timestep output)
            Print positions and velocities of particles;
#       pragma omp for
        for each particle q
            Compute total force on q;
#       pragma omp for
        for each particle q
            Compute position and velocity of q;
    }
```

This will have the desired effect on the two **for** each particle loops: the same team of threads will be used in both loops and for every iteration of the outer loop. However, we have a clear problem with the output statement. As it stands now, every thread will print all the positions and velocities, and we only want one thread to do the I/O. However, OpenMP provides the single directive for exactly this situation: we have a team of threads executing a block of code, but a part of the code should only be executed by one of the threads. Adding the `single` directive gives us the following pseudocode:

```
#   pragma omp parallel
    for each timestep {
        if (timestep output) {
#           pragma omp single
            Print positions and velocities of particles;
        }
#       pragma omp for
        for each particle q
            Compute total force on q;
#       pragma omp for
        for each particle q
            Compute position and velocity of q;
    }
```

There are still a few issues that we need to address. The most important has to do with possible race conditions introduced in the transition from one statement to another. For example, suppose thread 0 completes the first **for** each particle loop before thread 1, and it then starts updating the positions and velocities of its assigned particles in the second **for** each particle loop. Clearly, this could cause thread 1 to use an updated position in the first **for** each particle loop. However, recall that there is an implicit barrier at the end of each structured block that has been parallelized with a **for** directive. So, if thread 0 finishes the first inner loop before thread 1, it will block until thread 1 (and any other threads) finish the first inner loop, and it won't start the second inner loop until all the threads have finished the first. This will also prevent the possibility that a thread might rush ahead and print positions and velocities before they've all been updated by the second loop.

There's also an implicit barrier after the `single` directive, although in this program the barrier isn't necessary. Since the output statement won't update any memory locations, it's OK for some threads to go ahead and start executing the next iteration before output has been completed. Furthermore, the first inner **for** loop in the next iteration only updates the `forces` array, so it can't cause a thread executing the output statement to print incorrect values, and because of the barrier at the end of the first inner loop, no thread can race ahead and start updating positions and velocities in the second inner loop before the output has been completed. Thus, we could modify the `single` directive with a `nowait` clause. If the OpenMP implementation supports it, this simply eliminates the implied barrier associated with the `single` directive. It can also be used with **for**, `parallel` **for**, and `parallel` directives. Note that in this case, addition of the `nowait` clause is unlikely to have much effect on performance, since the two **for** each particle loops have implied barriers that will prevent any one thread from getting more than a few statements ahead of any other.

Finally, we may want to add a schedule clause to each of the **for** directives to ensure that the iterations have a block partition:

```
#       pragma omp for schedule(static, n/thread_count)
```

7.1.6 Parallelizing the reduced solver using OpenMP

The reduced solver has an additional inner loop: the initialization of the forces array to 0. If we try to use the same parallelization for the reduced solver, we should also parallelize this loop with a **for** directive. What happens if we try this? That is, what happens if we try to parallelize the reduced solver with the following pseudocode?

```
#   pragma omp parallel
    for each timestep {
        if (timestep output) {
#           pragma omp single
            Print positions and velocities of particles;
        }
#       pragma omp for
        for each particle q
            forces[q] = 0.0;
#       pragma omp for
        for each particle q
            Compute total force on q;
#       pragma omp for
        for each particle q
            Compute position and velocity of q;
    }
```

Parallelization of the initialization of the forces should be fine, since there's no dependence among the iterations. The updating of the positions and velocities is the same in both the basic and reduced solvers, so if the computation of the forces is OK, then this should also be OK.

How does parallelization affect the correctness of the loop for computing the forces? Recall that in the reduced version, this loop has the following form:

```
#   pragma omp for /* Can be faster than memset */
    for each particle q {
        force_qk[X] = force_qk[Y] = 0;
        for each particle k > q {
            x_diff = pos[q][X] - pos[k][X];
            y_diff = pos[q][Y] - pos[k][Y];
            dist = sqrt(x_diff*x_diff + y_diff*y_diff);
            dist_cubed = dist*dist*dist;
            force_qk[X]
                = G*masses[q]*masses[k]/dist_cubed * x_diff;
            force_qk[Y]
                = G*masses[q]*masses[k]/dist_cubed * y_diff;

            forces[q][X] += force_qk[X];
            forces[q][Y] += force_qk[Y];
            forces[k][X] -= force_qk[X];
            forces[k][Y] -= force_qk[Y];
        }
    }
```

As before, the variables of interest are pos, masses, and forces, since the values in the remaining variables are only used in a single iteration, and hence can be private. Also, as before, elements of the pos and masses arrays are only read, not updated. We therefore need to look at the elements of the forces array. In this version, unlike the basic version, a thread *may* update elements of the forces array other than those corresponding to its assigned particles. For example, suppose we have two threads and four particles and we're using a block partition of the particles. Then the total force on particle 3 is given by

$$\mathbf{F}_3 = -\mathbf{f}_{03} - \mathbf{f}_{13} - \mathbf{f}_{23}.$$

Furthermore, thread 0 will compute $\mathbf{f}_{03}$ and $\mathbf{f}_{13}$, while thread 1 will compute $\mathbf{f}_{23}$. Thus, the updates to forces[3] *do* create a race condition. In general, then, the updates to the elements of the forces array introduce race conditions into the code.

A seemingly obvious solution to this problem is to use a critical directive to limit access to the elements of the forces array. There are at least a couple of ways to do this. The simplest is to put a critical directive before all the updates to forces:

```
#   pragma omp critical
    {
        forces[q][X] += force_qk[X];
        forces[q][Y] += force_qk[Y];
        forces[k][X] -= force_qk[X];
        forces[k][Y] -= force_qk[Y];
    }
```

However, with this approach access to the elements of the forces array will be effectively serialized. Only one element of forces can be updated at a time, and contention for access to the critical section is actually likely to seriously degrade the performance of the program. (See Exercise 7.3.)

An alternative would be to have one critical section for each particle. However, as we've seen, OpenMP doesn't readily support varying numbers of critical sections, so we would need to use one lock for each particle instead and our updates would look something like this:

```
omp_set_lock(&locks[q]);
forces[q][X] += force_qk[X];
forces[q][Y] += force_qk[Y];
omp_unset_lock(&locks[q]);

omp_set_lock(&locks[k]);
forces[k][X] -= force_qk[X];
forces[k][Y] -= force_qk[Y];
omp_unset_lock(&locks[k]);
```

This assumes that the master thread will create a shared array of locks, one for each particle, and when we update an element of the forces array, we first set the lock

Table 7.1 First-phase computations for reduced algorithm with block partition.

Thread	Particle	Thread		
		0	**1**	**2**
0	0	$f_{01} + f_{02} + f_{03} + f_{04} + f_{05}$	0	0
	1	$-f_{01} + f_{12} + f_{13} + f_{14} + f_{15}$	0	0
1	2	$-f_{02} - f_{12}$	$f_{23} + f_{24} + f_{25}$	0
	3	$-f_{03} - f_{13}$	$-f_{23} + f_{34} + f_{35}$	0
2	4	$-f_{04} - f_{14}$	$-f_{24} - f_{34}$	f_{45}
	5	$-f_{05} - f_{15}$	$-f_{25} - f_{35}$	$-f_{45}$

Table 7.2 First-phase computations for reduced algorithm with cyclic partition.

Thread	Particle	Thread		
		0	**1**	**2**
0	0	$f_{01} + f_{02} + f_{03} + f_{04} + f_{05}$	0	0
1	1	$-f_{01}$	$f_{12} + f_{13} + f_{14} + f_{15}$	0
2	2	$-f_{02}$	$-f_{12}$	$f_{23} + f_{24} + f_{25}$
0	3	$-f_{03} + f_{34} + f_{35}$	$-f_{13}$	$-f_{23}$
1	4	$-f_{04} - f_{34}$	$-f_{14} + f_{45}$	$-f_{24}$
2	5	$-f_{05} - f_{35}$	$-f_{15} - f_{45}$	$-f_{25}$

corresponding to that particle. Although this approach performs much better than the single critical section, it still isn't competitive with the serial code. (See Exercise 7.4.)

Another possible solution is to carry out the computation of the forces in two phases. In the first phase, each thread carries out exactly the same calculations it carried out in the erroneous parallelization. However, now the calculations are stored in its *own* array of forces. Then, in the second phase, the thread that has been assigned particle q will add the contributions that have been computed by the different threads. In our example above, thread 0 would compute $-f_{03} - f_{13}$, while thread 1 would compute $-f_{23}$. After each thread was done computing its contributions to the forces, thread 1, which has been assigned particle 3, would find the total force on particle 3 by adding these two values.

Let's look at a slightly larger example. Suppose we have three threads and six particles. If we're using a block partition of the particles, then the computations in the first phase are shown in Table 7.1. The last three columns of the table show each thread's contribution to the computation of the total forces. In phase 2 of the computation, the thread specified in the first column of the table will add the contents of each of its assigned rows—that is, each of its assigned particles.

Note that there's nothing special about using a block partition of the particles. Table 7.2 shows the same computations if we use a cyclic partition of the particles. Note that if we compare this table with the table that shows the block partition, it's clear that the cyclic partition does a better job of balancing the load.

To implement this, during the first phase our revised algorithm proceeds as before, except that each thread adds the forces it computes into its own subarray of `loc_forces`:

```
#   pragma omp for
    for each particle q {
        force_qk[X] = force_qk[Y] = 0;
        for each particle k > q {
            x_diff = pos[q][X] - pos[k][X];
            y_diff = pos[q][Y] - pos[k][Y];
            dist = sqrt(x_diff*x_diff + y_diff*y_diff);
            dist_cubed = dist*dist*dist;
            force_qk[X]
                  = G*masses[q]*masses[k]/dist_cubed * x_diff;
            force_qk[Y]
                  = G*masses[q]*masses[k]/dist_cubed * y_diff;

            loc_forces[my_rank][q][X] += force_qk[X];
            loc_forces[my_rank][q][Y] += force_qk[Y];
            loc_forces[my_rank][k][X] -= force_qk[X];
            loc_forces[my_rank][k][Y] -= force_qk[Y];
        }
    }
```

During the second phase, each thread adds the forces computed by all the threads for its assigned particles:

```
#   pragma omp for
    for (q = 0; q < n; q++) {
        forces[q][X] = forces[q][Y] = 0;
        for (thread = 0; thread < thread_count; thread++) {
            forces[q][X] += loc_forces[thread][q][X];
            forces[q][Y] += loc_forces[thread][q][Y];
        }
    }
```

Before moving on, we should make sure that we haven't inadvertently introduced any new race conditions. During the first phase, since each thread writes to its own subarray, there isn't a race condition in the updates to `loc_forces`. Also, during the second phase, only the "owner" of thread q writes to forces[q], so there are no race conditions in the second phase. Finally, since there is an implied barrier after each of the parallelized **for** loops, we don't need to worry that some thread is going to race ahead and make use of a variable that hasn't been properly initialized, or that some slow thread is going to make use of a variable that has had its value changed by another thread.

Table 7.3 Run-times of the *n*-body solvers parallelized with OpenMP (times in seconds).

Threads	Basic	Reduced Default Sched	Reduced Forces Cyclic	Reduced All Cyclic
1	7.71	3.90	3.90	3.90
2	3.87	2.94	1.98	2.01
4	1.95	1.73	1.01	1.08
8	0.99	0.95	0.54	0.61

7.1.7 Evaluating the OpenMP codes

Before we can compare the basic and the reduced codes, we need to decide how to schedule the parallelized **for** loops. For the basic code, we've seen that any schedule that divides the iterations equally among the threads should do a good job of balancing the computational load. (As usual, we're assuming no more than one thread/core.) We also observed that a block partitioning of the iterations would result in fewer cache misses than a cyclic partition. Thus, we would expect that a block schedule would be the best option for the basic version.

In the reduced code, the amount of work done in the first phase of the computation of the forces decreases as the **for** loop proceeds. We've seen that a cyclic schedule should do a better job of assigning more or less equal amounts of work to each thread. In the remaining parallel **for** loops—the initialization of the loc_forces array, the second phase of the computation of the forces, and the updating of the positions and velocities—the work required is roughly the same for all the iterations. Therefore, *taken out of context*, each of these loops will probably perform best with a block schedule. However, the schedule of one loop can affect the performance of another (see Exercise 7.11), so it may be that choosing a cyclic schedule for one loop and block schedules for the others will degrade performance.

With these choices, Table 7.3 shows the performance of the *n*-body solvers when they're run on one of our systems with no I/O. The solver used 400 particles for 1000 timesteps. The column labeled "Default Sched" gives times for the OpenMP reduced solver when all of the inner loops use the default schedule, which, on our system, is a block schedule. The column labeled "Forces Cyclic" gives times when the first phase of the forces computation uses a cyclic schedule and the other inner loops use the default schedule. The last column, labeled "All Cyclic," gives times when all of the inner loops use a cyclic schedule. The run-times of the serial solvers differ from those of the single-threaded solvers by less than 1%, so we've omitted them from the table.

Notice that with more than one thread, the reduced solver, using all default schedules, takes anywhere from 50 to 75% longer than the reduced solver with the cyclic forces computation. Using the cyclic schedule is clearly superior to the default schedule in this case, and any loss in time resulting from cache issues is more than made up for by the improved load balance for the computations.

For only two threads, there is very little difference between the performance of the reduced solver with only the first forces loop cyclic and the reduced solver with all loops cyclic. However, as we increase the number of threads, the performance of the reduced solver that uses a cyclic schedule for all of the loops does start to degrade. In this particular case, when there are more threads, it appears that the overhead involved in changing distributions is less than the overhead incurred from false sharing.

Finally, notice that the basic solver takes about twice as long as the reduced solver with the cyclic scheduling of the forces computation. So if the extra memory is available, the reduced solver is clearly superior. However, the reduced solver increases the memory requirement for the storage of the forces by a factor of `thread_count`, so for very large numbers of particles, it may be impossible to use the reduced solver.

7.1.8 Parallelizing the solvers using Pthreads

Parallelizing the two *n*-body solvers using Pthreads is very similar to parallelizing them using OpenMP. The differences are only in implementation details, so rather than repeating the discussion, we will point out some of the principal differences between the Pthreads and the OpenMP implementations. We will also note some of the more important similarities.

- By default, local variables in Pthreads are private, so all shared variables are global in the Pthreads version.
- The principal data structures in the Pthreads version are identical to those in the OpenMP version: vectors are two-dimensional arrays of **double**s, and the mass, position, and velocity of a single particle are stored in a struct. The forces are stored in an array of vectors.
- Startup for Pthreads is basically the same as the startup for OpenMP: the main thread gets the command-line arguments, and allocates and initializes the principal data structures.
- The main difference between the Pthreads and the OpenMP implementations is in the details of parallelizing the inner loops. Since Pthreads has nothing analogous to a `parallel` **for** directive, we must explicitly determine which values of the loop variables correspond to each thread's calculations. To facilitate this, we've written a function `Loop_schedule`, which determines
  - the initial value of the loop variable,
  - the final value of the loop variable, and
  - the increment for the loop variable.
  The input to the function is
  - the calling thread's rank,
  - the number of threads,
  - the total number of iterations, and
  - an argument indicating whether the partitioning should be block or cyclic.
- Another difference between the Pthreads and the OpenMP versions has to do with barriers. Recall that the end of a `parallel` **for** directive OpenMP has an implied barrier. As we've seen, this is important. For example, we don't want a thread

to start updating its positions until all the forces have been calculated, because it could use an out-of-date force and another thread could use an out-of-date position. If we simply partition the loop iterations among the threads in the Pthreads versions, there won't be a barrier at the end of an inner **for** loop and we'll have a race condition. Thus, we need to add explicit barriers after the inner loops when a race condition can arise. The Pthreads standard includes a barrier. However, some systems don't implement it, so we've defined a function that uses a Pthreads condition variable to implement a barrier. See Subsection 4.8.3 for details.

7.1.9 Parallelizing the basic solver using MPI

With our composite tasks corresponding to the individual particles, it's fairly straightforward to parallelize the basic algorithm using MPI. The only communication among the tasks occurs when we're computing the forces, and, to compute the forces, each task/particle needs the position and mass of every other particle. MPI_Allgather is expressly designed for this situation, since it collects on each process the same information from every other process. We've already noted that a block distribution will probably have the best performance, so we should use a block mapping of the particles to the processes.

In the shared-memory MIMD implementations, we collected most of the data associated with a single particle (mass, position, and velocity) into a single struct. However, if we use this data structure in the MPI implementation, we'll need to use a derived datatype in the call to MPI_Allgather, and communications with derived datatypes tend to be slower than communications with basic MPI types. Thus, it will make more sense to use individual arrays for the masses, positions, and velocities. We'll also need an array for storing the positions of all the particles. If each process has sufficient memory, then each of these can be a separate array. In fact, if memory isn't a problem, each process can store the entire array of masses, since these will never be updated and their values only need to be communicated during the initial setup.

On the other hand, if memory is short, there is an "in-place" option that can be used with some MPI collective communications. For our situation, suppose that the array pos can store the positions of all *n* particles. Further suppose that vect_mpi_t is an MPI datatype that stores two contiguous **double**s. Also suppose that *n* is evenly divisible by comm_sz and loc_n = n/comm_sz. Then, if we store the local positions in a separate array, loc_pos, we can use the following call to collect all of the positions on each process:

```
MPI_Allgather(loc_pos, loc_n, vect_mpi_t,
        pos, loc_n, vect_mpi_t, comm);
```

If we can't afford the extra storage for loc_pos, then we can have each process *q* store its local positions in the *q*th block of pos. That is, the local positions of each process *P* should be stored in the appropriate block of each process' pos array:

```
P0: pos[0], pos[1], ... , pos[loc_n-1]
P1: pos[loc_n], pos[loc_n+1], ... , pos[loc_n + loc_n-1]
    ...
Pq: pos[q*loc_n], pos[q*loc_n+1], ... , pos[q*loc_n + loc_n-1]
    ...
```

With the pos array initialized this way on each process, we can use the following call to MPI_Allgather:

```
MPI_Allgather(MPI_IN_PLACE, loc_n, vect_mpi_t,
    pos, loc_n, vect_mpi_t, comm);
```

In this call, the first loc_n and vect_mpi_t arguments are ignored. However, it's not a bad idea to use arguments whose values correspond to the values that will be used, just to increase the readability of the program.

In the program we've written, we made the following choices with respect to the data structures:

- Each process stores the entire global array of particle masses.
- Each process only uses a single *n*-element array for the positions.
- Each process uses a pointer loc_pos that refers to the start of its block of pos. Thus, on process 0 local_pos = pos, on process 1 local_pos = pos + loc_n, and, so on.

With these choices, we can implement the basic algorithm with the pseudocode shown in Program 7.2. As usual, process 0 will read and broadcast the command

```
1   Get input data;
2   for each timestep {
3       if (timestep output)
4           Print positions and velocities of particles;
5       for each local particle loc_q
6           Compute total force on loc_q;
7       for each local particle loc_q
8           Compute position and velocity of loc_q;
9       Allgather local positions into global pos array;
10  }
11  Print positions and velocities of particles;
```

Program 7.2: Pseudocode for the MPI version of the basic *n*-body solver.

line arguments. It will also read the input and print the results. In Line 1, it will need to distribute the input data. Therefore, Get input data might be implemented as follows:

```
if (my_rank == 0) {
    for each particle
        Read masses[particle], pos[particle], vel[particle];
}
```

```
MPI_Bcast(masses, n, MPI_DOUBLE, 0, comm);
MPI_Bcast(pos, n, vect_mpi_t, 0, comm);
MPI_Scatter(vel, loc_n, vect_mpi_t,
        loc_vel, loc_n, vect_mpi_t,
        0, comm);
```

So process 0 reads all the initial conditions into three *n*-element arrays. Since we're storing all the masses on each process, we broadcast masses. Also, since each process will need the global array of positions for the first computation of forces in the main **for** loop, we just broadcast pos. However, velocities are only used locally for the updates to positions and velocities, so we scatter vel.

Notice that we gather the updated positions in Line 9 at the end of the body of the outer **for** loop of Program 7.2. This ensures that the positions will be available for output in both Line 4 and Line 11. If we're printing the results for each timestep, this placement allows us to eliminate an expensive collective communication call: if we simply gathered the positions onto process 0 before output, we'd have to call MPI_Allgather before the computation of the forces. With this organization of the body of the outer **for** loop, we can implement the output with the following pseudocode:

```
Gather velocities onto process 0;
if (my_rank == 0) {
    Print timestep;
    for each particle
        Print pos[particle] and vel[particle]
}
```

7.1.10 Parallelizing the reduced solver using MPI

The "obvious" implementation of the reduced algorithm is likely to be extremely complicated. Before computing the forces, each process will need to gather a subset of the positions, and after the computation of the forces, each process will need to scatter some of the individual forces it has computed and add the forces it receives. Fig. 7.6 shows the communications that would take place if we had three processes, six particles, and used a block partitioning of the particles among the processes. Not surprisingly, the communications are even more complex when we use a cyclic distribution (see Exercise 7.14). Certainly it would be possible to implement these communications. However, unless the implementation were *very* carefully done, it would probably be *very* slow.

Fortunately, there's a much simpler alternative that uses a communication structure that is sometimes called a *ring pass*. In a ring pass, we imagine the processes as being interconnected in a ring (see Fig. 7.7). Process 0 communicates directly with processes 1 and comm_sz − 1, process 1 communicates with processes 0 and 2, and so on. The communication in a ring pass takes place in phases, and during each phase each process sends data to its "lower-ranked" neighbor, and receives data from its "higher-ranked" neighbor. Thus, 0 will send to comm_sz − 1 and receive from 1.

Process 0
Particles   0, 1

Process 1
2, 3

Process 2
4, 5

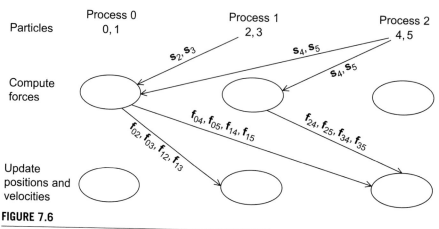

Compute
forces

s_2, s_3 s_4, s_5

s_4, s_5

$f_{04}, f_{05}, f_{14}, f_{15}$

$f_{24}, f_{25}, f_{34}, f_{35}$

$f_{02}, f_{03}, f_{12}, f_{13}$

Update
positions and
velocities

FIGURE 7.6

Communication in a possible MPI implementation of the reduced n-body solver.

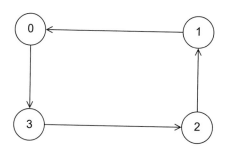

FIGURE 7.7

A ring of processes.

1 will send to 0 and receive from 2, and so on. In general, process q will send to process $(q - 1 + \text{comm\_sz}) \% \text{comm\_sz}$ and receive from process $(q + 1) \% \text{comm\_sz}$.

By repeatedly sending and receiving data using this ring structure, we can arrange that each process has access to the positions of all the particles. During the first phase, each process will send the positions of its assigned particles to its "lower-ranked" neighbor and receive the positions of the particles assigned to its higher-ranked neighbor. During the next phase, each process will forward the positions it received in the first phase. This process continues through $\text{comm\_sz} - 1$ phases until each process has received the positions of all of the particles. Fig. 7.8 shows the three phases if there are four processes and eight particles that have been cyclically distributed.

Of course, the virtue of the reduced algorithm is that we don't need to compute all of the inter-particle forces, since $\mathbf{f}_{kq} = -\mathbf{f}_{qk}$ for every pair of particles q and k. To see how to exploit this, first observe that using the reduced algorithm, the interparticle forces can be divided into those that are *added* into and those that are *subtracted* from the total forces on the particle. For example, if we have six particles, then the

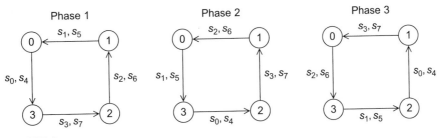

FIGURE 7.8

Ring pass of positions.

reduced algorithm will compute the force on particle 3 as

$$\mathbf{F}_3 = -\mathbf{f}_{03} - \mathbf{f}_{13} - \mathbf{f}_{23} + \mathbf{f}_{34} + \mathbf{f}_{35}.$$

The key to understanding the ring pass computation of the forces is to observe that the interparticle forces that are *subtracted* are computed by another task/particle, while the forces that are *added* are computed by the owning task/particle. Thus, the computations of the interparticle forces on particle 3 are assigned as follows:

Force	$\mathbf{f}_{03}$	$\mathbf{f}_{13}$	$\mathbf{f}_{23}$	$\mathbf{f}_{34}$	$\mathbf{f}_{35}$
Task/Particle	0	1	2	3	3

So, suppose that for our ring pass, instead of simply passing `loc_n = n/comm_sz` positions, we also pass `loc_n` forces. Then in each phase, a process can

1. compute interparticle forces resulting from interaction between its assigned particles and the particles whose positions it has received, and
2. once an interparticle force has been computed, the process can add the force into a local array of forces corresponding to its particles, *and* it can subtract the interparticle force from the received array of forces.

See, for example, [17,37] for further details and alternatives.

Let's take a look at how the computation would proceed when we have four particles, two processes, and we're using a cyclic distribution of the particles among the processes (see Table 7.4). We're calling the arrays that store the local positions and local forces `loc_pos` and `loc_forces`, respectively. These are not communicated among the processes. The arrays that are communicated among the processes are `tmp_pos` and `tmp_forces`.

Before the ring pass can begin, both arrays storing positions are initialized with the positions of the local particles, and the arrays storing the forces are set to 0. Before the ring pass begins, each process computes those forces that are due to interaction among its assigned particles. Process 0 computes $\mathbf{f}_{02}$ and process 1 computes $\mathbf{f}_{13}$. These values are added into the appropriate locations in `loc_forces` and subtracted from the appropriate locations in `tmp_forces`.

Table 7.4 Computation of forces in ring pass.

Time	Variable	Process 0	Process 1
Start	loc_pos	s_0, s_2	s_1, s_3
	loc_forces	$0, 0$	$0, 0$
	tmp_pos	s_0, s_2	s_1, s_3
	tmp_forces	$0, 0$	$0, 0$
After Comp of Forces	loc_pos	s_0, s_2	s_1, s_3
	loc_forces	$\mathbf{f}_{02}, 0$	$\mathbf{f}_{13}, 0$
	tmp_pos	s_0, s_2	s_1, s_3
	tmp_forces	$0, -\mathbf{f}_{02}$	$0, -\mathbf{f}_{13}$
After First Comm	loc_pos	s_0, s_2	s_1, s_3
	loc_forces	$\mathbf{f}_{02}, 0$	$\mathbf{f}_{13}, 0$
	tmp_pos	s_1, s_3	s_0, s_2
	tmp_forces	$0, -\mathbf{f}_{13}$	$0, -\mathbf{f}_{02}$
After Comp of Forces	loc_pos	s_0, s_2	s_1, s_3
	loc_forces	$\mathbf{f}_{01} + \mathbf{f}_{02} + \mathbf{f}_{03}, \mathbf{f}_{23}$	$\mathbf{f}_{12} + \mathbf{f}_{13}, 0$
	tmp_pos	s_1, s_3	s_0, s_2
	tmp_forces	$-\mathbf{f}_{01}, -\mathbf{f}_{03} - \mathbf{f}_{13} - \mathbf{f}_{23}$	$0, -\mathbf{f}_{02} - \mathbf{f}_{12}$
After Second Comm	loc_pos	s_0, s_2	s_1, s_3
	loc_forces	$\mathbf{f}_{01} + \mathbf{f}_{02} + \mathbf{f}_{03}, \mathbf{f}_{23}$	$\mathbf{f}_{12} + \mathbf{f}_{13}, 0$
	tmp_pos	s_0, s_2	s_1, s_3
	tmp_forces	$0, -\mathbf{f}_{02} - \mathbf{f}_{12}$	$-\mathbf{f}_{01}, -\mathbf{f}_{03} - \mathbf{f}_{13} - \mathbf{f}_{23}$
After Comp of Forces	loc_pos	s_0, s_2	s_1, s_3
	loc_forces	$\mathbf{f}_{01} + \mathbf{f}_{02} + \mathbf{f}_{03}, -\mathbf{f}_{02} - \mathbf{f}_{12} + \mathbf{f}_{23}$	$-\mathbf{f}_{01} + \mathbf{f}_{12} + \mathbf{f}_{13}, -\mathbf{f}_{03} - \mathbf{f}_{13} - \mathbf{f}_{23}$
	tmp_pos	s_0, s_2	s_1, s_3
	tmp_forces	$0, -\mathbf{f}_{02} - \mathbf{f}_{12}$	$-\mathbf{f}_{01}, -\mathbf{f}_{03} - \mathbf{f}13 - \mathbf{f}_{23}$

Now, the two processes exchange tmp_pos and tmp_forces and compute the forces due to interaction among their local particles and the received particles. In the reduced algorithm, the lower-ranked task/particle carries out the computation. Process 0 computes $\mathbf{f}_{01}, \mathbf{f}_{03}$, and $\mathbf{f}_{23}$, while process 1 computes $\mathbf{f}_{12}$. As before, the newly computed forces are added into the appropriate locations in loc_forces and subtracted from the appropriate locations in tmp_forces.

To complete the algorithm, we need to exchange the tmp arrays one final time.[1] Once each process has received the updated tmp_forces, it can carry out a simple vector sum

```
loc_forces += tmp_forces
```

to complete the algorithm.

---

[1] Actually, we only need to exchange tmp_forces for the final communication.

```
1   source = (my_rank + 1) % comm_sz;
2   dest = (my_rank - 1 + comm_sz) % comm_sz;
3   Copy loc_pos into tmp_pos;
4   loc_forces = tmp_forces = 0;
5
6   Compute forces due to interactions among local particles;
7   for (phase = 1; phase < comm_sz; phase++) {
8      Send current tmp_pos and tmp_forces to dest;
9      Receive new tmp_pos and tmp_forces from source;
10     /* Owner of the positions and forces we're receiving */
11     owner = (my_rank + phase) % comm_sz;
12     Compute forces due to interactions among my particles
13        and owner's particles;
14  }
15  Send current tmp_pos and tmp_forces to dest;
16  Receive new tmp_pos and tmp_forces from source;
```

Program 7.3: Pseudocode for the MPI implementation of the reduced *n*-body solver.

Thus, we can implement the computation of the forces in the reduced algorithm using a ring pass with the pseudocode shown in Program 7.3.

Recall that using `MPI_Send` and `MPI_Recv` for the send-receive pairs in Lines 8–9 and 15–16 is *unsafe* in MPI parlance, since they can hang if the system doesn't provide sufficient buffering. In this setting, recall that MPI provides `MPI_Sendrecv` and `MPI_Sendrecv_replace`. Since we're using the same memory for both the outgoing and the incoming data, we can use `MPI_Sendrecv_replace`.

Also recall that the time it takes to start up a message is substantial. We can probably reduce the cost of the communication by using a single array to store both `tmp_pos` and `tmp_forces`. For example, we could allocate storage for an array `tmp_data` that can store 2 × `loc_n` objects with type `vect_t` and use the first `loc_n` for `tmp_pos` and the last `loc_n` for `tmp_forces`. We can continue to use `tmp_pos` and `tmp_forces` by making these pointers to `tmp_data[0]` and `tmp_data[loc_n]`, respectively.

The principal difficulty in implementing the actual computation of the forces in Lines 12–13 lies in determining whether the current process should compute the force resulting from the interaction of a particle q assigned to it and a particle r, whose position it has received. If we recall the reduced algorithm (Program 7.1), we see that task/particle q is responsible for computing $\mathbf{f}_{qr}$ if and only if $q < r$. However, the arrays `loc_pos` and `tmp_pos` (or a larger array containing `tmp_pos` and `tmp_forces`) use *local* subscripts, not global subscripts. That is, when we access an element of (say) `loc_pos`, the subscript we use will lie in the range $0, 1, \ldots, \texttt{loc\_n} - 1$, not $0, 1, \ldots, n - 1$; so, if we try to implement the force interaction with the following pseudocode, we'll run into (at least) a couple of problems:

```
for (loc_part1 = 0; loc_part1 < loc_n-1; loc_part1++) {
    for (loc_part2 = loc_part1+1;
```

```
            loc_part2 < loc_n;
            loc_part2++) {
         Compute_force(loc_pos[loc_part1], masses[loc_part1],
               tmp_pos[loc_part2], masses[loc_part2],
               loc_forces[loc_part1], tmp_forces[loc_part2]);
      }
}
```

The first, and most obvious, is that `masses` is a global array and we're using local subscripts to access its elements. The second is that the relative sizes of `loc_part1` and `loc_part2` don't tell us whether we should compute the force due to their interaction. We need to use global subscripts to determine this. For example, if we have four particles and two processes, and the preceding code is being run by process 0, then when `loc_part1 = 0`, the inner loop will skip `loc_part2 = 0` and start with `loc_part2 = 1`; however, if we're using a cyclic distribution, `loc_part1 = 0` corresponds to global particle 0 and `loc_part2 = 0` corresponds to global particle 1, and we *should* compute the force resulting from interaction between these two particles.

Clearly, the problem is that we shouldn't be using local particle indexes, but rather we should be using *global* particle indexes. Thus, using a cyclic distribution of the particles, we could modify our code so that the loops also iterate through global particle indexes:

```
for (loc_part1 = 0, glb_part1 = my_rank;
      loc_part1 < loc_n-1;
      loc_part1++, glb_part1 += comm_sz) {
   for (glb_part2
         = First_index(glb_part1, my_rank, owner, comm_sz),
      loc_part2 = Global_to_local(glb_part2, owner, loc_n);
      loc_part2 < loc_n;
      loc_part2++, glb_part2 += comm_sz) {
         Compute_force(loc_pos[loc_part1], masses[glb_part1],
               tmp_pos[loc_part2], masses[glb_part2],
               loc_forces[loc_part1], tmp_forces[loc_part2]);
   }
}
```

The function `First_index` should determine a global index `glb_part2` with the following properties:

1. The particle `glb_part2` is assigned to the process with rank `owner`.
2. `glb_part1 < glb_part2 < glb_part1 + comm_sz`.

The function `Global_to_local` should convert a global particle index into a local particle index, and the function `Compute_force` should compute the force resulting from the interaction of two particles. We already know how to implement `Compute_force`. See Exercises 7.16 and 7.17 for the other two functions.

Table 7.5 Performance of the MPI
n-body solvers (in seconds).

Processes	Basic	Reduced
1	17.30	8.68
2	8.65	4.45
4	4.35	2.30
8	2.20	1.26
16	1.13	0.78

Table 7.6 Run-times for OpenMP and MPI *n*-body
solvers (in seconds).

Processes/ Threads	OpenMP		MPI	
	Basic	Reduced	Basic	Reduced
1	15.13	8.77	17.30	8.68
2	7.62	4.42	8.65	4.45
4	3.85	2.26	4.35	2.30

7.1.11 Performance of the MPI solvers

Table 7.5 shows the run-times of the two *n*-body solvers when they're run with 800
particles for 1000 timesteps on an Infiniband-connected cluster. All the timings were
taken with one process per cluster node.

The run-times of the serial solvers differed from the single-process MPI solvers
by less than 1%, so we haven't included them.

Clearly, the performance of the reduced solver is much superior to the performance of the basic solver, although the basic solver achieves higher efficiencies. For
example, the efficiency of the basic solver on 16 nodes is about 0.95, while the efficiency of the reduced solver on 16 nodes is only about 0.70.

A point to stress here is that the reduced MPI solver makes much more efficient
use of memory than the basic MPI solver; the basic solver must provide storage for
all *n* positions on each process, while the reduced solver only needs extra storage for
$n/$comm_sz positions and $n/$comm_sz forces. Thus, the extra storage needed on each
process for the basic solver is nearly comm_sz$/2$ times greater than the storage needed
for the reduced solver. When *n* and comm_sz are very large, this factor can easily make
the difference between being able to run a simulation only using the process' main
memory and having to use secondary storage.

The nodes of the cluster on which we took the timings have four cores, so we can
compare the performance of the OpenMP implementations with the performance of
the MPI implementations (see Table 7.6). We see that the basic OpenMP solver is a
good deal faster than the basic MPI solver. This isn't surprising, since MPI_Allgather
is such an expensive operation. Perhaps surprisingly, though, the reduced MPI solver
is quite competitive with the reduced OpenMP solver.

Let's take a brief look at the amount of memory required by the MPI and OpenMP reduced solvers. Say that there are n particles and p threads or processes. Then each solver will allocate the same amount of storage for the local velocities and the local positions. The MPI solver allocates n **doubles** per process for the masses. It also allocates $4n/p$ **doubles** for the tmp_pos and tmp_forces arrays, so in addition to the local velocities and positions, the MPI solver stores

$$n + 4n/p$$

doubles per process. The OpenMP solver allocates a total of $2pn + 2n$ **doubles** for the forces and n **doubles** for the masses, so in addition to the local velocities and positions, the OpenMP solver stores

$$3n/p + 2n$$

doubles per thread. Thus, the difference in the local storage required for the OpenMP version and the MPI version is

$$n - n/p$$

doubles. In other words, if n is large relative to p, the local storage required for the MPI version is substantially less than the local storage required for the OpenMP version. So, for a fixed number of processes or threads, we should be able to run much larger simulations with the MPI version than the OpenMP version. Of course, because of hardware considerations, we're likely to be able to use many more MPI processes than OpenMP threads, so the size of the largest possible MPI simulations should be *much* greater than the size of the largest possible OpenMP simulations. The MPI version of the reduced solver is much more scalable than any of the other versions, and the "ring pass" algorithm provides a genuine breakthrough in the design of n-body solvers.

7.1.12 Parallelizing the basic solver using CUDA

When we apply Foster's method to parallelizing the n-body solvers for GPUs, the first three steps are the same as they were for the general parallelization in Subsection 7.1.3. So our composite or agglomerated tasks correspond to the individual particles in the simulation. As before, this will greatly reduce the number of inter-task communications.

Now, however, the mapping stage is somewhat different. In the MIMD implementations of the n-body solvers, we expected the number of tasks or particles to be much greater than the number of available cores, and for a very large problem this will probably be the case for a GPU-solver as well. However, recall that for GPUs, we usually *want* many more threads than SIMD cores (or datapaths) so that we can hide the latency of slow operations, such as loads and stores. So when we implement the mapping phase for a CUDA solver, it may make sense to have just one task or particle per thread. Let's look at implementing this approach—one particle per thread—and we'll take a look at multiple particles per thread in the Programming Assignments. (See Programming Assignment 7.5.)

With this assumption each thread will compute something like this:

```
for each timestep t {
    Compute total force on my particle;
    Compute position and velocity of my particle;
}
```

However, there are a couple of race conditions in this pseudocode. Suppose threads *A* and *B* are executing the body of the loop for the same timestep. Also suppose that thread *A* updates the position of its particle *before* thread *B* starts the computation of the force on its particle. Then thread *B* will use the *wrong* value of the position of the particle assigned to thread *A* in the computation of the force on the particle assigned to *B*. So in any timestep no thread should proceed to the update of its particle's position and velocity until *after* all the threads have completed the updates to their particle's forces.

Now suppose that thread *A* has completed timestep *t*, and it has started computing the force on its particle in timestep *t* + 1. Also suppose that thread *B* is still executing timestep *t*, and it hasn't yet updated its particle's position. Then thread *A* will use the wrong value for thread *B*'s position in the computation of the force on its particle. So no thread should be able start timestep *t* + 1 until all the threads have completed timestep *t*.

Both of these race conditions can be prevented by the use of barriers. If we have a barrier across all the threads after the computation of the forces, then the first race condition can't occur, since thread *A* won't be able to proceed to the update of its particle's position and velocity until all the threads—including *B*—have completed the computation of the total force on their particles. Similarly, if we have a barrier after the updates to the positions and velocities, then no thread can proceed to timestep *t* + 1 until all the threads have completed their updates to the positions and velocities.

In Chapter 6 we saw that we could implement a barrier across *all* the threads by returning from one kernel and starting another. For example, suppose we wanted to place a barrier between calls to Function_x and Function_y in the following code:

```
/* Prototypes */
__device__ void Function_x (...);
__device__ void Function_y (...);
__global__ void My_kernel (...);

...
/* Host code */
...
My_kernel <<<blk_ct, th_per_blk>>> (...);

/* Device code */
__global__ void My_kernel (...) {
    Function_x (...);
    /* Want barrier here */
```

```
    Function_y (...);

    return;
}
```

Then we can rewrite our code so that `Function_x` and `Function_y` are kernels:

```
/* Prototypes */
__global__ void Function_x (...);
__global__ void Function_y (...);
...

/* Host code */
...
Function_x <<<blk_ct, th_per_blk>>> (...);
Function_y <<<blk_ct, th_per_blk>>> (...);
...

/* Device code */
__global__ void Function_x (...) {
    ...
}

__global__ void Function_y (...) {
    ...
}
```

In the second program the default behavior is for no thread to start executing `Function_y` until all the threads executing `Function_x` have returned.

Of course, in the general setting, there may be other code in `My_kernel`. In our case, the kernel with the race conditions might look something like this:

```
__global__ void Nbody_sim(vect_t forces[],
    struct particle_s curr[],
    double delta_t, int n, int n_steps) {
    int step;
    int my_particle = blkIdx.x*blkDim.x + threadIdx.x;

    for (step = 1; step <= n_steps; step++) {
        Compute_force(my_particle, forces, curr, n);
        Update_pos_vel(my_particle, forces, curr, n, delta_t);
    }
}
```

The only parts of the kernel that require the use of multiple threads are the calls to `Compute_force` and `Update_pos_vel`. So we can put the rest of the kernel in a *host* function[2]:

---

[2] In CUDA the __host__ designation is the default. So we could omit it for this function.

```
__host__ void Nbody_sim(vect_t forces[],
      struct particle_s curr[],
      double delta_t, int n, int n_steps,
      int blk_ct, int th_per_blk) {
   int step;

   for (step = 1; step <= n_steps; step++) {
      Compute_force <<<blk_ct, th_per_blk>>> (
            forces, curr, n);
      Update_pos_vel <<<blk_ct, th_per_blk>>> (
            forces, curr, n, delta_t);
   }
   cudaDeviceSynchronize();
}
```

Now, instead of being device functions, Compute_force and Update_pos_vel are kernels, and instead of computing the particle indexes in Nbody_sim, we compute them in the new kernels.

In this code there is an implicit barrier before each call to Update_pos_vel. The call won't be started until all the threads executing Compute_force have returned. Also each call to Compute_force (except the first) will wait for the call to Update_pos_vel in the preceding iteration to complete.

The call to cudaDeviceSynchronize isn't needed, except as protection against accidentally executing host code that depends on the completion of the final call to Update_pos_vel. Recall that by default host code does *not* wait for the completion of a kernel.

7.1.13 A note on cooperative groups in CUDA

With the introduction of CUDA 9 and the Nvidia Pascal processor, it became possible to synchronize multiple thread blocks without starting a new kernel. The idea is to use a construct that is similar to an MPI communicator. In CUDA the new construct is called a **cooperative group**. A cooperative group consists of a collection of threads that can be synchronized with a function call that's made in a kernel or a device function.

To use cooperative groups, you must be running CUDA SDK 9 or later, and to use cooperative groups consisting of threads from more than one thread block, you must be using a Pascal processor or later.

Programming assignment 7.6 outlines how we might modify the basic CUDA solver so that it uses cooperative groups.

7.1.14 Performance of the basic CUDA *n*-body solver

Let's take a look at the performance of our basic CUDA *n*-body solver. We ran the CUDA program on an Nvidia-Pascal system (CUDA capability 6.1) with 20 SMs, 128 SPs per SM, and a 1.73 GHz clock. Table 7.7 shows the run-times. Every run was

Table 7.7 Run-times (in seconds) of basic CUDA n-body solver.

Blocks	Particles	Serial	CUDA
1	1024	7.01e-2	1.93e-3
2	2048	2.82e-1	2.89e-3
32	32,768	5.07e+1	7.34e-2
64	65,536	2.02e+2	2.94e-1
256	262,144	3.23e+3	2.24e+0
1024	1,048,576	5.20e+4	4.81e+1

with 1024 threads per block using **float** as the datatype. The serial times were taken on the host machine, which had an Intel Xeon Silver processor running at 2.1 GHz. Clearly, the CUDA implementation is *much* faster than the serial implementation. Indeed, the CUDA implementation is at least an order-of-magnitude faster, and it can be more than three orders of magnitude faster. Also note that the CUDA system is *much* more scalable than the serial implementation. As we increase the problem size, the run-times of the CUDA implementation usually increase much more slowly than the run-times of the serial implementation.

7.1.15 Improving the performance of the CUDA n-body solver

While the basic CUDA n-body solver is both faster and more scalable than the serial solver, its run-time *does* increase dramatically as the number of particles increases. For example, the time required for simulating 1,048,576 particles is nearly 25,000 times greater than the time required for 1024 particles. So we would like to further improve its performance.

If you've read the section on the OpenMP or Pthreads implementation, you'll recall that for these implementations, we used temporary storage for the calculations of the forces involving particles assigned to a particular thread. This allowed us to improve the overall performance by halving the number of calculations carried out by each thread. However, the implementations allocated storage for

$$2 \times \texttt{thread\_count} \times n$$

floating point values, and with n threads, this would require storage for $2n^2$ values. So if we were simulating a million particles, this would require storage for 2 *trillion* floating point values. If we use `floats`, this is 8 trillion bytes, and if we use `doubles`, it is 16 trillion bytes. We could probably reduce the amount of temporary storage by observing that we don't need to store n values for each particle (see Tables 7.1 and 7.2), but this observation reduces the temporary storage by less than a factor of 2. (See Exercise 7.9.) So we would still need storage for more than 1 trillion bytes. At the time of this writing (January 2020), our most powerful Nvidia processor (Turing) has about 25 gigabytes of global storage. So we'd need storage that's about 40 times larger than what's available to us now.

Another possibility is to use shared memory.

7.1.16 **Using shared memory in the *n*-body solver**

Recall that in CUDA *global* memory is accessible to all the SMs and hence all the threads. However, the time required to load data from or store data to global memory is quite large compared to operations on data in registers (e.g., arithmetic operations). To compute the force on particle q, the basic solver must access the positions and masses of *all* of the particles. In the worst case, this will require that we access global memory for all n particles. Nvidia processors do have an L2 cache that is shared among all the SMs and a per-SM L1 cache. So some of these accesses will not require loads from global memory, but we can do better.

Recall that each SM has a block of on-chip memory called *shared memory*. The storage in the shared memory is shared among the threads in a thread block that's running on that SM. Since shared memory is on the same chip as an SM, it is relatively small, but since it is on-chip, it is *much* faster than global memory. So if we can arrange that many of our loads are from shared memory instead of global memory, we may be able to provide significant performance improvements.

Since shared memory is *much* smaller than global memory, we can't hope to use shared memory to store a copy of *all* the data we need from global memory. An alternative is to partition the data we need from global memory into logical "tiles" that are small enough to fit into shared memory. Then each thread block can iterate through the tiles. After the first tile is loaded into shared memory, a thread block carries out all of its computations that make use of the data in the first tile. Then the second tile is loaded into shared memory, and the thread block carries out all of its computations that make use of the data in the second tile, etc. This gives us the following pseudocode:

```
1    for each tile {
2        Load tile into shared memory;
3        Barrier across the thread block;
4        Each thread carries out its computations
5            making use of the data in this tile;
6        Barrier across the thread block;
7    }
```

Effectively, the tiles are being used as cache lines, except that accessing them is managed by our program instead of the hardware. The first barrier ensures that we don't start the calculations in Lines 4–5 until all of the data from the tile has been loaded. The second barrier ensures that we don't start overwriting the data in the current tile until all the threads have finished using it.

To apply this pseudocode to the *n*-body problem, we load the masses and positions of a collection of m particles in line 2, and then in Lines 4–5 a thread block will compute the forces between *its* assigned particles and the m particles the block has just loaded. So we need to determine m.

If each thread loads the position and mass of a single particle, we can execute the loads without thread divergence. Furthermore, CUDA currently limits the number of threads in a thread block to 1024, and if we load the positions and masses of 1024 particles, each thread will load three `floats` or three `doubles`, and we'll need shared memory storage for at most

$$3 \times 8 \times 1024 = 24 \text{ Kbytes.}$$

Since current-generation Nvidia processors have at least 48 Kbytes of shared memory, we can easily load the positions and masses of $m = $ `th_per_blk` particles from global memory into shared memory. Finally, it's easy and natural to choose the particles in a tile to have contiguous indexes. So our tiles will be

- Tile 0: positions and masses of particles 0 through (`th_per_blk` − 1)
- Tile 1: positions and masses of particles `th_per_blk` through (2 × `th_per_blk` − 1)
- ...
- Tile k: positions and masses of particles (k × `th_per_blk`) through ((k + 1) × `th_per_blk` − 1)
- ...
- Tile (`blk_ct` − 1): positions and masses of particles ((`blk_ct` − 1) × `th_per_blk`) through (`blk_ct` × `th_per_blk` − 1)

In other words, we're using a *block* partition of the positions and masses to form the tiles.

This gives us the following pseudocode:

```
1    // Thread q computes total force on particle q
2    Load position and mass of particle q
3    F_q = 0;   /* Total force on particle q */
4    for each tile t = 0, 1, ... , blk_ct-1 {
5       // my_loc_rk = rank of thread in thread block
6       Load position and mass of particle
7          t*tile_sz + my_loc_rk into shared memory;
8       __syncthreads ();
9
10      // All of the positions and masses for the particles in
11      //    tile t have been loaded into shared memory
12      for each particle k in tile t (except particle q) {
13         Load position and mass of k from shared memory;
14         Compute force f_qk of particle k on particle q;
15         F_q += f_qk;
16      }
17      // Don't start reading data from next tile until all
18      //    threads in this block are done with the current tile
19      __syncthreads ();
20   }
21   Store F_q;
```

The code outside the **for** `each tile` loop is the same as the code outside the **for** `each particle` loop in the original basic solver: before the loop we load a position and a mass, and after the loop we store a force. The outer loop iterates through the tiles, and, as we noted, each tile consists of th_per_blk particles. So each thread loads the position and mass of a single particle.

Before proceeding to the calculation of the forces, we call __syncthreads(). Recall that this is a barrier among the threads in a single thread block. So in any thread block, no thread will proceed to the computation of forces due to the particles in tile t until all of the data on these particles is in shared memory.

Once the data on the particles in tile t are loaded, each thread uses the inner **for** loop to compute the forces on its particle that are due to the particles in tile t. Note that if there are enough registers to store local variables, then all of these calculations will use data in registers or data in shared memory.

After each thread has computed the forces on its particle due to the particles in tile t, we have another call to __syncthreads(). This prevents any thread from continuing to the next iteration of the outer loop until all the threads in the thread block have completed the current iteration. This will ensure that the data on the particles in tile t isn't overwritten by data on particles in tile t+1 until all the threads in the block are done with the data on the particles in tile t.

Now let's look at how many words are loaded and stored by the original basic CUDA solver and the CUDA shared memory solver in the computation of the forces in a single timestep. Both solvers start by having each thread load the position and mass of a single particle, and both solvers complete the computation of the forces by storing the total force on each particle. So if there are *n* particles, both solvers will start by loading 3*n* words—two for each position, and one for each mass—and they'll store 2*n* words: two for each force.

Since these are the same for both solvers, we can just compare the number of words loaded and stored in the main loop of each of the solvers.

In the basic solver, each thread will iterate through *all* of the particles, except the thread's own particle. So each thread will load $3(n - 1)$ words, and since we're assuming there are *n* threads, this will result in a total of $3n(n - 1)$ words.

In the shared memory solver, each tile consists of th_per_blk particles, and there are blk_ct tiles. So each thread in each block will load 3 words in Lines 6–7. Thus each execution of these lines by each of the threads will result in loading 3*n* words. Since there are blk_ct tiles, we get a total of

$$3n \times \text{blk\_ct}$$

words loaded in the outer for loop. The loads in the inner for loop are from shared memory. So their cost is *much* less than the loads from global memory, and we have that the ratio of words loaded from global memory in the basic solver to the shared memory solver is

$$\frac{3n(n - 1)}{3n \times \text{blk\_ct}} \approx \frac{n}{\text{blk\_ct}} = \text{th\_per\_blk}.$$

So, for example, if we have 512 threads per block, then the original, basic solver will load approximately 512 times as many words from global memory as the shared memory solver.

Table 7.8 Run-times (in seconds) of CUDA basic and CUDA shared memory *n*-body solvers.

Blocks	Particles	Basic	Shared Mem	Speedup
1	1024	1.93e-3	1.59e-3	1.21
2	2048	2.89e-3	2.17e-3	1.33
32	32,768	7.34e-2	4.17e-2	1.76
64	65,536	2.94e-1	1.98e-1	1.48
256	262,144	2.24e+0	1.96e+0	1.14
1024	1,048,576	4.81e+1	2.91e+1	1.65

The actual performance of the basic and shared-memory solvers is shown in Table 7.8. We ran the shared-memory solver on the same system that we ran the basic solver on: an Nvidia Pascal-based system (CUDA capability 6.1) with 20 SMs, 128 SPs per SM, and a 1.73 GHz clock. Also, as with the basic solver, we always used 1024 threads per block. The shared memory implementation is always a good deal faster than the basic implementation, and on average, the speedup is 1.43.

7.2 Sample sort

7.2.1 Sample sort and bucket sort

Sample sort is a generalization of an algorithm known as *bucket sort*. In bucket sort, we're given two values $c < d$, and the keys are assumed to have a uniform distribution over the interval $[c, d)$. This means that if we take two subintervals $[p, q)$ and $[r, s)$ of the same length, then the number of keys in $[p, q)$ is approximately equal to the number of keys in $[r, s)$. So suppose we have n keys (often **floats** or **ints**), and we divide $[c, d)$ into b "buckets" or "bins" of equal length. Finally suppose that $h = (d - c)/b$. Then there should be approximately n/b keys in each of the buckets:

$$[c, c + h), [c + h, c + 2h), \ldots, [c + (b - 1)h, b).$$

In bucket sort we first assign each key to its bucket, and then sort the keys in each bucket.

For example, suppose $c = 1, d = 10$, the keys are

$$9, 8, 2, 1, 4, 6, 3, 7, 2,$$

and we're supposed to use 3 buckets. Then $h = (10 - 1)/3 = 3$, and the buckets are

$$[1, 4), [4, 7), [7, 10).$$

So the contents of the three buckets should be

$$\{2, 1, 3, 2\}, \{4, 6\}, \text{ and } \{9, 8, 7\},$$

respectively.

In sample sort, we don't know the distribution of the keys. So we take a "sample" and use this sample to estimate the distribution. The algorithm then proceeds in the same way as bucket sort. For example, suppose we have b buckets and we choose our sample by picking s keys from the n keys. Here $b \leq s \leq n$. Suppose that after sorting the sample, its keys are

$$a_0 \leq a_1 \leq \ldots \leq a_{s-1},$$

and we choose the keys in slots that are separated by s/b sample elements as "splitters." If the splitters are

$$e_1, e_2, \ldots, e_{b-1}$$

then we can use

$$[\min, e_1), [e_1, e_2), \ldots, [e_{b-1}, \max)$$

as our buckets.[3] Here min is any value less than or equal to the smallest element in the list, and max is a value greater than the largest element in the list.

For example, suppose that after being sorted our sample consists of the following 12 keys:

$$2, 10, 23, 25, 29, 38, 40, 49, 53, 60, 62, 67.$$

Also suppose we want $b = 4$ buckets. Then the first three keys in the sample should go in the first bucket, the next three into the second bucket, etc. So we can choose as the "splitters"

$$e_1 = (23 + 25)/2 = 24, \quad e_2 = (38 + 40)/2 = 39, \quad e_3 = (53 + 60)/2 = 57.$$

Note that we're using the convention that if the splitter is not a whole number, we round up.

After determining the splitters, we assign each of the keys in the list to one of the buckets

$$[\min, 24), [24, 39), [39, 57), [57, \max).$$

Thus, we can use the pseudocode in Program 7.4 as the basis for an implementation of serial sample sort. Note that we're distinguishing between "sublists" of the input list and "buckets" in the sorted list. The sublists partition the input list into b sublists, each of which has n/b elements. The buckets partition the list into b subsets of elements with the property that all the keys in one bucket lie between consecutive splitters. So after each of the buckets has been sorted, we can concatenate them to get a sorted list.

---

[3] If there are many duplicates in the list, we may want to allow the buckets to be closed intervals, and if duplicate elements are equal to a splitter, we can divide the duplicates between a pair of consecutive buckets. We won't consider this possibility in the programs we develop.

Clearly, to convert this pseudocode into a working program we must implement the sample generation function and the mapping function and any associated data structures. There are many possibilities. Let's take a look at some of them.

```
1    /*  Input:
2     *       list[n]:  the  input  list
3     *              n:  number  of  elements  in  list  (evenly
4     *                  divisible  by  s  and  b)
5     *              s:  number  of  elements  in  the  sample  (evenly
6     *                  divisible  by  b)
7     *              b:  number  of  buckets
8     *  Output:
9     *       list[n]:  the  original  input  data  in  sorted  order
10    */
11
12       /*  Take  the  sample  */
13       Gen_sample(list,  n,  s,  b,  sample);
14
15       Sort(sample,  s);
16
17       Find_splitters(sample,  s,  splitters,  b);
18
19       /*  Assign  each  element  of  each  sublist  to
20        *  its  destination  bucket  in  mapping,  a  data
21        *  structure  that  can  store  all  n  elements.
22        */
23       Map(list,  n,  splitters,  b,  mapping);
24
25       /*  Sort  each  bucket  and  copy  it  back  to  the  original  list  */
26       Sort_buckets(mapping,  n,  b,  list);
```

Program 7.4: Pseudocode for a serial sample sort.

7.2.2 Choosing the sample

An obvious method for generating the sample is to use a random number generator. For example,

```
// Seed the random number generator
srandom (...);
chosen = EMPTY_SET;   // subscripts of chosen elements
for (i = 0; i < s; i++) {
    // Get a random number between 0 and n
    sub = random() % n;
    // Don't choose the same subscript twice
    while (Is_an_element(sub, chosen))
```

```
      sub = random() % n;

   // Add sub to chosen elements
   Add(chosen, sub);
   sample[i] = list[sub]
}
```

The variable chosen is a set abstract data type that's used to store the subscripts of elements that have been added to the sample. The **while** loop is used to ensure that we don't choose the same subscript twice. For example, suppose random() % n is 5 when i = 2 and when i = 4. Then without the **while** loop the sample will include element list[5] twice. Checking whether sub has already been chosen can be expensive, but if the sample size, s, is small relative to n, the size of the list, then it's unlikely that we'll choose the same subscript twice, and we'll only need to check chosen once for each element in the sample.

An alternative takes a "deterministic" sample.

```
for each sublist {
   Sort(sublist, n/b);

   Choose s/b equally spaced elements
      from the sorted sublist and add
      these to the sample;
}
```

For example, suppose that after being sorted a sublist consists of the elements $\{21, 27, 49, 62, 86, 92\}$ and we're choosing two elements from each sublist for the sample. Then we could choose elements 49 and 92 from this sublist. Even if the sample size is large, the main cost of this method will be the calls to Sort. So this method will work best if there is a large number of relatively small sublists.

7.2.3 A simple implementation of the Map function

Perhaps the simplest implementation of the Map function allocates b arrays, each capable of storing at least n/b elements—for example, $2n/b$ elements. This is the mapping data structure in the pseudocode in Program 7.4. These arrays will be the buckets. After allocating the buckets, we can simply iterate through the elements of list determining for each element which bucket it should be mapped to and copying the element to the appropriate bucket.

```
Allocate storage for b buckets;
for each element x of list {
   Search splitters for x's bucket;
   if (bucket is full)
      Allocate additional storage for
         bucket b;
   Append x to bucket b;
}
```

The catch is that, in general, a bucket may need to store more than the initially allocated number of elements, and we'll need to increase its size. We can do this using the C library function `realloc`:

```
void* realloc(void* ptr, size_t size);
```

For us, `ptr` will be one of the buckets, and `size` is the amount of storage, in bytes, that we want for the resized bucket. If there is insufficient space immediately following the memory that has already been allocated, `realloc` will try to find another block that is large enough and copy the contents of the old block into the new block. When this happens, the pointer to the new block is returned by the call. If this isn't necessary, `realloc` returns the old pointer. If there isn't enough memory on the heap for the new allocation, `realloc` returns `NULL`. Note that any memory referred to by `ptr` before the call to `realloc` *must* have been allocated on the heap.

A final note here: `realloc` can be an expensive call. So in general, we don't allocate storage for just one additional element, since there's a good chance that we'll need to allocate more memory for the same bucket. Of course, in this setting, we don't know how much more memory we will ultimately need. If you'll be using the sample sort on a particular type of data, it may be useful to run some tests on data that is similar to the data encountered in the application that will be using the sort. However, for a fully general sort, we won't have this option. So we might try allocating another n/b elements when a bucket is full.

7.2.4 An alternative implementation of Map

An alternative implementation of `Map` uses a single n-element array, `temp`, for the buckets, but it builds some additional data structures that provide information about the buckets. These data structures might store the following information:

1. How many elements will be moved from each sublist to each bucket, `from_to_cts`;
2. The subscript of the first element in each sublist that goes to each bucket `bkt_starts_in_slists`;
3. The subscript of the first element in each bucket that comes from each sublist `slist_starts_in_bkts`, and the total number of elements in each bucket;
4. The location of the first element of each bucket in `temp`, `bkt_starts`.

The first two data structures are $b \times b$ arrays with rows corresponding to sublists and columns corresponding to buckets. The third data structure is a $(b+1) \times b$ array with rows $0, 1, \ldots, b-1$ corresponding to sublists, columns corresponding to buckets, and the last row storing the total number of elements in each bucket. The fourth data structure is a b-element array.

Note that we may not need all of these data structures. For example, in our serial implementation, we only use the first b rows of `slist_starts_in_bkts` to compute the last row. So we could simply sum each column of `from_to_cts` to get the final row and store it in a one-dimensional array.

We can build these data structures using the following pseudocode:

```
Sort each sublist;
for each sublist
    for each bucket
        Count elements in sublist
            going to bucket and store
            in from_to_cts[sublist,bucket];
for each sublist
    Build bkt_starts_in_slists
        by forming exclusive prefix sums
        of row sublist of from_to_cts;
for each bucket
    Build sublist_starts_in_bkts
        by forming exclusive prefix sums
        of column bucket of from_to_cts;
Build bkt_starts by forming
    exclusive prefix sums of last row
    of sublist_starts_in_bkts;
```

In spite of the apparent complexity of this code, it only uses **for** loops and three basic algorithms:

1. A sort for the sublists,
2. Counting elements of a sublist that fall in a range determined by the splitters, and
3. Exclusive prefix sums of a row or column of a two-dimensional array. (See paragraph 'Prefix sums' on p. 404.)

For the sort, we should probably choose a sorting algorithm that has relatively little overhead, so that it will be quite fast on short lists. So let's look at some details of the second and third algorithms.

To illustrate the second algorithm, we need to specify the details of the organization of the array of splitters. We'll assume that there are $b + 1$ elements in the array splitters and they're stored as shown in Table 7.9. The constants MINUS_INFTY and INFTY have been chosen so that they are, respectively, less than or equal to and greater than any possible element of the input list.

Since the sublists have been sorted, we can implement the second item by iterating sequentially through a sublist:

```
int curr_row[b];  /* row of from_to_cts     */
dest_bkt = 0;     /* subscript in splitters */

Assign(curr_row, 0);  /* Set elements to 0 */
for (j = 0; j < n/b; j++) {
    while (sublist[j] >= splitters[dest_bkt+1])
        dest_bkt++;
    curr_row[dest_bkt]++;
}
```

Table 7.9 The splitters in sample sort.

Bucket	Elements
0	MINUS_INFTY = splitters[0] ≤ key < splitters[1]
1	splitters[1] ≤ key < splitters[2]
⋮	⋮
i	splitters[i] ≤ key < splitters[i+1]
⋮	⋮
b-1	splitters[b-1] ≤ key < splitters[b] = INFTY

So we iterate through the elements of the current sublist in the **for** statement. For each of these elements, we iterate through the remaining splitters until we find a splitter that is strictly greater than the current element.

Suppose, for example, that $n = 24$, $b = 4$, and the splitters are $\{-\infty, 45, 75, 91, \infty\}$. Now suppose one of the sublists contains the elements $\{21, 27, 49, 62, 86, 92\}$. Then both 21 and 27 are less than 45. So the first two elements will go to bucket 0. Similarly, 49 and 62 are both less than 75 (and greater than 45). So they should go to bucket 1. However $86 > 75$, but less than 91; so it should go to bucket 2. Finally, $92 > 91$, but less than ∞; and it should go to bucket 3. This gives us two elements going to bucket 0, two elements going to bucket 1, one element going to bucket 2, and one element going to bucket 3.

Note that since the splitters are sorted, the linear search in the while loop could be replaced with binary search. Alternatively, since the sublists are sorted, we could use a binary search to find the location of each splitter in the sublist. (See Exercise 7.20.)

Prefix sums

A **prefix sum** is an algorithm that is applied to the elements of an n-element array of numbers, and its output is an n-element array that's obtained by adding up the elements "preceding" the current element. For example, suppose our array contains the elements

$$5, 1, 6, 2, 4$$

Then an *inclusive* prefix sum of this array is

$$5, 6, 12, 14, 18$$

So

```
incl_pref_sum[0] = array[0];
```

and for $i > 0$

```
incl_pref_sum[i] = array[i] + incl_pref_sum[i-1];
```

An *exclusive* prefix sum of our array is

$$0, 5, 6, 12, 14$$

So in an exclusive prefix sum, element 0 is 0, and the other elements are obtained by adding the *preceding* element of the array to the most recently computed element of the prefix sum array:

```
excl_pref_sum[0] = 0;
```

and for $i > 0$

```
excl_pref_sum[i] = array[i-1] + excl_pref_sum[i-1];
```

Observe that since the sublists have been sorted, the exclusive prefix sums of the rows of `from_to_cts` tell us which element of a sublist should be the first element from the sublist in each bucket. In our preceding example, we have $b = 4$ buckets and for the sublist $\{21, 27, 49, 62, 86, 92\}$ the row corresponding to it in `from_to_cts` is

$$2, 2, 1, 1.$$

The exclusive prefix sums of this array are

$$0, 2, 4, 5$$

So the subscripts of the first element in the sublist going to each of the four buckets are shown in the following table:

Bucket	Subscript of First Element Going to Bucket	First Element Going to Bucket
0	0	21
1	2	49
2	4	86
3	5	92

In fact, the elements of each of the remaining data structures can be computed using exclusive prefix sums. Here's a complete example with $n = 24$, $b = 4$, $s = 8$, and the following list:

```
Sublist  0:    15, 35, 77, 83, 86, 93
Sublist  1:    21, 27, 49, 62, 86, 92
Sublist  2:    26, 26, 40, 59, 63, 90
Sublist  3:    11, 29, 36, 67, 68, 72
```

As before, the splitters are $\{-\infty, 45, 75, 91, \infty\}$. Using this information, the various data structures are

from_to_cts				
	bucket			
sublist	**0**	**1**	**2**	**3**
0	2	0	3	1
1	2	2	1	1
2	3	2	1	0
3	3	3	0	0

bkt_starts_in_slists				
sublist	**bucket**			
0	0	2	2	5
1	0	2	4	5
2	0	3	5	6
3	0	3	6	6

slist_starts_in_bkts				
sublist	**bucket**			
0	0	0	0	0
1	2	0	3	1
2	4	2	4	2
3	7	4	5	2
bkt sizes	10	7	5	2

bkt_starts				
bucket	0	1	2	3
1st elt	0	10	17	22

Note that the last row of slist_starts_in_bkts is the number of elements in each bucket, and it is obtained by continuing the prefix sums down the columns of from_to_cts. Finally, note that bkt_starts is obtained by taking the exclusive prefix sums of the last row of slist_starts_in_bkts, and it tells us where each bucket should start in the array temp. So we don't need to do any reallocating in this version, and there's no wasted space.

Finishing up

Now we can assign elements of list to temp, the array that will store the buckets. Since bkt_starts tells us where the first element of each bucket is (e.g., the first element in bucket 1 should go to temp[10]), we can finish by iterating through the sublists, and in each sublist iterating through the buckets:

```
1   for each slist of list {
2       sublist = pointer to start of next block
3                of n/b elements in list;
```

```
4      for each bkt in temp {
5          bucket = pointer to start of bkt in temp;
6          dest = 0;
7
8          /* Now iterate through the elements of sublist */
9          first_src_sub = bkt_starts_in_slist[slist, bkt];
10         if (bkt < b)
11             last_src_sub = bkt_starts_in_slist[slist, bkt+1];
12         else
13             last_src_sub = n/b;
14         for (src = first_src_sub; src < last_src_sub; src++)
15             bucket[dest++] = sublist[src];
16     }
17  }
```

The assignments in Lines 2 and 4 are just defining pointers to the beginning of the current sublist and the beginning of the current bucket. The innermost loop iterates through the elements of the current sublist that go to the current bucket. In general, the limits of this iteration can be obtained from bkt_starts_in_slist. However, since there are only b columns in this array, the last element has to be treated as a special case—hence the else clause.

7.2.5 Parallelizing sample sort

Serial sample sort can be divided into a sequence of well-defined phases:

1. Take a sample from the input list.
2. Find the splitters from the sample.
3. Build any needed data structures from the list and the splitters.
4. Map keys in the list into buckets.
5. Sort the buckets.

There are other possible phases—e.g., sorts at various points in the algorithm—and, as we've already seen, there are a variety of possible implementations of each phase.

Serial sample sort can be naturally parallelized for MIMD systems by assigning sublists and buckets to processes/threads. For example, if each process/thread is assigned a sublist, then Steps 1 and 2 can have embarrassingly parallel implementations. First each process/thread chooses a "subsample" from its sublist, and then the splitters are chosen by each process/thread from its subsample.

Since CUDA thread blocks can operate independently of each other, in CUDA it is natural to assign sublists and buckets to thread blocks instead of individual threads.

We'll go into detail for the various APIs below, but we'll assume that each MIMD process/thread is assigned a sublist and a bucket. For the CUDA implementations, we'll look at two different assignments. The first, basic, implementation will also assign a bucket and a sublist to each thread. We'll use a more natural, but more complex, approach in the second implementation, where we'll assign buckets and sublists to thread blocks.

Note also that in our serial implementations, we used a generic sort function several times. So for the MIMD parallel implementations, we'll assume that the sorts are either single-threaded or that we have a parallel sort function that performs well on a relatively small list. For the basic CUDA implementation, we'll use single-threaded sorts on the host and each thread. For the second implementation, we'll assume that we have a parallel sort function that performs well when implemented with a single thread block and one that works well with multiple thread blocks on relatively small lists.

Finally, for our "generic" parallelizations, we'll focus on the MIMD APIs, and we'll address the details of CUDA implementations in Section 7.2.9.

Basic MIMD implementation

The first implementation uses many of the ideas from our first serial implementation:

1. Use a random number generator to choose a sample from the input list.
2. Sort the sample.
3. Choose splitters from the sample.
4. Use the splitters to map keys to buckets.
5. Sort buckets.

As we noted earlier, Step 1 can be embarrassingly parallel if we assign a sublist to each MIMD process/thread. For Step 2 we can either have a single process/thread carry out the sort, or we can use a parallel sort that is reasonably efficient on short lists (e.g., parallel odd-even transposition sort).

If the sample is distributed among the processes/threads, then Step 3 can be implemented by having each process/thread choose a single splitter. So if we used a single process/thread for Step 2, we can first distribute the sample among the processes/threads. Finally, to determine which elements of the list go to which buckets, each process/thread will need access to all of the splitters. So in a shared memory system, the splitters should be shared among the threads, and in a distributed memory system, the splitters should be gathered to each process.

For Step 4, one task might be to determine which bucket an element of the list should be mapped to. A second might be to assign the element to its destination bucket. Clearly these two tasks can be aggregated, since the second can't proceed until the first is completed. The obvious problem here is that we need some sort of synchronization among tasks with the same destination bucket. For example, on a shared memory system, we might use a temporary array to record for each element its destination bucket. Then a process/thread can iterate through the list, and when it finds an element going to its assigned bucket, it can copy from the element its destination bucket. This will require an extra array to record destinations, but if the list isn't huge, this should be possible.

For distributed memory, this approach could be very expensive: we might gather both the original list and the list of destination buckets to each process. A less costly alternative would be to construct some of the data structures outlined in the second serial implementation and use these with an MPI collective communication called

`MPI_Alltoallv` to redistribute the elements of the list from each process to its correct destination. We'll go into detail in the MPI section (Section 7.2.8), but the idea is that each process has a block of elements going to every other process. For example, process q has some elements that should go to process 0, some that should go to process 1, etc., and this is the case for each process. So we would like for each process to scatter its original elements to the correct processes, *and* we would like for each process to gather its "final" elements from every process.

In either the shared- or the distributed-memory implementation, we can finish by having each process/thread sort its elements using a serial sort.

A second implementation

As you might have guessed our second implementation uses many of the ideas from the second serial implementation.

1. Sort the elements of each sublist.
2. Choose a random or a "deterministic" sample as discussed in our serial implementation.
3. Sort the sample.
4. Choose equally spaced splitters from the sample.
5. Use the input list and the splitters to build the data structures needed to implement the assignment of elements to buckets.
6. Assign elements of the list to the correct buckets.
7. Sort each bucket.

As with the first implementation, we'll assign a single sublist and a single bucket to each thread/process.

As we noted near the beginning of this section, for steps 1, 3, and 7 we assume each sort is either done by a single thread/process, or that we already have a good parallel sort for "small" lists.

Step 2 is embarrassingly parallel for both the random and deterministic samples. For the deterministic sample, a task is choosing an element of the sample, and since they're equally spaced, a single thread/process can choose a subset of the sample elements from a single sublist.

Step 4 is the same as choosing the splitters in the first parallel implementation.

For Step 5, there are two basic parts. First we need to count the number of elements going from each sublist to each bucket. So as before, one task might be determining for an element of the list which bucket it should go to. Now a second task could be incrementing the appropriate element of the data structure. Clearly these tasks can be aggregated. Since the number of processes/threads is the same as the number of buckets, we can map the aggregate tasks for the elements of a sublist to a single process/thread.

The second part of Step 5 is implementing one or more prefix sums. Since the number of processes/threads is the same as the number of buckets, one process/thread can be responsible for each set of prefix sums. See below and, for example, [27].

In Step 6, a single task can be to assign an element of a sublist to its bucket. We aggregate the tasks so that a single process/thread is responsible for a sublist or a bucket. So a process/thread can "push" the keys in its sublist to their destination buckets. Or, alternatively, since a process/thread is responsible for a single bucket, it can "pull" all the keys from the original list that go to its bucket.

7.2.6 Implementing sample sort with OpenMP

Our OpenMP implementation of the sample sort will assume that the list to be sorted is stored in memory as a contiguous array. After the sort is complete, the values will have been moved to their appropriate locations within the array. The number of elements in the list, n, and the sample size, s, must be evenly divisible by the number of threads, thread_count (as specified by OMP_NUM_THREADS).

First implementation

For our initial implementation, the main thread will be responsible for setting up data structures and performing any serial steps before the sort begins. This will include sampling the list, sorting the sample, and calculating the splitters. We'll use the first approach described in Section 7.2.2 to determine which values to sample from the list by selecting a set of unique random numbers that correspond to the list indexes of the samples. Once the sample has been taken, we will sort it with the C library qsort function, since it is reasonably efficient with small lists. With the sample sorted, we calculate thread_count $+1$ splitters and use them to define a range of values that will be assigned to each thread's bucket. The buckets are assigned sequentially across the threads, with thread 0 managing the smallest values (MINUS_INFTY to the first splitter) and the last thread managing the largest values (from the last splitter to INFTY). Once these steps are complete, the remainder of the algorithm runs in an OpenMP parallel region.

Each thread's bucket is represented as an array of elements that fall within its splitter range and a global *bucket count* to track bucket sizes. We can use a slightly modified version of the simple approach outlined in Section 7.2.3 to populate the buckets. Following is the pseudocode:

```
1    for each element x of list {
2        if (x in bucket range) {
3            if (bucket is full) {
4                Resize bucket;
5            }
6
7            Append x to bucket;
8            Increment global bucket count;
9        }
10   }
```

Since resizing a bucket requires an expensive call to realloc, buckets are initially allocated with enough space for n/thread_count elements. If a resize operation is required, the bucket size is doubled. This helps reduce calls to `realloc` at the expense of somewhat higher overall memory consumption.

Once the buckets are populated, threads sort their individual buckets via the C library `qsort` function. Since the bucket sizes vary from thread to thread, a barrier ensures that the bucket counts have all been calculated before proceeding to the final step: copying each thread's bucket back to the appropriate positions in the original list. Since the list is shared among all the threads, this final operation can be carried out in parallel. Here's complete pseudocode for the multi-threaded portion of the algorithm:

```
 1   #   pragma omp parallel
 2       {
 3           bucket_counts[rank] = 0;
 4           my_bucket_size = n / thread_count;
 5           my_bucket = malloc(my_bucket_size * sizeof(int));
 6
 7           // Retrieve bucket range from the global splitter list
 8           bucket_low, bucket_high = Calc_splitter_range(rank);
 9
10           for each element x of list {
11               if (x >= bucket_low && x < bucket_high) {
12                   if (bucket is full) {
13                       my_bucket_size = my_bucket_size * 2;
14                       my_bucket = realloc(
15                           my_bucket, my_bucket_size);
16                   }
17
18                   // Append x to the bucket, then
19                   //         increment global bucket count
20                   my_bucket[bucket_counts[rank]++] = x;
21               }
22           }
23
24           // 0 = subscript of first element
25           Sort(my_bucket, 0, my_bucket_size);
26
27           // Wait for all threads to sort
28   #       pragma omp barrier
29
30           // Determine bucket positions using
31           //         global bucket counts
32           list_position = Calc_list_position(
33               rank, bucket_counts[rank])
34
35           // Copy sorted sublist from bucket
36           //         back to the global list
37           Copy(list + list_position, my_bucket, my_bucket_size)
38       }
```

Table 7.10 shows run-times (in seconds) of our initial implementation on a system with dual AMD EPYC 7281 16-Core Processors. The input list has $2^{24} = 16,777,216$ ints, and the sample has 65,536 ints. The run-time of the C library qsort is also provided for comparison. The program was compiled with compiler optimizations enabled (gcc -O3), 1000 trials were run for each test configuration, and the minimum time is reported.

Table 7.10 Run-times of the first OpenMP implementation of sample sort (times are in seconds).

Threads	Run-time
1	3.54
2	2.29
4	1.69
8	1.34
16	1.40
32	1.36
64	1.33
qsort	2.59

While the run-times reported in Table 7.10 demonstrate an improvement over the serial version of the algorithm, the implementation has some inefficiencies and clearly does not scale. One of the most significant issues with this approach is that every thread inspects the entire input list, resulting in a large amount of duplicated effort. As described previously in Section 7.2.4, we can avoid this overhead by modifying our Map implementation. Additionally, our sampling approach is limited to a single thread and may be inefficient with larger sample sizes due to repeated iterations through the list of "chosen" elements (see Section 7.2.2).

Second implementation

To improve the performance of our initial implementation, we take additional preparatory steps to populate the data structures described in Section 7.2.4. This ultimately decreases the amount of duplicated work performed by the threads and eliminates the need to reallocate space for buckets. While each thread in our first implementation "pulled" relevant values from the global list into its bucket based on its splitters, the second implementation "pushes" values from each bucket to their appropriate positions in the global list. Consequently, our second approach should be faster, and the difference in performance between the two approaches will increase as the list size increases.

Our first task is to divide the global list evenly among the threads, and then sort the values under their purview in place. We can determine the sublist indexes as follows:

```
my_sublist_size = n / thread_count;
my_first = my_rank * my_sublist_size;
my_last  = my_first + my_sublist_size;
```

And then sort the elements from `my_first` through `my_last` in parallel.

To accelerate the sampling process, we use the "deterministic" sampling algorithm described in Section 7.2.2. We begin by calculating individual "subsample" ranges for each thread to populate:

```
subsample_size = s / thread_count;
subsample_idx = subsample_size * my_rank;
```

Each thread samples from the original list by deterministically choosing elements at a fixed interval. A barrier ensures all the threads have generated their portion of the sample, and then the first thread is responsible for sorting it and choosing the splitters. Note that this process is the only portion of the second implementation that is completed by a single thread.

```
// Use the deterministic sampling approach
Choose_sample(
        my_first, my_last, subsample_idx, subsample_size);

#   pragma omp barrier

    if (my_rank == 0) {
        Sort(sample, 0, s)
        splitters = Choose_splitters(sample, s, thread_count);
    }

#   pragma omp barrier
```

Once the sublists are sorted and the splitters have been chosen, each thread can determine the number of elements in its bucket that belong to each of the other threads (based on their particular splitters). These counts are stored in a global $n \times n$ count matrix, with each thread responsible for initializing one row of the matrix. Once all of the threads have completed this step, we can determine the number of elements that will be placed in each thread's bucket by summing the corresponding column in the count matrix. Next, we allocate memory for the buckets and calculate the exclusive prefix sums for each row of the count matrix; this provides the indices for the start and end of each thread's bucket within the global list.

Once the preparatory work is complete and our data structures have been populated, each thread will "pull" in values from the other threads by merging them into its bucket. After the process is complete, the buckets are copied to their appropriate locations within the global list to conclude the parallel sort. The full pseudocode for the parallel portion of this approach follows:

```
1       my_sublist_size = n / thread_count;
2       my_first = my_rank * my_sublist_size;
3       my_last = my_first + my_sublist_size;
4       subsample_size = s / thread_count;
5       subsample_idx = subsample_size * my_rank;
6
```

```
 7      // Sort this thread's portion of the global list
 8      Sort(list, my_first, my_last);
 9
10      // Use the deterministic sampling approach
11      Choose_sample(my_first, my_last,
12             subsample_idx, subsample_size);
13
14  #   pragma omp barrier
15
16      // This portion is handled by a single thread
17      if (my_rank == 0) {
18         Sort(sample, 0, s);
19         splitters = Choose_splitters(sample, s, thread_count);
20      }
21
22  #   pragma omp barrier
23
24      // Count the number of elements in this thread's sublist
25      // that will be pushed to each thread, stored as a row
26      // of global count_matrix
27      Find_counts(my_rank);
28  #   pragma omp barrier
29
30      // Find the bucket size for this thread
31      bucket_size = 0;
32      for each thread_rank {
33         idx = thread_rank * thread_count + my_rank;
34         bucket_size += count_matrix[idx];
35      }
36  #   pragma omp barrier
37
38      // Each thread calculates exclusive prefix sums for its row
39      Calc_prefix_sums();
40  #   pragma omp barrier
41
42      // Preparation complete. Begin merging:
43      for each thread_rank {
44         Merge(my_bucket, thread_rank);
45      }
46  #   pragma omp barrier
47
48      // Determine bucket positions using global bucket counts
49      list_low, list_high = Calc_list_positions(
50             rank, bucket_counts[rank]);
51
52      // Copy sorted sublist from bucket back to the global list
53      Copy(global_list, my_bucket, list_low, list_high);
```

Here, the `Merge` process is more or less the same as one would find in a standard merge sort: assuming both lists are sorted in ascending order, compare the "heads" of each list and move the smaller of the two heads into a temporary array. Once the end of either list is reached, the remaining values of the other list can be copied directly into the array.

Table 7.11 shows run-times (in seconds) of our second implementation on the same testbed with input list of $2^{24} = 16,777,216$ **int**s and sample of 65,536 **int**s. Again, the run-time of the C library `qsort` is also provided for comparison, and the program was compiled with compiler optimizations enabled (`gcc -03`); 1000 trials were run for each test configuration, and the minimum time is reported.

Table 7.11 Run-times of the second OpenMP implementation of sample sort (times are in seconds).

Threads	Run-time
1	2.65
2	1.33
4	1.01
8	0.35
16	0.21
32	0.15
64	0.13
qsort	2.59

While the second approach requires a bit more upfront effort (both computationally and programmatically), it provides better run-times across the board, as each thread no longer iterates through the entire list of elements.

7.2.7 Implementing sample sort with Pthreads

As was the case with the two n-body solvers, the Pthreads implementation of sample sort is very similar to the OpenMP implementation. To avoid repeating the previous discussion, we will focus on the differences between the two approaches.

For our first iteration of the algorithm, we can replace OpenMP parallel regions with thread procedures. After the initial steps are completed by the main thread (gathering parameters, allocating data structures, and choosing the sample and splitters), we use `pthread_create` to start the parallel portion of the algorithm. The pseudocode follows:

```
1    bucket_counts[rank] = 0;
2    my_bucket_size = n / thread_count;
3    my_bucket = malloc(my_bucket_size * sizeof(int));
4
5    // Retrieve bucket range from the global splitter list
6    bucket_low, bucket_high = Calc_splitter_range(rank);
```

```
7
8     for each element x of list {
9        if (x >= bucket_low && x < bucket_high) {
10          if (bucket is full) {
11              my_bucket_size = my_bucket_size * 2;
12              my_bucket = realloc(my_bucket, my_bucket_size);
13          }
14
15          // Append x to bucket, increment global bucket count
16          my_bucket[bucket_counts[rank]++] = x;
17        }
18     }
19
20     Sort(my_bucket, 0, my_bucket_size);
21
22     // Wait for all threads to sort
23     pthread_barrier_wait(&bar);
24
25     // Determine bucket positions using global bucket counts
26     list_position = Calc_list_position(
27            rank, bucket_counts[rank])
28
29     // Copy sorted sublist from bucket back to the global list
30     Copy(list + list_position, my_bucket, my_bucket_size)
```

Note that the primary difference here is we are using a barrier from the Pthreads library. Usage is typical of Pthreads objects:

```
1     // Shared barrier variable
2     pthread_barrier_t bar;
3
4     ...
5
6     // Initialize barrier 'bar' with n threads
7     // (done once during the serial portion of the algorithm):
8     pthread_barrier_init(&bar, NULL, n);
9
10    ...
11
12    // Wait for all threads to reach the barrier:
13    pthread_barrier_wait(&bar);
14
15    ...
16
17    // Destroy the barrier in the main thread, when it's
18    // no longer needed:
19    pthread_barrier_destroy(&bar);
20    //
```

See Exercise 4.10 in Chapter 4.

However, since Pthread barriers are not available on all platforms, we can also build our own implementation with a condition variable:

```
1    void Barrier(void) {
2        pthread_mutex_lock(&bar_mut);
3        bar_count++;
4        if (bar_count == thread_count) {
5            bar_count = 0;
6            pthread_cond_broadcast(&bar_cond);
7        } else {
8            while (pthread_cond_wait(&bar_cond, &bar_mut) != 0);
9        }
10       pthread_mutex_unlock(&bar_mut);
11   }
```

Here `bar_mut` and `bar_cond` are shared variables. See Subsection 4.8.3 in Chapter 4 for further details.

Table 7.12 shows run-times (in seconds) of our initial implementation on a system with dual AMD EPYC 7281 16-Core Processors. The input list has $2^{24} = 16,777,216$ ints, and the sample has 65,536 ints. The run-time of the C library qsort is also provided for comparison. The program was compiled with compiler optimizations enabled (gcc -03); 1000 trials were run for each test configuration, and the minimum time is reported.

Table 7.12 Run-times of the first Pthreads implementation of sample sort (times are in seconds).

Threads	Run-time
1	3.55
2	2.31
4	1.71
8	1.48
16	1.45
32	1.28
64	1.24
qsort	2.59

Just as with the OpenMP version, this implementation provides some improvement but is clearly not scalable. We applied the same optimizations to the second implementation (parallel deterministic sample, determining bucket sizes in advance, pulling values into the buckets and merging). Table 7.13 shows the run-times (in seconds) on the same testbed with input list of $2^{24} = 16,777,216$ ints and sample of 65,536 ints. Again, the run-time of the C library qsort is also provided for comparison and the program was compiled with compiler optimizations enabled (gcc -03).

Table 7.13 Run-times of the second Pthreads implementation of sample sort (times are in seconds).

Threads	Run-time
1	2.65
2	1.36
4	1.08
8	0.35
16	0.20
32	0.14
64	0.12
qsort	2.59

We can observe that performance is very similar for both the OpenMP and Pthreads approaches. This is not surprising; apart from some minor syntactical differences, their implementations are also quite similar.

7.2.8 Implementing sample sort with MPI

For the MPI implementation, we assume that the initial list is distributed among the processes, but the final, sorted list is gathered to process 0. We also assume that both the total number of elements n and the sample size s are evenly divisible by the number of processes p. So each process will start with a sublist of n/p elements, and each process can choose a "subsample" of s/p elements.

First implementation

Each process can choose a sample of its sublist using the method outlined above for the first serial algorithm. Then these can be gathered onto process 0. Process 0 will sort the sample, choose the splitters, and broadcast the splitters to all the processes. Since there will be p buckets, there will be $p + 1$ splitters: the smallest less than or equal to any element in the list (MINUS_INFTY), and the largest greater than any element in the list (INFTY).

Then each process can determine to which bucket each of its elements belongs, and we can use the MPI function MPI_Alltoallv to distribute the contents of each sublist among the buckets. The syntax is

```
int MPI_Alltoallv(
        const void      *send_data       /* in  */,
        const int       *send_counts     /* in  */,
        const int       *send_displs     /* in  */,
        MPI_Datatype    send_type        /* in  */,
        void            *recv_data       /* out */,
        const int       *recv_counts     /* in  */,
        const int       *recv_displs     /* in  */,
```

```
MPI_Datatype    recv_type          /* in   */,
MPI_Comm        comm               /* in   */);
```

In our setting, `send_data` is the calling process' sublist, and `recv_data` is the calling process' bucket. Furthermore, the function assumes that the elements going to any one process occupy contiguous locations in `send_data`, and the elements coming from any one process should occupy contiguous locations in `recv_data`.[4] The communicator is `comm`, which for us will be `MPI_COMM_WORLD`. For us, both `send_type` and `recv_type` are `MPI_INT`. The arrays `send_counts` and `send_displs` specify where the elements of the sublist should go, and the arrays `recv_counts` and `recv_displs` specify where the elements of the bucket should come from.

As an example, suppose we have three processes, the splitters are $-\infty, 4, 10, \infty$ and the sublists and buckets are shown in the following table:

	Processes		
	0	**1**	**2**
sublists	3, 5, 6, 9	1, 2, 10, 11	4, 7, 8, 12
buckets	3, 1, 2	5, 6, 9, 4, 7, 8	10, 11, 12

Then we see that processes should send the following numbers of elements to each process:

From	To Process		
Process	**0**	**1**	**2**
0	1	3	0
1	2	0	2
2	0	3	1

So `send_counts` on each process should be

$$\text{Process } 0 : \{1, 3, 0\}$$
$$\text{Process } 1 : \{2, 0, 2\}$$
$$\text{Process } 2 : \{0, 3, 1\}$$

and `recv_counts` should be

$$\text{Process } 0 : \{1, 2, 0\}$$
$$\text{Process } 1 : \{3, 0, 3\}$$
$$\text{Process } 2 : \{0, 2, 1\}$$

So `recv_counts` can be obtained by taking the *columns* of the matrix whose rows are the `send_counts`.

---

[4] These assumptions apply when the datatype arguments occupy contiguous locations in memory.

The array `send_counts` tells `MPI_Alltoallv` how many **int**s should be sent from a given process to any other process, and the `recv_counts` array tells the function how many elements come from each process to a given process. However, to determine where the elements of a sublist going to a given process begin, `MPI_Alltoallv` also needs an offset or **displacement** for the first element that goes from the process to each process, and the first element that comes to a process from each process. In our setting, these displacements can be obtained from the exclusive prefix sums of `send_counts`. Thus, in our example, the `send_displs` arrays on the different processes are

$$\text{Process } 0 : \{0, 1, 4\}$$
$$\text{Process } 1 : \{0, 2, 2\}$$
$$\text{Process } 2 : \{0, 0, 3\}$$

and the `recv_displs` arrays are

$$\text{Process } 0 : \{0, 1, 3\}$$
$$\text{Process } 1 : \{0, 3, 3\}$$
$$\text{Process } 2 : \{0, 0, 2\}$$

Note also that by adding in the final elements of `recv_counts` to the corresponding final elements `recv_displs` we get the total number of elements needed for `recv_data`:

$$\text{Process } 0 : \texttt{recv\_data[]}, \; 3 + 0 \text{ ints}$$
$$\text{Process } 1 : \texttt{recv\_data[]}, \; 3 + 3 \text{ ints}$$
$$\text{Process } 2 : \texttt{recv\_data[]}, \; 2 + 1 \text{ ints}$$

We'll just use a *large n*-element array for `recv_data`. So we won't need to worry about allocating the exact amount of needed storage for each process' `recv_data`. After the call to `MPI_Alltoallv`, each process sorts its local list.

Note that, in general, we'll need to first get the *numbers* of elements coming from each process. We could use `MPI_Alltoallv` to do this, but since each process is sending a single int to every process, we can use a somewhat simpler function: `MPI_Alltoall`:

```
int  MPI_Alltoall (
       const  void      *send_data      /*  in   */,
       int               send_count     /*  in   */,
       MPI_Datatype      send_type      /*  in   */,
       void             *recv_data      /*  out  */,
       int               recv_count     /*  in   */,
       MPI_Datatype      recv_type      /*  in   */,
       MPI_Comm          comm           /*  in   */);
```

For us, each process should put in `send_data` the number of elements going to each process; `send_count` and `recv_count` should be 1, and the types should both be `MPI_INT`.

At this point our convention is that we need to gather the entire list onto process 0. At first, it might seem that we can simply use `MPI_Gather` to collect the distributed list onto process 0. However, `MPI_Gather` expects each process to contribute the same number of elements to the global list, and, in general, this won't be the case. We encountered another version of `MPI_Gather`, `MPI_Gatherv`, back in Chapter 3. (See Exercise 3.13.) This has the following syntax:

```
int  MPI_Gatherv(
        void*          loc_list              /*  in   */,
        int            loc_list_count        /*  in   */,
        MPI_Datatype   send_elt_type         /*  in   */,
        void*          list                  /*  out  */,
        const int*     recv_counts           /*  in   */,
        const int*     recv_displs           /*  in   */,
        MPI_Datatype   recv_elt_type         /*  in   */,
        int            root                  /*  in   */,
        MPI_Comm       comm                  /*  in   */);
```

This is very similar to the syntax for `MPI_Gather`:

```
int  MPI_Gather(
        void*          loc_list              /*  in   */,
        int            loc_list_count        /*  in   */,
        MPI_Datatype   send_elt_type         /*  in   */,
        void*          list                  /*  out  */,
        const int      recv_count            /*  in   */,
        MPI_Datatype   recv_elt_type         /*  in   */,
        int            root                  /*  in   */,
        MPI_Comm       comm                  /*  in   */);
```

The only differences are that in `MPI_Gatherv`, there is an *array* of `recv_counts` instead of a single int, and there is an array of displacements. These differences are completely natural: each process is contributing a collection `loc_list` of `loc_list_count`, each of which has type `send_elt_type`. In `MPI_Gather`, each process contributes the same number of elements, so `loc_list_count` is the same for each process. In `MPI_Gatherv`, the processes contributed different numbers of elements. So instead of a single `recv_count`, we need an array of `recv_counts`, one for each process. Also, by analogy with `MPI_Alltoallv`, we need an array of *displacements*, `recv_displs`, that indicates where in `list` each process' contribution begins.

So we need to know how many elements each process is contributing, and we need to know the offset in `list` of the first element of each process' contribution.

Now recall that each process has built two arrays that are used in the call to `MPI_Alltoallv`. The first one specifies the number of elements the process receives from every process, and the second one specifies the offset (or displacement) of the first element received from each process. If we call these arrays `my_fr_counts` and `my_fr_offsets`, respectively, we can get the total number of elements received from each process by adding the last elements

```
my_new_count  =   my_fr_counts[p−1]  +  my_fr_offsets[p−1];
```

We can "gather" these counts onto process 0 to get the elements of recv_counts, and we can form the exclusive prefix sums of recv_counts to get recv_displs.

To summarize, then, Program 7.5 shows pseudocode for this version of sample sort.

```
/* s = global sample size */
loc_s = s/p;
Gen_sample(my_rank, loc_list, loc_n, loc_samp, loc_s);
Gather_to_0(loc_samp, global_sample);
if (my_rank == 0) Find_splitters(global_sample, splitters);
Broadcast_from_0(splitters);

/* Each process has all the splitters */
Sort(loc_list, loc_n);
/* my_to_counts has storage for p ints */
Count_elts_going_to_procs(
       loc_list, splitters, my_to_counts);

/* Get number of elements coming from each process
 *          in my_fr_counts */
MPI_Alltoall(my_to_counts, 1, MPI_INT,
       my_fr_counts, 1, MPI_INT, comm);

/* Construct displs or offset arrays for each process */
Excl_prefix_sums(my_to_counts, my_to_offsets, p);
Excl_prefix_sums(my_fr_counts, my_fr_offsets, p);

/* Redistribute the elements of the list            */
/* Each process receives its elements into tlist */
MPI_Alltoallv(
       loc_list, my_to_counts, my_to_offsets, MPI_INT,
       tlist, my_fr_counts, my_fr_offsets, MPI_INT, comm);
my_new_count = my_fr_offsets[p-1] + my_fr_counts[p-1];

Sort(tlist, my_new_count);

/* Gather tlists to process 0 */
MPI_Gather(&my_new_count, 1, MPI_INT, bkt_counts, 1,
       MPI_INT, 0, comm);
if (my_rank == 0)
       Excl_prefix_sums(bkt_counts, bkt_offsets, p);
MPI_Gatherv(tlist, my_new_count, MPI_INT, list, bkt_counts,
       bkt_offsets, MPI_INT, 0, comm);
```

Program 7.5: Pseudocode for first MPI implementation of sample sort.

Table 7.14 Run-times of first
MPI implementation of sample sort
(times are in seconds).

Processes	Run-time
1	9.94e-1
2	4.77e-1
4	2.53e-1
8	1.32e-1
16	7.04e-2
32	5.98e-2
qsort	7.39e-1

Table 7.14 shows run-times of this implementation in seconds. The input list has $2^{22} = 4,194,304$ **int**s, and the sample has 16,384 **int**s. The last row of the table shows the run-time of the C library qsort function. We're using a system running Linux with two Xeon Silvers 4116 running at 2.1 GHz. Each of the Xeons has 24 cores, and the MPI implementation is MPICH 3.0.4. Compiler optimization made little difference in overall run-times. So we're reporting run-times with no optimization.

Second implementation

Our first MPI implementation seems to be fairly good. Except for the 32 core run, the efficiency was at least 0.88, and with 2 or more cores, it is always faster than the C library qsort function. So we would like to improve the efficiency when we're using 32 cores.

There are several modifications we can make to the code that may improve performance:

1. Replace the serial sort of the sample with a parallel sort.
2. Use a parallel algorithm to find the splitters.
3. Write our own implementation of the Alltoallv function.

The first two modifications are straightforward. For the first, we can replace the gather of the sample to process zero and the serial sort of the sample, by a parallel odd-even transposition sort. (See Section 3.7.)

After the parallel sort of the sample, the sample will be distributed among the processes. So there's no need to scatter the sample, and the keys on process $q - 1$ are all $\le$ the keys on process q, for $q = 1, 2, \ldots, p - 1$. Thus, since there are p splitters, we can compute splitter[i] by sending the maximum element of the sample on each process to the next higher-ranked process (if it exists), and the next higher-ranked process can average this value with its minimum value. (See Fig. 7.9.)

So let's take a look at implementing Alltoallv. When p is a power of 2, we can implement this with a butterfly structure (Fig. 3.9). First we'll use a small example to

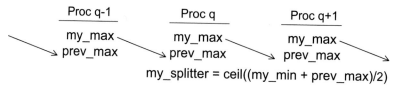

FIGURE 7.9

Finding the splitters.

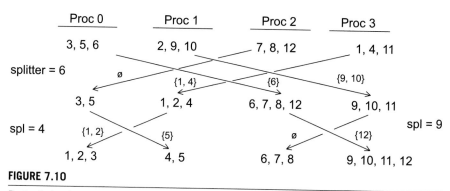

FIGURE 7.10

Butterfly implementation of `Alltoallv`.

illustrate how this is done. We'll then talk about the general case. So suppose we have four processes, and the splitters are $\{-\infty, 4, 6, 9, \infty\}$. Also suppose that the sublists are shown in the following table:

		Processes		
	0	**1**	**2**	**3**
sublists	3, 5, 6	2, 9, 10	7, 8, 12	1, 4, 11

Then a butterfly implementation of `Alltoallv` can be illustrated as shown in Fig. 7.10.

The algorithm begins with the sorted sublists distributed among the processors. The first stage uses the "middle" splitter, `splitter[2]` = 6 to divide the keys. Keys that are less than 6 will go to the first half of the distributed list, and keys greater than or equal to 6 will go to the second half of the distributed list. Processes 0 and 2 exchange keys, and processes 1 and 3 exchange keys. After the exchanges take place what remains of the old sublist is merged with the received keys.

Now that the distributed list has been split into two halves, there is no need for further communication between the processes that are less than $p/2$ and the processes that are greater than or equal to $p/2$. So we can now split the first half of the list between processes 0 and 1, and we can split the second half of the list between processes 2 and 3.

To split the first half of the list among processes 0 and 1, we'll use splitter[1] = 4, and to split the second half, we'll use splitter[3] = 9. As in the first "phase," the paired processes (0–1 and 2–3) will exchange keys, and each process will merge its remaining keys with its new keys.

This gives us a distributed, sorted list, and we can use MPI_Gather and MPI_Gatherv to gather the list to process 0: this is the same as the final step in the algorithm we used for the first MPI implementation.

So let's take a look at some of the details in the implementation of Alltoallv. (See Program 7.6.)

```
1    unsigned bitmask = which_splitter = p >> 1;
2
3    while (bitmask >= 1) {
4        partner = my_rank ^ bitmask;
5        if (my_rank < partner) {
6            Get_send_args(loc_list, loc_n, &offset, &count,
7                splitter[which_splitter], SEND_TO_UPPER);
8            new_loc_n = offset
9            bitmask >>= 1;
10           which_splitter -= bitmask /* splitter for next pass */
11       } else {
12           Get_send_args(loc_list, loc_n, &offset, &count,
13               splitter[which_splitter], SEND_TO_LOWER);
14           new_loc_n = loc_n - count;
15           bitmask >>= 1;
16           which_splitter += bitmask;
17       }
18       MPI_Sendrecv(loc_list + offset, count, MPI_INT, partner,
19               0, rcv_buf, n, MPI_INT, partner, 0, comm, &status);
20       MPI_Get_count(&status, MPI_INT, &rcv_count);
21
22       loc_n = new_loc_n;
23       if (my_rank < partner) {
24           offset = 0;
25       } else {
26           offset = count;
27       }
28
29       /* Merges loc_list and rcv_buf into tmp_buf and swaps
30        *      loc_list and tmp_buf */
31       Merge(&loc_list, &loc_n, offset,
32               rcv_buf, rcv_count, &tmp_buf);
33   } /* while bitmask >= 1 */
```

Program 7.6: Pseudocode for the butterfly algorithm in the second MPI sample sort.

The key to understanding how it works is to look at the binary or base 2 representations of the various scalar values. The variable bitmask determines which processes are paired: process my_rank is paired with the process with rank

```
partner = my_rank ^ bitmask;
```

(Recall that ^ takes the bitwise exclusive or of its operands.) The initial value of bitmask is p >> 1, and after each phase of the butterfly, it's updated by a right shift bitmask >= 1. When it's equal to 0, the iterations stop.

In our example, when $p = 4 = 100_2$, bitmask will take on the values 10_2, 01_2, and 00_2. The following table shows which processes are paired:

bitmask	Processes			
	$0 = 00_2$	$1 = 01_2$	$2 = 10_2$	$3 = 11_2$
10_2	$2 = 10_2$	$3 = 11_2$	$0 = 00_2$	$0 = 01_2$
01_2	$1 = 01_2$	$0 = 00_2$	$3 = 11_2$	$2 = 10_2$

Note that all the processes will have the same value of bitmask.

The variable which_splitter is the subscript of the element in the splitter array that's used to determine which elements should be sent from one process to its partner. So, in general, the processes will have different values of which_splitter. It is also initialized to p >> 1, but it is updated using its value in the previous iteration and the updated bitmask. If the process' current partner has a higher rank, then, in the next iteration, it should choose the splitter

```
which_splitter = which_splitter - bitmask
```

If the partner has a lower rank, it should choose the splitter

```
which_splitter = which_splitter + bitmask
```

The following table shows the values of which_splitter for each process when $p = 8$. (bitm is an abbreviation of bitmask.)

New bitm	Processes							
	$0 = 000_2$	$1 = 001_2$	$2 = 010_2$	$3 = 011_2$	$4 = 100_2$	$5 = 101_2$	$6 = 110_2$	$7 = 111_2$
XXX	$4 = 100_2$	$4 = 100_2$	$4 = 100_2$	$4 = 100_2$	$4 = 100_2$	$4 = 100_2$	$4 = 100_2$	$4 = 100_2$
010_2	$2 = 010_2$	$2 = 010_2$	$2 = 010_2$	$2 = 010_2$	$6 = 110_2$	$6 = 110_2$	$6 = 110_2$	$6 = 110_2$
001_2	$1 = 001_2$	$1 = 001_2$	$3 = 011_2$	$3 = 011_2$	$5 = 101_2$	$5 = 101_2$	$7 = 111_2$	$7 = 111_2$

The function Get_send_args determines the arguments offset and count that are used in the call to MPI_Sendrecv. The variable offset gives the subscript of the first element in loc_list that will be sent to the partner, and the variable count is the number of elements that will be sent. They are determined by a search of the local list for the location of the splitter. This can be carried out with a linear or a binary search. We used linear search.

After the call to MPI_Sendrecv, we use the status argument and a call to MPI_Get_count to determine the number of elements that were received, and then call a merge function that merges the old local list and the received list into `tmp_buf`. At the end of the merge, the new list is swapped with the old `loc_list`.

An important difference between the implementation that uses `MPI_Alltoallv` and this implementation is that to use `MPI_Alltoallv`, we must determine *before* the call

- How many elements go *from* each process *to* every other process, and
- The offsets or displacements of the first element going from each process to every other process.

As we've already seen, this requires communication, so it's relatively time consuming. With the hand-coded implementation, we can determine these values as the algorithm proceeds, and the *amount* of the data is, effectively, included in the communication of the actual data. We can access the amount by using the status and the (local) MPI function `MPI_Get_count` after receiving the data.

In the approach outlined above for implementing the data exchange, we use `MPI_Sendrecv` and a receive buffer that's large enough to receive any list (e.g., an *n*-element buffer). A more memory-efficient implementation can split the send and the receive, and before the receive is executed, it can call `MPI_Probe`:

```
int MPI_Probe(
        int            source     /* in  */,
        int            tag        /* in  */,
        MPI_Comm       comm       /* in  */,
        MPI_Status*    status_p   /* out */);
```

This function will block until it is notified that there is a message from process `source` with the tag `tag` that has been sent with communicator `comm`. When the function returns, the fields referred to by `status_p` will be initialized, and the calling process can use `MPI_Get_count` to determine the size of the incoming message. With this information, the process can, if necessary, call `realloc` to enlarge the receive buffer before calling `MPI_Recv`.

To ensure safety in the MPI sense (see Section 3.7.3) of this pair of communications, we can use a **nonblocking** send instead of the standard send. The syntax is

```
int MPI_Isend(
        void*          buf        /* in  */,
        int            count      /* in  */,
        MPI_Datatype   datatype   /* in  */,
        int            dest       /* in  */,
        int            tag        /* in  */,
        MPI_Comm       comm       /* in  */,
        MPI_Request*   request_p  /* out */);
```

The first six arguments are the same as the arguments to `MPI_Send`. However, unlike `MPI_Send`, `MPI_Isend` only *starts* the communication. In other words, it notifies the system that it should send a message to `dest`. After notifying the system, `MPI_Isend`

Table 7.15 Run-times of the two MPI implementation of sample sort (times are in seconds).

Processes	Version 1 Run-time	Version 2 Run-time
1	9.94e-1	8.39e-1
2	4.77e-1	4.03e-1
4	2.53e-1	2.10e-1
8	1.32e-1	1.10e-1
16	7.04e-2	6.36e-2
32	5.98e-2	4.67e-2
qsort	7.39e-1	7.39e-1

returns: the "I" stands for "immediate," since the function returns (more or less) immediately. Because of this, the user code *cannot* modify the send buffer `buf` until after the communication has been explicitly *completed*.

There are several MPI functions that can be used for completing a nonblocking operation, but the only one we'll need is `MPI_Wait`:

```
int MPI_Wait(
        MPI_Request*    request_p    /* in  */,
        MPI_Status*     status_p     /* out */);
```

The `request_p` argument is the argument that was returned by the call to `MPI_Isend`. It is an opaque object that is used by the system to identify the operation. When our program calls `MPI_Wait`, the program blocks until the send is completed. For this reason, nonblocking communications are sometimes called "split" or "two-phase" communications. We won't make use of the `status_p` argument, so we can pass in `MPI_STATUS_IGNORE`.

To summarize, if we want to minimize memory use, we can replace the call to `MPI_Sendrecv` in Program 7.6 with the function shown in Program 7.7. The function

```
int MPI_Abort(
        MPI_Comm    comm         /* in  */,
        int         errorcode    /* in  */);
```

will terminate all the processes in `comm` and return `errorcode` to the invoking environment. So we need to define the constant `REALLOC_FAILED`. Note that current implementations of MPI will terminate *all* processes, not just those in `comm`.

Let's take a look at the performance of our the version of sample sort that uses a hand-coded butterfly implementation of `Alltoallv` and `MPI_Sendrecv` (with large buffers). Table 7.15 shows run-times of this version and our original version of sample sort that uses `MPI_Alltoallv`. All the times were taken on the same system (see page 423). Both versions used no compiler optimization, and the input data was identical. We see that the second version is always faster than the first, and it is usually more efficient. The sole exception is that with $p = 16$, the first version has an ef-

```
 1   void Send_recv(int snd_buf[], int count, int** rcv_buf_p,
 2         int* rcv_buf_sz_p, int partner, MPI_Comm comm) {
 3      int rcv_count, *rcv_ptr;
 4      MPI_Status status;
 5      MPI_Request req;
 6
 7      /* Start the send */
 8      MPI_Isend(snd_buf, count, MPI_INT, partner, 0, comm, &req);
 9
10      /* Wait for info on storage needed for rcv_buf */
11      MPI_Probe(partner, 0, comm, &status);
12      MPI_Get_count(&status, MPI_INT, &rcv_count);
13      if (rcv_count > *rcv_buf_sz_p) {
14         rcv_ptr = realloc(*rcv_buf_p, rcv_count);
15         if (rcv_ptr == NULL)
16            /* Call to realloc failed, quit */
17            MPI_Abort(comm, REALLOC_FAILED);
18         else
19            *rcv_buf_p = rcv_ptr;
20      }
21
22      /* Now do the receive */
23      *rcv_buf_sz = rcv_count;
24      MPI_Recv(*rcv_buf_p, rcv_count, MPI_INT,
25            partner, 0, comm, &status);
26
27      /* Complete the send */
28      MPI_Wait(&req, MPI_STATUS_IGNORE);
29   } /* Send_recv */
```

Program 7.7: Send_recv function for minimizing memory requirements.

ficiency of 0.88, while the second has an efficiency of 0.82. This is because the run-time of the first version is so much greater. Note that neither version has a very high efficiency with $p = 32$: the first version has an efficiency of 0.52 and the second 0.56. So we'll look at a couple of further improvements to the second version in Exercise 7.22.

7.2.9 Implementing sample sort with CUDA

As usual, we'll look at two implementations. The first splits the implementation between the host and the device. The second has the same basic structure as our second serial implementation. Unlike the first implementation, however, the entire algorithm, except for synchronization, is executed on the device.

First implementation

For the first implementation, we'll avoid issues with interblock synchronization by restricting ourselves to a single thread block. We'll start the implementation by using the host to choose the sample, sort the sample, and choose the splitters. Then a kernel will build the data structures needed to assign elements to the correct buckets. As with the MIMD implementations, each thread will do the work associated with one sublist and one bucket.

The host code is identical to the first serial implementation: it uses a random number generator to choose the sample, the qsort library function to sort the sample, and a simple **for** loop to choose the splitters, which are computed from equally spaced elements of the sample.

The device code starts by finding out the number of elements from the list that will be mapped to each bucket. These numbers are stored in an array counts. Then we implement parallel exclusive prefix sums of the elements of counts. This will tell us the subscript of the first element of each thread's bucket. When we have the prefix sums, each thread can copy the elements of its bucket to contiguous locations in a temporary array. The device code finishes with each thread sorting its bucket and copying the elements of the bucket back to the original list. (See Program 7.8.)

Let's take a look at some of the details. First, the list being sorted is the argument list, so it has n elements. As in our other examples, splitters has $b + 1$ or th_per_blk+1 elements, since we explicitly store MINUS_INFTY and INFTY as the first and last elements, respectively. The counts array also has storage for th_per_blk+1 elements. This will allow us to find the first and last elements in a bucket without treating the last bucket as a special case.

We'll use the shared memory array shmem to store and sort the individual buckets. It will be partitioned into subarrays: the ith subarray will store the ith bucket, for $i = 0, 1, \ldots, b - 1$. This will limit the number of elements we can sort. For example, in our system, the maximum amount of shared memory available to a thread block is 49,152 bytes or $49,152/4 = 12,288$ **ints**.

Note that after most of the function calls, we need a barrier (i.e., a call to _syncthreads), because a subsequent function call will have a thread that is making use of the output from the earlier call that may have been computed by another thread. For example, the output of Find_counts is the array counts, and counts[i] is computed by the thread with my_rank = i, for $0 \le i \le$ th_per_blk. But in the call to Excl_prefix_sums, a different thread will, in general, make use of this entry. So there would be a race condition if there weren't a barrier between the two calls.

Except for Excl_prefix_sums, the device functions called in the kernel are pretty straightforward to implement. In both Find_count and Copy_to_my_bkt, each thread iterates through all the elements of list. In Find_counts, each thread counts the number of elements belonging to its bucket. This information is used to determine both the locations and sizes of the buckets. In Copy_to_my_bkt, each thread copies the elements belonging to its bucket into the appropriate subarray of shmem.

```
1   __global__ void Dev_ssort(int list[], int splitters[],
2          int counts[], int n) {
3     int my_rank = threadIdx.x;    // Only one thread block
4     int th_per_blk = blockDim.x;
5     int my_first, my_last, my_count, *my_bkt;
6     __shared__ int shmem[MAX_SHMEM];    // n <= MAX_SHMEM
7
8     Find_counts(list, n, splitters, counts,
9            th_per_blk, my_rank);
10    __syncthreads();
11
12    my_count = counts[my_rank];
13    // Now counts has number of elements going to each bucket
14    // counts has th_per_blk+1 elements.
15    // Last element is uninitialized
16    Excl_prefix_sums(counts, th_per_blk+1, my_rank);
17    if (my_rank == 0) counts[th_per_blk] = n;
18    __syncthreads();
19
20    int* my_bkt = shmem + counts[my_rank];
21    Copy_to_my_bkt(list, n, splitters,
22           my_bkt, my_count, my_rank);
23
24    Serial_sort(my_bkt, my_count);
25    __syncthreads();
26
27    my_first = counts[my_rank];
28    my_last = counts[my_rank+1];
29    Copy_fr_my_bkt(list, my_bkt, my_rank, counts[my_rank],
30           counts[my_rank+1]);
31 }  /*  Dev_ssort */
```

Program 7.8: Kernel code for the first CUDA implementation of sample sort.

The Serial_sort function can be any single-threaded sort. We used heap sort. Finally, in Copy_fr_my_bkt, each thread copies its sorted bucket back to the appropriate subarray of list.

Prefix sums

Recall that when we were discussing implementations of serial sample sort, we talked about prefix sums. (See paragraph 'Prefix sums' on p. 404.) To complete our implementation of this sample sort, we can use *exclusive* prefix sums of the elements of the counts array to determine *where* each bucket should be located in the shmem array. Recall that if x is a *b*-element array, then the exclusive prefix sums of x are the sums

$$0, \ x[0], \ x[0]+x[1], \ \dots \ , \ x[0] \ + \ x[1] \ + \ \dots \ + \ x[b-2].$$

So a serial algorithm for computing exclusive prefix sums is

```
ps[0] = 0;
for (i = 1; i < b; i++)
    ps[i] = ps[i-1] + x[i-1];
```

As in the serial implementation, the exclusive prefix sums tell us where each bucket begins. This can be done "in place," i.e., the prefix sums are stored in x with the following algorithm:

```
told = x[0];
x[0] = 0;
for (i = 1; i < b; i++) {
    tnew = x[i];
    x[i] = told + x[i-1];
    told = tnew;
}
```

Since we were originally interested in serial prefix sums, this was all we needed. In fact, given the dependence of the calculation of the ith prefix sum on the $i - 1$st, it might at first seem that this is inherently serial. However, there is a relatively simple parallel algorithm that is also a good deal faster than the serial algorithm:

```
if (my_rank == 0)
    tmp = my_rank;
else
    tmp = x[my_rank-1];
__syncthreads();
x[my_rank] = tmp;
__syncthreads();

for (shift = 1; shift < b; shift <<= 1) {
    if (my_rank >= shift) tmp = x[my_rank-shift];
    __syncthreads();
    if (my_rank >= shift) x[my_rank] += tmp;
    __syncthreads();
}
```

For example, suppose x has six elements that are initialized as shown in Table 7.16. Note that since the thread block can have multiple warps, we need the calls to __syncthreads. Without them one thread can "race ahead" of the other threads and use results from an earlier iteration.

This is not the most efficient parallel implementation, but it's easy to understand and code. See, for example, [27] for a more efficient implementation.

Performance

How does our CUDA sample sort perform? Table 7.17 shows results taken on an Nvidia system with an Intel Xeon Silver 4116 host running at 2.1 GHz. The devices are Pascal GPUs with a maximum clock rate of 1.73 GHz, 20 SMs, and 128 SPs

Table 7.16 Computation of exclusive prefix sums using the parallel algorithm.

	Subscripts in x					
Time	**0**	**1**	**2**	**3**	**4**	**5**
Start	3	6	7	5	3	5
Initialization	0	3	6	7	5	3
shift = 1	0	3	9	13	12	8
shift = 2	0	3	9	16	21	21
shift = 4	0	3	9	16	21	24

Table 7.17 First version of CUDA sample sort, with $n = 12,288$, $s = 1024$. Times are in seconds.

Threads	Run-times
32	7.48e-3
64	6.25e-3
128	5.14e-3
256	5.18e-3
512	5.19e-3
1024	5.34e-3
qsort	2.69e-3

per SM. For the performance tests, $n = 12,288$, $s = 1024$, the sample sort was run at least 30 times with each set of input, and minimum run-times are reported. The briefest examination shows that the performance is *not* very good. The C library qsort function running on a single core of the host is almost twice as fast as the fastest run of the sample sort. Also, since the program is *slower* with 512 and 1024 threads than it is with 256, the program isn't at all scalable.

Table 7.18 shows the fastest run-times of the program for a given sample size. For each sample size, we ran the program with the number of threads ranging from 64 up to the sample size, except for sample size 2048, where we only ran up to 1024 threads, since the maximum number of threads that can be run on a single block in our system is 1024. These data show that for larger sample sizes, the host code (choosing the sample, sorting it, and finding the splitters) formed a huge part of the overall run-time. It seems clear that if the sample size is large relative to the list size, then our method for choosing the sample is *far* too slow. The data also show that the time spent on the device decreases as the sample size increases. There two possible reasons for this: as the sample size increases, on average, the amount of work per thread decreases. Also, as the sample size increases, it's probable that the buckets will become more uniform in size, and the total work per thread will decrease.

Finally, observe that when two SIMD threads run a serial sorting algorithm, such as heap sort, on two different lists, we would expect that there would be large amounts

Table 7.18 First version of CUDA sample sort. Fastest times for various sample sizes. Times are in milliseconds.

Sample Size	Number of Threads	Host Time	Device Time	Total time
128	128	0.00572	3.29	3.30
256	256	0.176	3.02	3.20
512	512	0.614	2.95	3.56
1024	256	2.93	2.25	5.18
2048	512	8.93	1.27	10.2

Table 7.19 Run-times for 32 CUDA threads running heap sort. Times are in milliseconds.

Number of Thread Blocks	Run-time
1	1.16
2	1.05
4	1.04
8	1.02
16	0.814
32	0.764
qsort	1.82

of thread divergence, and hence very little parallelism. We looked at this by timing a CUDA program in which a list with n elements is shared among a collection of threads, and each thread must sort its sublist using heap sort. We sorted each list using 32 CUDA threads. However, we varied the number of thread blocks. When there is a single block of 32 threads, all the threads are in the same warp, and hence they are SIMD threads. At the other extreme, each thread belonged to its own thread block, and to all intents and purposes, they are MIMD threads. Table 7.19 shows the results when $n = 12,288$, and the program is run on the same Pascal system used for the sample sort. The C library function qsort was run on a core of the host system on the same list. So we see that even with 32 SIMD threads, the CUDA heap sort is faster than the C library qsort. Of more interest is that the speedup of 32 "MIMD" threads—i.e., 1 thread/block—is better than 1.5. So for this problem, thread divergence increased the run-time by more than 50%.

We'll make use of our observations on the performance of the first CUDA sample sort in the design of our second CUDA implementation.

A second CUDA implementation of sample sort

It is easy to see how we might make a number of improvements to the first CUDA implementation. For example, the data in Table 7.18 show that for large sample sizes, the overall run-time of the program is dominated by the time spent in the serial code:

choosing the sample, sorting it, and finding the splitters. Clearly, then, we should use algorithms that parallelize these steps.

Another obvious issue is that the first implementation can't sort a list with more than 12,288 elements, and it may be very difficult to outperform a good serial code on lists this small. So we should write a program that will sort relatively large lists, and to bring the full power of the GPU to bear, we should write a program that makes good use of many, large thread blocks.

A final issue with the first implementation is that the threads aren't making use of the power of SIMD execution. When each thread is running a complex algorithm with many conditional branches—e.g., an algorithm like heap sort—we have little reason to think that the threads will be executing the same instruction at the same time. On a MIMD system, this is generally not a problem, but the ideal SIMD application has all of the threads executing the same instruction simultaneously. When threads need to take different branches, other threads are necessarily idled, and their computational power is wasted.

Several researchers have developed implementations of sample sort for GPUs. Our solution is a simplified version of the implementation discussed in [13]. In this implementation, each bucket corresponds to a thread block. Here's an outline; we'll go into more detail shortly.

1. **Choose the sample.** Each thread block operates independently in choosing the sample. First each block's sublist is copied into shared memory and sorted using a slightly modified version of our first bitonic sort algorithm (see Exercise 6.12). Then threads choose equally spaced elements from the sorted sublist. These are copied into a `sample` array in global memory.
2. **Sort the sample.** The global sample is sorted using our second bitonic sort algorithm (see Section 6.14.4).
3. **Choose the splitters.** A collection of threads chooses equally spaced splitters from `sample`.
4. **Determine destination buckets.** For each element of the list, determine its destination bucket by using a binary search of the splitters. The number of elements going from sublist j to bucket i is stored in `mat_counts[i][j]`. After determining the destination bucket of an element, we use `atomicAdd` to increment the appropriate entry in `mat_counts`.
5. **Parallel prefix sums.** Form the exclusive prefix sums of the rows of `mat_counts`. After the completion of the prefix sums, `mat_counts[i][j]` will be the index in sublist j of its first element going to bucket i. This step also determines the total number of elements going to each bucket i, which is the sum of the elements in row i of the original `mat_counts`.
6. **Serial prefix sums.** To determine the location of the start of each bucket in a temporary array, we form the exclusive prefix sums of the bucket sizes. Unlike the preceding step, this is just a serial calculation.
7. **Send elements to buckets.** Now we map the elements of the list to their destination buckets in a temporary array.

8. **Sort buckets.** To finish up, we want to sort each bucket. However, to use the single block bitonic sort, each bucket needs to store 2^k elements, for some nonnegative integer k. To arrange this, we copy each bucket into temporary storage, pad it with copies of ∞, sort the padded list, and copy the elements of the sorted, padded list that are $< \infty$, back into the original list.

We specify `blk_ct`, `th_per_blk`, the list size `n`, and the sample size `s`, on the command-line. We make the following assumptions:

- All four of the command-line arguments are powers of 2.
- The number of elements in the list `n = 2*blk_ct*th_per_blk`. This is the assumption that's made on the relation between these values in our implementation of bitonic sort (see Section 6.14.4).
- `s < n`. If `s = n`, there's no point in using sample sort: we may as well use bitonic sort for the entire list.
- `blk_ct <= s`. The number of splitters is the same as the number of buckets, which is the same as the number of thread blocks. Since we choose the splitters from the sample, and we need one splitter for each thread block, we can't have more splitters than sample elements.

Note that except for the serial prefix sums (Step 6), all of the steps are parallel, but unlike our other CUDA programs, the different steps may use differing numbers of thread blocks and differing numbers of threads per block.

Let's take a look at some of the details.

1. **Choose the sample.** The number of blocks and the number of threads per block are the same as the command-line arguments.
2. **Sort the sample.** Since we're using our bitonic sort to sort the sample, we want

$$s = 2 \times \texttt{blk\_ct} \times \texttt{th\_per\_blk}$$

So the number of threads per block in this kernel is

$$\texttt{sth\_per\_blk} = \min\{s/2, \texttt{th\_per\_blk}\},$$

and the number of blocks is

$$\texttt{sblk\_ct} = \frac{s}{2 \times \texttt{sth\_per\_blk}}.$$

3. **Choose the splitters.** We use one thread to compute one splitter by simply finding the gap between splitters and multiplying by the thread's global rank. This is ideal for SIMD parallelism. So we use as many threads per block as possible. This is a constant defined in the program, `MAX_TH_PER_BLK`. (On current generation systems, it's 1024.) In case there are more splitters than this, we define the number of blocks to be

$$\texttt{spl\_blk\_ct} = \texttt{ceil}(\texttt{blk\_ct}/\texttt{MAX\_TH\_PER\_BLK})$$

4. **Determine destination buckets.** This uses the `blk_ct` blocks and `th_per_blk` threads per block, so each thread determines the destination buckets of two elements of the list.
5. **Parallel prefix sums.** Each bucket requires a prefix sum, and each row has `blk_ct` elements. So, in general, we want one thread block for each row, and `blk_ct` threads per row to carry out a parallel prefix sum. However, if there are more blocks than `MAX_TH_PER_BLK`, then the number of threads per block must be set to `MAX_TH_PER_BLK` instead of `blk_ct`.

   The algorithm is a modified version of an efficient prefix sum described in [27]. Note that we want to compute the subscript of the first element in each sublist that's being mapped to a given bucket. So using rows that correspond to a bucket, rather than the transposed array with rows that correspond to a sublist, results in a more efficient use of cache: we are using contiguous entries in a row rather than strided elements in a column.
6. **Serial prefix sums.** If `blk_ct` ≤ `MAX_TH_PER_BLK`, this could be implemented with a single thread block, but in the interests of simplifying the program we opted for a serial prefix sum.
7. **Send elements to buckets.** This uses the command-line arguments `blk_ct` and `th_per_blk`, so each thread is responsible for two elements. Given the kth element of the current sublist, we determine the subscript of the first element in the sublist that maps to the same bucket, `first_elt_sub`. The difference

$$k - \texttt{first\_elt\_sub}$$

   gives the offset of the element into the destination bucket. The previously calculated prefix sums allow us to determine the subscript of the first element in the destination bucket. Adding the offset and the start of the bucket gives the subscript of the destination of the element.
8. **Sort buckets.** Here, we want to use a single thread block to sort each bucket using the single-block implementation of bitonic sort. However, we expect that in general the buckets will have more than $2 \times$`th_per_blk` elements. So we may be able to make use of large thread blocks. When this is the case, we use `MAX_TH_PER_BLK` instead of the command line value `th_per_blk`. Note that the single thread block bitonic sort has been modified so that it can handle arbitrary powers of 2 for both the number of threads and the number of elements being sorted.

   An additional issue is the possibility that the size of the list being sorted by a thread block is larger than the maximum available storage for shared memory. This depends both on the size of the bucket being stored in the list and the padding that may be required to adjust the list size so that it's a power of two. So before calling the kernel that sorts each thread block, we find the padded size of each bucket and allocate storage in global memory for these buckets. Before the kernel calls the bitonic sort device function, it checks whether it needs to use global memory, and if it does, the padded bucket is copied into global memory. If it doesn't, the padded bucket is copied into shared memory.

Table 7.20 Average run-times and maximum bucket sizes for second CUDA sample sort. Times are in seconds.

blk_ct	th_per_blk	n	s	Run-time	qsort	Max Bkt Sz
256	256	131,072	16,384	1.58e-3	2.90e-2	2048
512	256	262,144	32,768	4.47e-3	5.49e-2	4096
256	512	262,144	32,768	2.14e-3	5.49e-2	2048
1024	256	524,288	65,536	7.30e-3	1.22e-1	8192
512	512	524,288	65,536	5.28e-3	1.22e-1	4096
256	1024	524,288	65,536	3.28e-3	1.22e-1	4096
1024	512	1,048,576	131,072	9.53e-3	2.46e-1	8192
512	1024	1,048,576	131,072	6.25e-3	2.46e-1	4096
1024	1024	2,097,152	262,144	1.42e-2	5.20e-1	8192

Performance of the second CUDA sample sort

We used the same Pascal-based system that we used for the first implementation. (See Section 'Performance' on p. 432.) Since there was considerable variation in the run-times for the individual executions, we're reporting *average* run-times over 30 runs. The sample size didn't seem to make a very large difference in the overall run-times, so we are only reporting on sample size = (list size)/8. (See Table 7.20.) The column labeled "qsort" is the run-time (in seconds) of the C library qsort function. It is run on the same list that the sample sort was run on, but it runs on a single core of the Intel Xeon host. The column labeled "Max Bkt Sz" is the size of the largest bucket after it has been padded with elements to make its size a power of two. For all of the problems in the table, the CUDA sample sort is more than 15 times faster than the qsort library functions. However, for very small problems ($n < 4096$), the qsort library function *was* faster. Sample sizes that were between 4 and 16 times smaller than the list size didn't make much difference in the overall run-times. However, sample sizes that were much smaller than a factor of 16 less than the list size could make a large difference. For example, with 1024 blocks, 1024 threads per block, a list of size 2,097,152, and a sample size of 16,384, the program was 40% slower than the program with the same input, but a sample of size 262,144. For large lists, sample size also had a noticeable effect on the maximum bucket size. Fortunately, the use of global memory for temporary storage of the buckets didn't seem to have a very significant effect on the overall run-time.

Perhaps, the most striking feature of the table is the decrease in run-times when the number of blocks was decreased, but the number of threads per block was increased. For example, with 262,144 threads and a list containing 524,288 ints, using 256 blocks of 1024 threads was more than twice as fast as using 1024 blocks of 256 threads. If we time the various kernels called by the sample sort, it's clear that most of the difference is due to the increased cost of the kernel that builds the matrix of counts. With 256 blocks of 1024 threads, we need to compute $256 \times 256 = 65,536$ entries, while with 1024 blocks of 256 threads, we need to compute $1024 \times 1024 =$

1,048,576 entries, and there is relatively little difference in the average cost of computing an entry. (See Exercise 7.24.)

7.3 A word of caution

In developing our solutions to the n-body problem and the problem of sorting using sample sort, we chose our serial algorithms because they were easy to understand, and their parallelizations were relatively straightforward. In no case did we choose a serial algorithm because it was the fastest or because it could solve the largest problem. Thus, it should not be assumed that either the serial or the parallel solutions are the best available. For information on "state-of-the-art" algorithms, see the bibliography, especially [14] for the n-body problem and [2] and [22] for sorting.

7.4 Which API?

How can we decide which API, MPI, Pthreads, OpenMP, or CUDA is best for our application? In general, there are many factors to consider, and the answer may not be at all clear-cut. However, here are a few points to consider.

As a first step, decide whether the application is suitable for use with CUDA. If a serial solution involves using more or less the same operations on different data items, then CUDA is a very good candidate. On the other hand, if a serial solution involves lots of conditional branches that depend on the data, then CUDA may not be a good choice: you may have to rework the serial solution so that it's more amenable to a SIMD parallelization.

As a second step, decide whether to use distributed-memory MIMD or shared-memory MIMD. To do this, first consider the amount of memory the application will need. In general, distributed-memory systems can provide considerably more main memory than shared-memory systems, so if the memory requirements are very large, you may need to use MPI.

If the problem will fit into the main memory of your shared-memory system, you may still want to consider using MPI. Since the total available cache on a distributed-memory system will probably be much greater than that available on a shared-memory system, it's conceivable that a problem that requires lots of main memory accesses on a shared-memory system will mostly access cache on a distributed-memory system, and, consequently, have much better overall performance.

However, even if you'll get a big performance improvement from the large aggregate cache on a distributed-memory system, if you already have a large and complex serial program, it often makes sense to write a shared-memory program. It's often possible to reuse considerably more serial code in a shared-memory program than a distributed-memory program. It's more likely that the serial data structures can be easily adapted to a shared-memory system. If this is the case, the development effort for the shared-memory program will probably be much less. This is especially true for OpenMP programs, since some serial programs can be parallelized by simply inserting some OpenMP directives.

Another consideration is the communication requirements of the parallel algorithm. If the processes/threads do little communication, an MPI program should be fairly easy to develop, and very scalable. At the other extreme, if the processes/threads need to be very closely coordinated, a distributed-memory program will probably have problems scaling to large numbers of processes, and the performance of a shared-memory program should be better.

If you decided that shared-memory is preferable, you will need to think about the details of parallelizing the program. As we noted earlier, if you already have a large, complex serial program, you should see if it lends itself to OpenMP. For example, if large parts of the program can be parallelized with `parallel` **for** directives, OpenMP will be much easier to use than Pthreads. On the other hand, if the program involves complex synchronization among the threads—for example, read-write locks or threads waiting on signals from other threads—then Pthreads will be much easier to use.

7.5 Summary

In this chapter, we've looked at serial and parallel solutions to two very different problems: the n-body problem and sorting using sample sort. In each case we began by studying the problem and looking at serial algorithms. We continued by using Foster's methodology for devising a parallel solution, and then, using the designs developed with Foster's methodology, we implemented parallel solutions using Pthreads, OpenMP, MPI, and CUDA.

In developing the reduced MPI solution to the n-body problem, we determined that the "obvious" solution would be difficult to implement correctly, and it would require a huge amount of communication. We therefore turned to an alternative "ring pass" algorithm, which proved to be much easier to implement and is probably more scalable.

We considered implementing a CUDA n-body solver that used some of the same ideas we used in developing the reduced Pthreads and OpenMP solvers. However, we found that the amount of memory required would be too great. So we looked at an alternative approach to improving the performance of the basic CUDA n-body solver: we reduced the number of accesses to global memory by loading from global memory into fast, on-chip shared memory. Then, when we carried out the calculations, we accessed shared memory rather than global memory. We accomplished this by dividing the arrays into "virtual" tiles. Then each thread block loaded a tile, did all the required calculations involving the particles in that tile, and, when it was done with the particles in the current tile, proceeded to the next tile. Effectively, then, we used shared memory as a programmer-managed cache.

We looked at two serial implementations of sample sort. In the first, we chose the sample using a random number generator, sorted the sample, and chose "splitters" to be equally spaced elements of the sample. Then the elements of the list were mapped to their destination "buckets" and each bucket was sorted.

The second serial implementation used a deterministic scheme to choose the sample, and the contents of the buckets were determined using *exclusive prefix sums*. An *inclusive* prefix sum of the elements of a list $\{a_0, a_1, \ldots, a_{n-1}\}$ is the new list

$$a_0, a_0 + a_1, \ldots, a_0 + a_1 + \cdots + a_{n-1}.$$

An exclusive prefix sum starts with 0. So it is

$$0, a_0, a_0 + a_1, \ldots, a_0 + a_1 + \cdots + a_{n-2}.$$

We parallelized both implementations using data parallelism. For MIMD systems, the "sublists" were contiguous blocks of n/p elements of the original list and the "buckets" were contiguous blocks elements of the final list—which, in general, did not have exactly n/p elements. For CUDA, we used the same terminology, but the definitions were different. In the first implementation, we used a single thread block. So the sublists had $n/$th_per_blk elements and the buckets had approximately $n/$th_per_blk elements. In the second implementation, we used multiple thread blocks, and sublists had $n/$blk_ct elements and buckets had approximately this many elements.

All of the implementations used similar ideas. In general, first implementations did the sampling, and identification of splitters using a single process/thread in the MIMD systems and the host processor with CUDA. The building of data structures, mapping of elements to buckets, and final sorting was done in parallel. The second implementation used parallelism at all stages, except that the OpenMP and Pthreads implementation used a serial sorting algorithm to sort the sample.

In closing, we looked briefly at the problem of deciding which API to use. Our first consideration was whether the problem was suitable for parallelization using CUDA: serial programs that use more or less the same operations on different data items are usually well suited to a CUDA parallelization. The second consideration is whether to use shared memory or distributed memory. To decide this, we should look at the memory requirements of the application and the amount of communication among the processes/threads. If the memory requirements are great or the distributed-memory version can work mainly with cache, then a distributed-memory program is likely to be much faster. On the other hand, if there is considerable communication, a shared-memory program will probably be faster.

When you're choosing between OpenMP and Pthreads, if there's an existing serial program and it can be parallelized by the insertion of OpenMP directives, then OpenMP is probably the clear choice. However, if complex thread synchronization is needed—for example, read-write locks or thread signaling—then Pthreads will be easier to use.

7.5.1 MPI

In the course of developing these programs, we learned about several additional MPI collective communication functions. In the first MPI implementation of sample sort, we recalled a generalization of MPI_Gather, MPI_Gatherv, which allows us to gather different numbers of elements from one process onto another:

```
int MPI_Gatherv(
        void*         loc_list          /* in  */,
        int           loc_list_count    /* in  */,
        MPI_Datatype  send_elt_type     /* in  */,
        void*         list              /* out */,
        const int*    recv_counts       /* in  */,
        const int*    recv_displs       /* in  */,
        MPI_Datatype  recv_elt_type     /* in  */,
        int           root              /* in  */,
        MPI_Comm      comm              /* in  */);
```

This gathers the contents of each process' loc_list into list on the process with rank root root in comm. However, unlike MPI_Gather, the number of elements in loc_list may differ from one process to another. So the argument recv_counts tells us the number of elements each process is sending, and the argument recv_displs tells us where each process' first element goes in list.

We also used two MPI implementations of "all-to-all scatter-gather." These combine the features of both scatter and gather: each process scatters its own collection to all the processes in the communicator, and each process gathers data from all the processes. The first implementation is MPI_Alltoall, which sends the same amount of data from each process to each process. The second implementation is MPI_Alltoallv, which can send different amounts of data to different processes. Their syntaxes are

```
int MPI_Alltoall(
        const void    *send_data        /* in  */,
        int           send_count        /* in  */,
        MPI_Datatype  send_type         /* in  */,
        void          *recv_data        /* out */,
        int           recv_count        /* in  */,
        MPI_Datatype  recv_type         /* in  */,
        MPI_Comm      comm              /* in  */);


int MPI_Alltoallv(
        const void    *send_data        /* in  */,
        const int     *send_counts      /* in  */,
        const int     *send_displs      /* in  */,
        MPI_Datatype  send_type         /* in  */,
        void          *recv_data        /* out */,
        const int     *recv_counts      /* in  */,
        const int     *recv_displs      /* in  */,
```

```
        MPI_Datatype   recv_type      /*  in  */,
        MPI_Comm       comm           /*  in  */);
```

We also learned about the nonblocking send function MPI_Isend:

```
int MPI_Isend(
        void*          buf            /*  in  */,
        int            count          /*  in  */,
        MPI_Datatype   datatype       /*  in  */,
        int            dest           /*  in  */,
        int            tag            /*  in  */,
        MPI_Comm       comm           /*  in  */,
        MPI_Request*   request_p      /*  out */);
```

We used this to implement a memory-efficient send-receive. The first six arguments are the same as the arguments to MPI_Send. When it's called, it *starts* a send, but when it returns, the send will not, in general, be completed. The output argument, request_p, is an opaque object. It's used by the program and MPI system to identify the communication. When we want to finish the communication, we can call the function MPI_Wait:

```
int MPI_Wait(
        MPI_Request*   request_p      /*  in  */,
        MPI_Status*    status_p       /*  out */);
```

When this function is called, it will block until the call associated with request_p (in our case, the call to MPI_Send) has completed. The status_p argument will return information on the call. In our setting we didn't use it. So we just passed in MPI_STATUS_IGNORE.

To determine whether the process with which we were communicating had sent its data, we used the function MPI_Probe.

```
int MPI_Probe{
        int            source         /*  in  */,
        int            tag            /*  in  */,
        MPI_Comm       comm           /*  in  */,
        MPI_Status*    status_p       /*  out */);
```

This blocks until it is notified that there is a message from process source with tag tag in communicator comm. The status_p will return the usual status fields. In particular, we can use the function MPI_Get_count to determine how big the message is, and we used this to be sure sufficient storage was allocated for the receive buffer for the message. If we're unable to allocate sufficient storage for the receive buffer, we can call MPI_Abort:

```
int MPI_Abort(
        MPI_Comm   comm           /*  in */,
        int        errorcode      /*  in */);
```

This is supposed to terminate all the processes in comm and return errorcode to the invoking environment, although most implementations terminate all the processes in MPI_COMM_WORLD.

7.6 Exercises

7.1 In each iteration of the serial *n*-body solver, we first compute the total force on each particle, and then we compute the position and velocity of each particle. Would it be possible to reorganize the calculations so that in each iteration we did all of the calculations for each particle before proceeding to the next particle? That is, could we use the following pseudocode?

```
for each timestep
    for each particle {
        Compute total force on particle;
        Find position and velocity of particle;
        Print position and velocity of particle;
    }
```

If so, what other modifications would we need to make to the solver? If not, why not?

7.2 Run the basic serial *n*-body solver for 1000 timesteps with a stepsize of 0.05, no output, and internally generated initial conditions. Let the number of particles range from 500 to 2000. How does the run-time change as the number of particles increases? Can you extrapolate and predict how many particles the solver could handle if it ran for 24 hours?

7.3 Parallelize the reduced version of the *n*-body solver with OpenMP or Pthreads and a single `critical` directive (OpenMP) or a single mutex (Pthreads) to protect access to the `forces` array. Parallelize the rest of the solver by parallelizing the inner **for** loops. How does the performance of this code compare with the performance of the serial solver? Explain your answer.

7.4 Parallelize the reduced version of the *n*-body solver with OpenMP or Pthreads and a lock/mutex for each particle. The locks/mutexes should be used to protect access to updates to the `forces` array. Parallelize the rest of the solver by parallelizing the inner **for** loops. How does the performance compare with the performance of the serial solver? Explain your answer.

7.5 In the shared-memory MIMD reduced *n*-body solver, if we use a block partition in both phases of the calculation of the forces, the loop in the second phase can be changed so that the **for** `thread` loop only goes up to `my_rank` instead of `thread_count`. That is, the code

```
#   pragma omp for
    for (part = 0; part < n; part++) {
        forces[part][X] = forces[part][Y] = 0.0;
        for (thread = 0; thread < thread_count; thread++){
            forces[part][X] += loc_forces[thread][part][X];
            forces[part][Y] += loc_forces[thread][part][Y];
        }
    }
```

can be changed to

```
#    pragma omp for
     for (part = 0; part < n; part++) {
        forces[part][X] = forces[part][Y] = 0.0;
        for (thread = 0; thread < my_rank; thread++) {
            forces[part][X] += loc_forces[thread][part][X];
            forces[part][Y] += loc_forces[thread][part][Y];
        }
     }
```

Explain why this change is OK. Run the program with this modification and compare its performance with the original code with block partitioning and the code with a cyclic partitioning of the first phase of the forces calculation. What conclusions can you draw?

7.6 In our discussion of the OpenMP implementation of the basic n-body solver, we observed that the implied barrier after the output statement wasn't necessary. We could therefore modify the `single` directive with a `nowait` clause. It's possible to also eliminate the implied barriers at the ends of the two **for** each particle q loops by modifying **for** directives with `nowait` clauses. Would doing this cause any problems? Explain your answer.

7.7 For the shared-memory implementation of the reduced n-body solver, we saw that a cyclic schedule for the computation of the forces outperformed a block schedule, in spite of the reduced cache performance. By experimenting with the OpenMP or the Pthreads implementation, determine the performance of various block-cyclic schedules. Is there an optimal block size for your system?

7.8 If **x** and **y** are double-precision n-dimensional vectors and α is a double-precision scalar, the assignment

$$\mathbf{y} \longleftarrow \alpha \mathbf{x} + \mathbf{y}$$

is called a DAXPY. DAXPY is an abbreviation of "Double precision Alpha times **X** Plus **Y**." Write a Pthreads or OpenMP program in which the master thread generates two large, random n-dimensional arrays and a random scalar, all of which are **double**s. The threads should then carry out a DAXPY on the randomly generated values. For large values of n and various numbers of threads, compare the performance of the program with a block partition and a cyclic partition of the arrays. Which partitioning scheme performs better? Why?

7.9 a. Sketch a table similar to Table 7.1 for a reduced solver using 4 particles and 4 threads. How many entries in the table can be nonzero?

b. Write a general formula for the number of entries in the table that can be nonzero if there are n particles and n threads.

7.10 Write an MPI program in which each process generates a large, initialized, m-dimensional array of **double**s. Your program should then repeatedly call `MPI_Allgather` on the m-dimensional arrays. Compare the performance of the calls to `MPI_Allgather` when the global array (the array that's initialized by the call to `MPI_Allgather`) has

a. a block distribution, and

b. a cyclic distribution.

To use a cyclic distribution, download the code `cyclic_derived.c` from the book's web site, and use the MPI datatype created by this code for the *destination* in the calls to `MPI_Allgather`. For example, we might call

```
MPI_Allgather(sendbuf, m, MPI_DOUBLE, recvbuf, 1,
        cyclic_mpi_t, comm);
```

if the new MPI datatype were called `cyclic_mpi_t`.

Which distribution performs better? Why? Don't include the overhead involved in building the derived datatype.

7.11 Consider the following code:

```
int n, thread_count, i, chunksize;
double x[n], y[n], a;
    . . .
#      pragma omp parallel num_threads(thread_count) \
           default(none) private(i) \
           shared(x, y, a, n, thread_count, chunksize)
       {
#        pragma omp for schedule(static, n/thread_count)
         for (i = 0; i < n; i++) {
             x[i] = f(i);   /* f is a function */
             y[i] = g(i);   /* g is a function */
         }
#        pragma omp for schedule(static, chunksize)
         for (i = 0; i < n; i++)
             y[i] += a*x[i];
       }  /* omp parallel */
```

Suppose $n = 64$, `thread_count = 2`, the cache-line size is 8 **doubles**, and each core has an L2 cache that can store 131,072 **doubles**. If `chunksize = n/thread_count`, how many L2 cache misses do you expect in the second loop? If `chunksize = 8`, how many L2 misses do you expect in the second loop? You can assume that both x and y are aligned on a cache-line boundary. That is, both $x[0]$ and $y[0]$ are the first elements in their respective cache lines.

7.12 Write an MPI program that compares the performance of `MPI_Allgather` using `MPI_IN_PLACE` with the performance of `MPI_Allgather` when each process uses separate send and receive buffers. Which call to `MPI_Allgather` is faster when run with a single process? What if you use multiple processes?

7.13 a. Modify the basic MPI implementation of the *n*-body solver so that it uses a separate array for the local positions. How does its performance compare with the performance of the original *n*-body solver? (Look at performance with I/O turned off.)

b. Modify the basic MPI implementation of the *n*-body solver so that it distributes the masses. What changes need to be made to the communications in the program? How does the performance compare with the original solver?

7.14 Using Fig. 7.6 as a guide, sketch the communications that would be needed in an "obvious" MPI implementation of the reduced n-body solver if there were three processes, six particles, and the solver used a cyclic distribution of the particles.

7.15 Modify the MPI version of the reduced n-body solver so that it uses two calls to `MPI_Sendrecv_replace` for each phase of the ring pass. How does the performance of this implementation compare to the implementation that uses a single call to `MPI_Sendrecv_replace`?

7.16 A common problem in MPI programs is converting global array indexes to local array indexes and vice-versa.
  a. Find a formula for determining a global index from a local index if the array has a block distribution.
  b. Find a formula for determining a local index from a global index if the array has a block distribution.
  c. Find a formula for determining a global index from a local index if the array has a cyclic distribution.
  d. Find a formula for determining a local index from a global index if the array has cyclic distribution.

  You can assume that the number of processes evenly divides the number of elements in the global array. Your solutions should only use basic arithmetic operators ($+$, $-$, $*$, $/$ and remainders). They shouldn't use any loops or branches.

7.17 In our implementation of the reduced n-body solver, we make use of a function `First_index`, which, given a global index of a particle assigned to one process, determines the "next higher" global index of a particle assigned to another process. The input arguments to the function are the following:
  a. The global index of the particle assigned to the first process
  b. The rank of the first process
  c. The rank of the second process
  d. The number of processes

  The return value is the global index of the second particle. The function assumes that the particles have a cyclic distribution among the processes. Write C-code for `First_index`. (*Hint*: Consider two cases: the rank of the first process is less than the rank of the second, and the rank of the first is greater than or equal to the rank of the second.)

7.18 Our CUDA implementations of the n-body solver assume that the number of threads is at least as large as the number of particles. To modify the implementation so that it handles the case in which the number of threads is less than the number of particles, we can use a **grid-stride loop** [26].

  Recall that the **stride** of a loop is the amount by which the loop index is incremented after each iteration:

```
int start = ...;
int stride = ...;
for (int i = start; i < max; i += stride)
```

```
Iteration code depending on i;
```

In a grid-stride loop, the stride is the number of elements in the grid. If both the grid and the thread blocks are one-dimensional, we have

```
/* gridDim.x = number of thread blocks */
int stride = gridDim.x*blockDim.x;
```

In a grid-stride loop, each thread starts at its "global" index:

```
int start = blockIdx.x*blockDim.x;
for (int i = start; i < max; i += stride)
    Iteration code depending on i;
```

In our basic n-body solver, `max` is just the number of particles:

```
for (int i = start; i < n; i += stride)
    Compute total force on particle i;
```

Modify the basic n-body solver so that it uses a grid-stride loop. What are the advantages to using grid-stride loops in this code? Are there any disadvantages? How would you modify the shared-memory n-body solver so that it uses a "grid-stride" loop?

7.19 Write a program that uses a random number generator to choose a sample:

```
chosen = empty set;
for (i = 0; i < s; i++) {
    // Get a random number between 0 and n
    sub = random() % n;
    // Don't choose the same subscript twice
    while (Is_an_element (sub, chosen))
        sub = random() % n;

    // Add sub to chosen elements
    Add(chosen, sub);
    // Add list[sub] to sample
    sample[i] = list[sub]
}
```

Just use an array for `chosen`, add a new subscript to the end of `chosen` and use linear search to implement `Is_an_element`.

What happens to the run-time of your program as the sample size gets large? Can you significantly improve the performance of the program by modifying your implementation of the set ADT `chosen`?

7.20 When we discussed a second algorithm for implementing serial sample sort (Section 7.2.4), we outlined an algorithm for determining the number of elements in a sorted sublist that are mapped to each bucket. Since both the splitters and sublists are sorted, we could have used binary search to implement this algorithm.

Suppose that inputs to the algorithm are n, the total number of elements in the list, b, the number of buckets (and the number of sublists), a sorted list of splitters (as shown in Table 7.9), and a sorted sublist. The output of the algorithm is a list of b ints, `curr_row`, where `curr_row[dest_bkt]` is the number of elements in the sublist that should be mapped to the bucket with index `dest_bkt`. Write pseudocode for algorithms that have these inputs, but use binary search to find the number of elements in the sublist mapped to each bucket.

a. For each element of the sublist this algorithm uses a binary search of the splitters to determine which bucket the element should be mapped to. When it finds the bucket that `sublist[j]` should be mapped to, the element `curr_row[dest_bkt]` should be incremented.

b. For each splitter this algorithm uses a binary search of the sublist to determine the location of the splitter in the sublist. That is, if `splj = splitters[j]` and `spljp1 = splitters[j+1]`, then the algorithm finds the index i of the element of the sublist such that

$$\text{sublist[i-1]} < \text{splj} \le \text{sublist[i]} < \text{spljp1} \le \text{sublist[i+1]}.$$

In other words, it finds the index i of the smallest element of `sublist` such that `splj` $\le$ `sublist[i]`. It uses these indexes to determine the number of elements in the sublist that should be mapped to each bucket. What does your algorithm do if no element of the sublist is between `splj` and `spljp1`?

7.21 In our second OpenMP implementation of sample sort, we use serial quicksort to sort the sample. If we replace this with parallel odd-even transposition sort (Subsection 5.6.2), how does this affect the performance of the program?

7.22 a. How would you modify the binary search in Exercise 7.20b so that it could be used in the second version of the MPI implementation of sample sort?

b. Modify the online code in `mpi−ssort−bfly.c` so that it uses your binary search. Your modification should reduce the range of later searches by making use of the fact that a process has already found the location of some of the other splitters.

c. How does the performance of this modification affect the overall runtime?

7.23 In the kernel code in Program 7.8, give examples showing how the kernel could produce incorrect results if any one of the calls to `__syncthreads()` were omitted. Hint: Choose two threads that belong to different warps, and show how one of the threads uses an incorrect value because the other thread finishes the earlier function at a later time.

7.24 In our second CUDA implementation of sample sort, we compute the number of elements in each sublist that should be mapped to each bucket. These values are stored in a (logical) two-dimensional array `mat_counts`. The columns in `mat_counts` correspond to sublists, and the rows of `mat_counts` correspond to buckets. So `mat_counts[i][j]` is the number of elements of the list that are in sublist j and that will be mapped to bucket i. Since the number of sublists and the number of buckets are both equal to the number of thread blocks, $blk\_ct = b$, `mat_counts` has b rows and b columns.

So the kernel that computes the entries in `mat_counts` must compute b^2 entries. The entries are computed from the elements of the list:

1. Determine which bucket the element should be mapped to by executing a binary search of the splitters.
2. If an element is in sublist j and should be mapped to bucket i, increment `mat_counts[i][j]` by 1 by using `atomicAdd`.

Recall that if the number of elements in the list is n, then

$$n = 2b \times \texttt{th\_per\_blk},$$

where this kernel is called with `th_per_blk` threads per block. Use this information to determine

a. An upper bound on the number of operations (comparisons and branches) required for a thread to execute statement 1, above. You can assume all of the splitters are distinct.

b. The average number of increments to each element of `mat_counts` in statement 2.

If `atomicAdd` serializes the increments to `mat_counts[i][j]`, and the thread that calls `atomicAdd` gets its increment executed last, estimate an upper bound on the number of operations (comparisons, branches, additions) required for the thread to complete statements 1 and 2, above.

7.25 Since the sublists have been sorted, an alternative to the preceding approach is to carry out a binary search of a sublist for the location of each splitter. Then the number of elements going to a particular bucket from a particular sublist, can be computed by taking the difference of the largest subscript of an element in the sublist that is less than a splitter and the smallest subscript of an element in the list that is greater than or equal to the previous splitter.

1. Write pseudocode for the binary search. You should not assume that all of the elements of the sublist are distinct. Also do not assume that each sublist has an element between consecutive splitters.
2. Write pseudocode for the computation of the number of elements in the sublist that are between successive splitters.

After executing the two operations above, the number of elements between successive splitters can be assigned to the appropriate entry in `mat_counts`.

a. How many blocks and how many threads per block should be used in a kernel that implements this new method of computing the elements of `Mat_counts`?

b. How do you think the run-time would compare to the kernel in the preceding problem? Explain your answer.

7.7 Programming assignments

7.1 Look up the classical fourth-order Runge Kutta method for solving an ordinary differential equation. Use this method instead of Euler's method to estimate the values of $s_q(t)$ and $s'_q(t)$. Modify the reduced version of the serial n-body solver, either the Pthreads or the OpenMP reduced n-body solver, the MPI basic n-body solver, and the first CUDA n-body solver. How does the output compare to the output using Euler's method? How does the performance of the two methods compare?

7.2 Modify the basic MPI n-body solver so that it uses a ring pass instead of a call to MPI_Allgather. When a process receives the positions of particles assigned to another process, it computes *all* the forces resulting from interactions between its assigned particles and the received particles. After receiving comm_sz − 1 sets of positions, each process should be able to compute the total force on each of its particles. How does the performance of this solver compare with the original basic MPI solver? How does its performance compare with the reduced MPI solver?

7.3 We can simulate a ring pass using shared-memory:

```
Compute loc_forces and tmp_forces
        due to my particle interactions;
Notify dest that tmp_forces are available;
for (phase = 1; phase < thread_count; phase++) {
    Wait for source to notify tmp_forces are available;
    Compute forces due to my particle interactions with
            ''received'' particles;
    Notify dest that tmp_forces are available;
}
Add my tmp_forces into my loc_forces;
```

To implement this, the main thread can allocate n storage locations for the total forces and n locations for the "temp" forces. Each thread will operate on the appropriate subset of locations in the two arrays. It's easiest to implement "notify" and "wait" using semaphores. The main thread can allocate a semaphore for each source-dest pair and initialize each semaphore to 0 (or "locked"). After a thread has computed the forces, it can call sem_post to notify the dest thread, and a thread can block in a call to sem_wait to wait for the availability of the next set of forces. Implement this scheme in Pthreads. How does its performance compare with the performance of the original reduced OpenMP/Pthreads solver? How does its performance compare with the reduced MPI solver? How does its memory usage compare with the reduced OpenMP/Pthreads solver? How does its memory usage compare with the reduced MPI solver?

7.4 The storage used in the second MPI n-body solver can be further reduced by having each process store only its $n/\text{comm\_sz}$ masses and communicating masses as well as positions and forces. This can be implemented by adding storage for an additional $n/\text{comm\_sz}$ **doubles** to the tmp_data array. How does

this change affect the performance of the solver? How does the memory required by this program compare with the memory required for the original MPI
n-body solver? How does its memory usage compare with the reduced OpenMP
solver?

7.5 We can easily modify the first CUDA n-body solver so that each thread is responsible for multiple particles. The idea is that if there are g threads in the
grid, then we assign particles to threads by iterating through a **for** loop that uses
stride g (see Exercise 7.18):

```
int g = gridDim.x*blockDim.x;
int my_rk = blockIdx.x*blockDim.x + threadIdx.x;
for (int particle = my_rk; particle < n; particle += g)
    Carry out computations involving particle;
```

Here, we're assuming there are n particles. Note that if $g = n$, then the body of
the loop will only be executed once, but if $g < n$, some or all of the threads will
execute the body multiple times for particles

$$\text{my\_rk, my\_rk} + g, \text{my\_rk} + 2g, \dots.$$

Edit the first n-body solver so that it can simulate a system in which the number
of particles is an even multiple of the number of threads. How does its performance compare to the system with one particle per thread?

What changes (if any) need to be made to this solver if the number of particles is
greater than the number of threads, but it is not an even multiple of the number
of threads?

What additional changes (if any) need to be made to the basic shared memory
n-body solver?

7.6 There are several changes that must be made to our first CUDA n-body solver
if we want to use cooperative groups instead of multiple kernels:

a. We need to include the header `cooperative_groups.h`

b. C++ uses a construct called *namespaces*. It allows us to use two libraries
with the same identifier(s) without conflict, so we need to declare this with
something like

```
using namespace cooperative_groups;
/* This is a shorthand for cooperative_groups: */
namespace cg = cooperative_groups;
```

c. We need to call the function `cudaLaunchCooperativeKernel` to launch the
kernel `Dev_sim` in `Dev_sim_Driver`. Its syntax is

```
__host__ cudaError_t cudaLaunchCooperativeKernel(
        const void*      func,
        dim3             gridDim,
        dim3             blockDim,
        void**           args,
        size_t           sharedMem,
        cudaStream_t     stream )
```

For us, `func` should be the kernel `Dev_sim`, and `gridDim` and `blockDim` are the usual arguments in triple angle brackets. The argument `args` should be initialized to contain pointers to the arguments to `Dev_sim` cast to type **void**∗:

```
void *dev_sim_args []
     = {(void *) &forces ,
        (void *) &curr ,
        (void *) &n ,
        (void *) &delta_t ,
        (void *) &n_steps ,
        (void *) &output_freq };
```

The last two arguments can be 0 and NULL, respectively.

d. `Dev_sim` should create the `grid` cooperative group:

```
cg :: grid_group  grid  =  cg :: this_grid ();
```

e. Make `Compute_force` and `Update_pos_vel` device functions instead of kernels.

f. Add barriers between calls to `Compute_force` and `Update_pos_vel` using calls to `sync`.

```
cg :: sync ( grid );
```

g. Compilation needs to be done in stages:

```
nvcc -dc -gencode \
       --gencode arch=compute_61 , code=sm_61 \
       -c nbody_basic_cg.cu
nvcc -c stats.c
nvcc -c set_device.c
nvcc -o nbody_basic_cg \
         nbody_basic_cg.o stats.o set_device.o
```

See code on the book website for further details.

How large a problem can you solve with this version? How does this compare to the original basic CUDA solver? How does the performance of the two solvers compare?

Where to go from here

Now that you know the basics of writing parallel programs using MPI, Pthreads, OpenMP, and CUDA, you may be wondering if there are any new worlds left to conquer. The answer is a resounding yes. Listed below are a few topics for further study. With each topic we've listed several references. Keep in mind, though, that this list is necessarily brief, and parallel computing is a fast-changing field. So you may want to do a little searching on the internet before settling down to do more study.

1. **MPI.** MPI is a large and evolving standard. We've only discussed a part of the original standard, MPI-1. We've learned about some of the point-to-point and collective communications and some of the facilities it provides for building derived datatypes. MPI-1 also provides standards for creating and managing communicators and topologies. We briefly discussed communicators in the text: recall that, roughly speaking, a communicator is a collection of processes that can send messages to each other. Topologies provide a means for imposing a logical organization on the processes in a communicator. For example, we talked about partitioning a matrix among a collection of MPI processes by assigning blocks of rows to each process. In many applications, it's more convenient to assign block submatrices to the processes. In such an application, it would be very useful to think of our processes as a rectangular grid in which a process is identified with the submatrix it's working with. Topologies provide a means for us to make this identification, so we may refer to the process in (say) the second row and fourth column instead of process thirteen. The texts [23,48,50] provide more in-depth introductions to MPI-1.

   MPI-2 added dynamic process management, one-sided communications, and parallel I/O to MPI-1. We briefly discussed one-sided communications and parallel I/O in Chapter 2. Also, when we discussed Pthreads, we mentioned that many Pthreads programs create threads as they're needed, rather than creating all the threads at the beginning of execution. Dynamic process management adds this and other capabilities to MPI process management. [24] is an introduction to MPI-2.

   MPI-3 added nonblocking collective communications and extensions to the one-sided communications in MPI-2. (See [40].) MPI-4 is currently under development, and it should be released in the near future. For its current status, see [39].

2. **Pthreads and semaphores.** We've discussed some of the functions Pthreads provides for starting and terminating threads, protecting critical sections, and synchronizing threads. There are a number of other functions, and it's also possible

to change the behavior of most of the functions we've looked at. Recall that the various object initialization functions take an "attribute" argument. We always simply passed in NULL for these arguments so the object functions would have the "default" behavior. Passing in other values for these arguments changes this. For example, if a thread obtains a mutex and it then attempts to relock the mutex (e.g., when it makes a recursive function call), the default behavior is undefined. However, if the mutex was created with the attribute PTHREAD_MUTEX_ERRORCHECK, the second call to pthread_mutex_lock will return an error. On the other hand, if the mutex was created with the attribute PTHREAD_MUTEX_RECURSIVE, then the attempt to relock will succeed.

Another subject is the use of nonblocking operations. For example, if a thread calls pthread_mutex_lock and the mutex has been locked by another thread, then the calling thread blocks until the mutex is unlocked. In many circumstances, if a thread finds that a mutex is locked, it makes sense for the thread to do some other work rather than simply blocking. The function pthread_mutex_trylock allows threads to do just this. There are also nonblocking versions of the read-write lock functions and sem_wait. These functions all provide the opportunity for a thread to continue working when a lock or semaphore is owned by another thread. So they have the potential to greatly increase the parallelism in an application. For further information on Pthreads, see [7,34].

3. **OpenMP.** We've learned about some of the most important directives, clauses, and functions in OpenMP: we've learned about how to start multiple threads, how to parallelize **for** loops, how to protect critical sections, how to schedule loops, how to modify the scope of variables, and how to implement task-parallelism. However, there is still a good deal more to learn. One of the most important features we haven't discussed is the capability of OpenMP (since version 4.0) to offload computations to a GPU. Another very important feature is thread affinity: if a thread is suspended and restarted on a core that doesn't share cache with the original core, then all of the cached data and instructions are lost. The addition of thread affinity to OpenMP (version 4.0) allows OpenMP programs to control thread placement.

The OpenMP Architecture Review Board is continuing development of the OpenMP standard. The latest documents are available at [47]. For some texts, see [9,10,50].

4. **CUDA and GPUs.** GPUs have very complex architectures, and, as a consequence, the CUDA API has *many* features—many more than we've discussed. Two of the most important are dynamic parallelism and streams. With dynamic parallelism, device code can start kernels. So a CUDA thread can start other threads, and codes can use recursion or adapt to changing situations. For example, a code (such as an *n*-body solver) can focus on short-range particle interactions by allocating more threads where there are clusters of nearby particles. Streams provide the infrastructure to start multiple kernels on a single device or multiple kernels on multiple devices. This capability is crucial for solving very large problems.

The Nvidia website has a vast quantity of documentation [43]. Their CUDA C++ Programming Guide [11] and their Developer Blog [42] are especially useful. The article [19] and the book [33] provide overviews of GPUs. [33] also provides an introduction to CUDA programming.

5. **Parallel hardware.** Parallel hardware tends to be a fast-moving target. Fortunately, the texts by Hennessy and Patterson [28,49] are updated fairly frequently. They provide a comprehensive overview on topics such as instruction-level parallelism, shared memory systems, and interconnects. For a book that focuses exclusively on parallel systems, see [12].

6. **General parallel programming.** There are a number of books on parallel programming that don't focus on a particular API. [52] provides a relatively elementary discussion of both distributed and shared memory programming. [30] has an extensive discussion of parallel algorithms and a priori analysis of program performance. [35] provides an overview of current parallel programming languages and some directions that may be taken in the near future.

   For discussions of shared-memory programming, see [4,5,29]. The text [5] discusses techniques for designing and developing shared-memory programs. In addition, it develops a number of parallel programs for solving problems such as searching and sorting. Both [4] and [29] provide in-depth discussions of determining whether an algorithm is correct and mechanisms for ensuring correctness.

7. **History of parallel computing.** It's surprising to many programmers that parallel computing has a long and venerable history—well, as long and venerable as most topics in computer science. The article [16] provides a very brief survey and some references. The website [53] is a timeline listing milestones in the development of parallel systems.

Now go out and conquer new worlds.

Bibliography

[1] Eduard Ayguadé, Alejandro Duran, Jay Hoeflinger, Federico Massaioli, Xavier Teruel, An experimental evaluation of the new OpenMP tasking model, in: International Workshop on Languages and Compilers for Parallel Computing, 2007, pp. 63–77.

[2] Selim G. Akl, Parallel Sorting Algorithms, Academic Press, Orlando, FL, 1985.

[3] Gene M. Amdahl, Validity of the single processor approach to achieving large scale computing capabilities, in: AFIPS Conference Proceedings, 1967, pp. 483–485.

[4] Gregory R. Andrews, Multithreaded, Parallel, and Distributed Programming, Addison-Wesley, Reading, MA, 2000.

[5] Clay Breshears, The Art of Concurrency: A Thread Monkey's Guide to Writing Parallel Applications, O'Reilly, Sebastopol, CA, 2009.

[6] Randal E. Bryant, David R. O'Hallaron, Computer Systems: A Programmer's Perspective, 3rd ed., Pearson, Boston, MA, 2016.

[7] David R. Butenhof, Programming with Posix Threads, Addison-Wesley, Boston, 1997.

[8] The Chapel Programming Language, https://chapel-lang.org/.

[9] Rohit Chandra, et al., Parallel Programming in OpenMP, Morgan Kaufmann Publishers, San Francisco, CA, 2001.

[10] Barbara Chapman, Gabriele Jost, Ruud van der Pas, Using OpenMP: Portable Shared Memory Parallel Programming, The MIT Press, Cambridge, MA, 2008.

[11] CUDA C++ Programming Guide, https://docs.nvidia.com/cuda/cuda-c-programming-guide/index.html.

[12] David Culler, J.P. Singh, Anoop Gupta, Parallel Computer Architecture: A Hardware/Software Approach, Morgan Kaufmann, San Francisco, 1998.

[13] Frank Dehne, Hamidreza Zaboli, Deterministic sample sort for GPUs, Parallel Processing Letters 22 (3) (2012).

[14] James Demmel, et al., UC Berkeley CS267 Home Page: Spring 2020, Hierarchical Methods for the N-Body Problem, https://drive.google.com/file/d/14cg0j1lGGY4gqTkMErsJm4tEsQ1LlUyU/view, 2020.

[15] Edsger W. Dijkstra, Cooperating sequential processes, in: F. Genuys (Ed.), Programming Languages: NATO Advanced Study Institute, 1968, pp. 43–112. A draft version of this paper is available online at https://www.cs.utexas.edu/users/EWD/transcriptions/EWD01xx/EWD123.html.

[16] Peter J. Denning, Jack B. Dennis, The resurgence of parallelism, Communications of the ACM 53 (6) (June 2010) 30–32.

[17] Ernst Nils Dorband, Marc Hemsendorf, David Merritt, Systolic and hyper-systolic algorithms for the gravitational N-body problem, with an application to Brownian motion, Journal of Computational Physics 185 (2003) 484–511.

[18] Eclipse Parallel Tools Platform, https://www.eclipse.org/ptp/.

[19] Kayvon Fatahalian, Mike Houston, A closer look at GPUs, Communications of the ACM 51 (10) (October 2008) 50–57.

[20] Michael Flynn, Very high-speed computing systems, Proceedings of the IEEE 54 (1966) 1901–1909.

[21] Ian Foster, Designing and Building Parallel Programs, Addison-Wesley, Boston, MA, 1995. This is also available online at https://www.mcs.anl.gov/~itf/dbpp/.

[22] M. Gopi, GPU sorting algorithms, in: Hamid Sarbazi-Azad (Ed.), Advances in GPU Research and Practice, Morgan Kaufmann, Cambridge, MA, 2016.

[23] William Gropp, Ewing Lusk, Anthony Skjellum, Using MPI, 2nd ed., MIT Press, Cambridge, MA, 1999.

[24] William Gropp, Ewing Lusk, Rajeev Thakur, Using MPI-2, MIT Press, Cambridge, MA, 1999.

[25] John L. Gustafson, Reevaluating Amdahl's law, Communications of the ACM 31 (5) (1988) 532–533.

[26] Mark Harris, CUDA Pro Tip: Write Flexible Kernels with Grid-Stride Loops, Nvidia Developer Blog, Apr. 22, 2013, https://devblogs.nvidia.com/cuda-pro-tip-write-flexible-kernels-grid-stride-loops/.

[27] Mark Harris, Shubhabrata Sengupta, John D. Owens, Parallel prefix sum (Scan) with CUDA, in: Hubert Nguyen (Ed.), GPU Gems 3, Addison-Wesley Professional, Boston, MA, 2007. Available online at https://developer.nvidia.com/gpugems/gpugems3/part-vi-gpu-computing/chapter-39-parallel-prefix-sum-scan-cuda.

[28] John Hennessy, David Patterson, Computer Architecture: A Quantitative Approach, 6th ed., Morgan Kaufmann, Burlington, MA, 2019.

[29] Maurice Herlihy, Nir Shavit, The Art of Multiprocessor Programming, Morgan Kaufmann, Burlington, MA, 2008.

[30] Ananth Grama, et al., Introduction to Parallel Computing, 2nd ed., Addison-Wesley, Harlow, Essex, UK, 2003.

[31] IBM, IBM InfoSphere Streams v1.2.0 supports highly complex heterogeneous data analysis, IBM United States Software Announcement 210-037, Feb. 23, 2010, http://www.ibm.com/common/ssi/rep_ca/7/897/ENUS210-037/ENUS210-037.PDF.

[32] Brian W. Kernighan, Dennis M. Ritchie, The C Programming Language, 2nd ed., Prentice Hall, Upper Saddle River, NJ, 1988.

[33] David B. Kirk, Wen-mei W. Hwu, Programming Massively Parallel Processors: A Hands-on Approach, 3rd ed., Morgan Kaufmann, 2016.

[34] Bill Lewis, Daniel J. Berg, Multithreaded Programming with Pthreads, Prentice-Hall, Upper Saddle River, NJ, 1998.

[35] Calvin Lin, Lawrence Snyder, Principles of Parallel Programming, Addison-Wesley, Boston, 2009.

[36] John Loeffler, No more transistors: the end of Moore's Law, Interesting Engineering, Nov 29, 2018. See https://interestingengineering.com/no-more-transistors-the-end-of-moores-law.

[37] Junichiro Makino, An efficient parallel algorithm for $O(N^2)$ direct summation method and its variations on distributed-memory parallel machines, New Astronomy 7 (2002) 373–384.

[38] John M. May, Parallel I/O for High Performance Computing, Morgan Kaufmann, San Francisco, 2000.

[39] Message-Passing Interface Forum, http://www.mpi-forum.org.

[40] Message-Passing Interface Forum, MPI: A Message-Passing Interface Standard, Version 3.1. Available as https://www.mpi-forum.org/docs/mpi-3.1/mpi31-report.pdf, 2015.

[41] Microsoft Visual Studio, https://visualstudio.microsoft.com.

[42] Nvidia Developer Blog, https://devblogs.nvidia.com/.

[43] Nvidia Developer Documentation, https://docs.nvidia.com/.

[44] Nvidia Nsight Eclipse, https://developer.nvidia.com/nsight-eclipse-edition.

[45] Leonid Oliker, et al., Effects of ordering strategies and programming paradigms on sparse matrix computations, SIAM Review 44 (3) (2002) 373–393.

[46] The Open Group, www.opengroup.org.

[47] OpenMP Architecture Review Board, OpenMP Application Program Interface, Version 5.0, https://www.openmp.org/specifications/, November 2018.

[48] Peter Pacheco, Parallel Programming with MPI, Morgan Kaufmann, San Francisco, 1997.

[49] David A. Patterson, John L. Hennessy, Computer Organization and Design: The Hardware/Software Interface, 5th ed., Morgan Kaufmann, Burlington, MA, 2013.

[50] Michael Quinn, Parallel Programming in C with MPI and OpenMP, McGraw-Hill Higher Education, Boston, 2004.

[51] Valgrind Home, https://valgrind.org.

[52] Barry Wilkinson, Michael Allen, Parallel Programming: Techniques and Applications Using Networked Workstations and Parallel Computers, 2nd ed., Prentice-Hall, Upper Saddle River, NJ, 2004.

[53] Gregory Wilson, The History of the Development of Parallel Computing, https://parallel.ru/history/wilson_history.html, 1994.

[54] X10: Performance and Productivity at Scale, http://x10-lang.org/.

Index